APPLIED MYC
AND BIOTECH

VOLUME 6

BIOINFORMATICS

APPLIED MYCOLOGY AND BIOTECHNOLOGY

VOLUME 6

BIOINFORMATICS

Edited by

Dilip K. Arora
Centre of Advanced Study in Botany
Banaras Hindu University
Varanasi, India

Randy M. Berka
Novozymes Biotech, Inc.
1445 Drew Avenue
Davis, CA 95616-4880, USA

Gautam B. Singh
Center for Bioinformatics
Department of Computer Science
& Engineering, Oakland University
Rochester, MI 48309, USA

ELSEVIER

Amsterdam – Boston – Heidelberg – London – New York – New Delhi
Oxford – Paris – San Diego – San Francisco – Singapore – Sydney – Tokyo

Elsevier
Radarweg 29, PO Box 211, 1000 AE Amsterdam, The Netherlands
The Boulevard, Langford Lane, Kidlington, Oxford OX5 1GB, UK

First edition 2006

Library of Congress Cataloging-in-Publication Data
A catalog record for this book is available from the Library of Congress

British Library Cataloguing in Publication Data
A catalogue record for this book is available from the British Library

ISBN-13: 978-0-444-51807-1
ISBN-10: 0-444-51807-X

Transferred to digital print, 2007
Printed and bound by CPI Antony Rowe, Eastbourne

Editors

Dilip K. Arora*
Centre of Advanced Study in Botany
Banaras Hindu University
Varanasi, 221005 India
E-mail: aroradilip@yahoo.co.in

Randy M. Berka
Novozymes Biotech, Inc.
1445 Drew Avenue
Davis, CA 95616-4880, USA
E-mail: rambo@novozymesbiotech.com

Gautam B. Singh
Center for Bioinformatics
Department of Computer Science
& Engineering, Oakland University
Rochester, MI 48309, USA
E-mail: singh@oakland.edu

*Present affiliation: National Bureau of Agriculturally Important Microorganisms, Kusmaur
P. O. Box 6, Mau Nath Bhanjan, Uttar Pradesh 275 101, India

Contents

Contributors

Catherine A. Abbott	School of Biological Sciences, Flinders University, PO Box 2100, Adelaide, South Australia.
David L. Baillie	Department of Molecular Biology and Biochemistry, Simon Fraser University, Burnaby, B.C., V5A 1S6, Canada.
Manoj Bhasin	Institute of Microbial Technology, Sector 39 A, Chandigarh, India.
Gregory Butler	Department of Computer Science, Concordia University, Montreal, Quebec, H3G 1M8, Canada.
Sangdun Choi	Department of Biological Sciences, College of Natural Sciences, Ajou University, Suwon, 443-749, Korea; Department of Neurobiology and Anatomy, The University of Texas Medical School at Houston, Houston, TX77030, USA.
Jacques Cohen	Volen Center for Complex Systems, Brandeis University, Waltham, MA 02454, USA.
Brian Fayz	Bioinformatics Facility, Applied Genomics Facility, Wayne State University, Detroit, MI 48201, USA.
Michelle Green	Bioinformatics Research Group, SRI International, Ravenswood Ave, EK207, Menlo Park, CA 94025, USA.
Igor V. Grigoriev	US Department of Energy Joint Genome Institute, Walnut Creek, CA 94598, USA.
Tom Hsiang	Department of Environmental Biology, University of Guelph, Guelph, Ontario, N1G 2W1, Canada.
Peter D. Karp	Bioinformatics Research Group, SRI International 333 Ravenswood Ave, EK207, Menlo Park, CA 94025, USA.
Dmytro Kevorkov	Département d'informatique, Université du Québec à Montréal, C.P. 8888, succ. Centre- Ville, Montréal, Canada.
Stephen A. Krawetz	Department of Obstetrics and Gynecology, Center for Molecular Medicine and Genetics, Institute for Scientific Computing, Wayne State University, Detroit, MI 48201, USA.
Markus Krummenacker	Bioinformatics Research Group, SRI International, Ravenswood Ave, EK207, Menlo Park, CA 94025, USA.

Pierre Legendre	Département de Sciences Biologiques, Université de Montréal, C.P. 6128, succ. Centre-ville, Montréal, Canada.
Robert Lennie	Bioinformatics Facility, Applied Genomics Facility, Wayne State University, Detroit, MI 48201, USA.
Diego A. Martinez	Los Alamos National Laboratory Joint Genome Institute, P.O. Box 1663 Los Alamos, NM 87545.
Vladimir Makarenkov	Département d'informatique, Université du Québec à Montréal, C.P. 8888, succ. Centre-Ville, Montréal, Canada.
R. Ian Menz	School of Biological Sciences, Flinders University, South Australia.
Audrius Meškauskas	Alte Gfennstr. 22, CH-8600 Dubendorf, Switzerland.
Xiang Jia Min	Centre for Structural and Functional Genomics, Concordia University, Montreal, Quebec H4B 1R6, Canada.
Helena Nevalainen	Department of Chemistry and Biomolecular Sciences, Macquarie University, Sydney, Australia
Suzanne M. Paley	Bioinformatics Research Group, SRI International 333 Ravenswood Ave, EK207 Menlo Park, CA 94025, USA.
Shobha Ranganathan	Department of Chemistry and Biomolecular Sciences, Macquarie University, Sydney, Australia
Melissa R. Pitman	School of Biological Sciences, Flinders University, South Australia.
Adrian E. Platts	Department of Obstetrics and Gynecology, Wayne State University, Detroit, MI 48201, USA.
G. P. S. Raghava	Institute of Microbial Technology, Sector 39 A, Chandigarh, India.
Asaf A. Salamov	US Department of Energy Joint Genome Institute, Walnut Creek, CA 94598.
Leah A. Santat	Division of Biology, California Institute of Technology, Pasadena, CA 91125, USA.
Gautam B. Singh	Center for Bioinformatics, Department of Computer Science & Engineering, Oakland University, Rochester, MI 48309, USA.

Reginald Storms	Department of Biology, Concordia University, Montreal, Quebec H4B 1R6, Canada.
Nicholas O'Toole	Centre for Structural and Functional Genomics, Concordia University, Montreal, Quebec, H4B 1R6, Canada.
Brian Tjaden	Computer Science Department, Wellesley College, Wellesley, MA 02481, USA.
Adrian Tsang	Centre for Structural and Functional Genomics, Concordia University, Montreal, Quebec, H4B 1R6, Canada.
Anna Tsykin	Hanson Institute, Adelaide, South Australia.
Dawei Wang	Bioinformatics Facility, Applied Genomics Facility, Wayne State University, Detroit, MI 48201, USA.
Christopher R. Wilkinson	Child Health Research Institute, Adelaide, South Australia.
Claire H. Wilson	School of Biological Sciences, Flinders University, PO Box 2100, Adelaide, South Australia.
Sangdao Wongsai	Department of Chemistry and Biomolecular Sciences, Macquarie University, Sydney, Australia
Bin Yao	Bioinformatics Facility, Applied Genomics Facility, Wayne State University, Detroit, MI 48201, USA.
Yeisoo Yu	Arizona Genomics Institute, University of Arizona, Tucson, AZ 85721, USA.

Preface

With the completion of the sequencing of human genome, our next challenge in this post-genomic era is the acquisition of knowledge underlying the function and coordination of genes and proteins. This will be accomplished through an increase in the bandwidth and processing capabilities of genome data analysis pipelines by integration of in-vitro and in-vivo data sets to develop computational models that are adaptive in nature and complex enough to capture the characteristics of living systems. True progress is being made by the amalgamation of skills of researchers from computer science, mathematics and physics to the experimental expertise of scientists in biochemistry and molecular genetics. While these multidisciplinary teams are an exciting new phenomenon of the post-genomic era, the problems posited by the discipline would not afford justice any other way. Our objective in putting together this volume is to provide an insight into the principles, tools and applications in bioinformatics. This volume is compiled in a manner that would appeal to the professionals working in the area of mycology; however, several chapters are not specialized in nature and would be of interest to bioinformaticians in general.

This volume of *Applied Mycology and Biotechnology* entitled *Bioinformatics* is a logical extension of the previous issues on Fungal Genomics. Our use of the term bioinformatics is based on a broad interpretation that involves the implementation of mathematics, statistics, computer science and information technology to address questions relating to the biology of fungal organisms. As with the preceding books in this series, we are mindful of the challenges faced in developing a comprehensive volume on fungal bioinformatics because of the breadth and complexity of the information that is being generated by an international community of fungal biologists. Nevertheless, we have embarked on a mission to offer contributions by authors who typify the diversity of scientific approaches and thought processes in the current climate in order to provide readers with a reference point from which to embark for future investigations.

The volume is divided into three sections. The first section, *Principles,* focuses on providing a survey of theoretical underpinnings on the technological tools and applications. The section begins by describing the experimental design, analysis, processing of microarrays, a high throughput technology for biological analysis and functional determination that has significantly changed the way we can quantitatively measure and observe gene expression. The following chapter on protein homology modeling reviews the significance of this technique for a mycologist and discusses the advantages and limitations of creating a structural model. This chapter is followed by review of the methodologies for constructing phylogeny which provide significant information about the structure of genes and their sometimes convergent evolution. In particular, the chapter discusses reticulate evolution including horizontal gene transfer between taxa, hybridization events and homoplasy. The final chapter in this section provides a higher level view on the volume and integration of biological data within the context of fungal genomics research and raises some significant questions on the future of mycology within a broader context of comparative genomics and drug discovery.

The second section entitled *Tools,* begins by providing an overview of the tools utilized for the annotation of fungal genomes and addresses issues related to automated annotation generation in a high throughput biotechnology environment. This is followed by a detailed description of the various bioinformatics packages utilized for sequence analysis and in a sense provides a basis and the background information for the tools utilized for annotation as well as analysis of biological data. The following chapter describes the tools, particularly the statistical programs, needed for analysis of microarray data. These are significant for characterizing expression levels observed and simplify the enormous task of interpreting the expression levels of tens of thousands of genes. This final chapter in this section provides a comprehensive summary of the tools available for genome annotation, comparative genomics, protein structure prediction, functional classification of proteins, and the identification of potential vaccine candidates.

The third section focuses on describing the *Applications* of the concepts and methodologies presented in the first two sections. This section begins by describing a tool that utilizes a hierarchical controlled vocabulary for data mining cDNA and microarray expression data. The following chapter discusses the specific area of secreted proteins, or secretome, in fungal species. As secreted proteins are very important in fungal species, an analysis of the secretome of a number of fungal genomes is presented. An automated agent based for data mining fungal genomes is described in the next chapter. The software tool uses the internet for capturing and downloading information from fungal database servers and requires minimal programming efforts. The final chapter in this section reviews the burgeoning field of biolinguistics, which aims at applying the theory of human natural languages to the problem of interpreting biological data. The approach's capabilities extend to the comparison of biological sequences using phylogenetic and bio-chemical properties providing higher sensitivity in genomic and proteomic data analysis.

It is our hope that the bioinformatics scientists and biotechnologists, particularly in the area of mycology, would use this material covering the principles, methodologies and applications to leverage their quest for knowledge and move ahead by acquiring the understanding necessary to forge novel discoveries in the future. We are grateful to the authors who generously contributed chapters and to the editorial board for their help in assembling this volume. We sincerely thank Lisa Tickner, Senior Editor of Elsevier Life Sciences for her expert technical assistance.

Dilip K. Arora
Randy A. Berka
Gautam B. Singh

Applied Mycology and Biotechnology
An International Series
Volume 6. Bioinformatics

Experimental Design and Analysis of Microarray Data

Claire H. Wilson[1], Anna Tsykin[2,4], Christopher R. Wilkinson[3,4], Catherine A. Abbott[1]
[1]School of Biological Sciences, Flinders University, PO Box 2100, Adelaide, South Australia, 5001, Australia; [2]Hanson Institute, Adelaide, SA, Australia; [3]Child Health Research Institute, Adelaide, SA, Australia; [4]School of Mathematical Sciences, University of Adelaide, Adelaide, SA, Australia (claire.wilson@flinders.edu.au, anna.tsykin@adelaide.edu.au, christopher.wilkinson@adelaide.edu.au; cathy.abbott@flinders.edu.au)

The advent of microarray technology has significantly changed the way we can quantitatively measure and observe gene expression at the mRNA level within a given biological sample of interest, allowing for the monitoring of tens to hundreds of thousands of genes within a single experiment. The two main array platforms are spotted two-colour arrays and one-colour *in situ*-synthesized arrays. Microarrays are used for a wide range of applications including gene annotation, investigation of gene-gene interactions, elucidation of gene regulatory networks and gene-expression profiling of *Saccharomyces cerevisiae* and other fungal organisms. Academic researchers and both the pharmaceutical and agricultural industries have an enormous interest in developing microarrays both as diagnostic tools and for use in basic research into how pathogens, such as fungi, interact with their host. Microarray experiments generate vast quantities of raw gene expression data, therefore good experimental design and statistical analysis is required for the extraction of accurate and useful information regarding the expression of genes. In this review we firstly provide an overview of the arrival and development of microarray technology. We then focus on the issues surrounding experimental design and the processing of microarray images, followed by a discussion on methods for cleaning and normalizing raw gene expression data and a final discussion of the importance statistical analysis plays in identifying differentially expressed genes.

1. INTRODUCTION

Since the late 1990's, following the successful sequencing of the *Eschericia coli* genome, there has been a rapid advancement in genome-scale sequencing of both prokaryotic and eukaryotic organisms. At present the publicly accessible National

Corresponding author: Catherine A. Abbott

Center for Biotechnology Information (NCBI) (http://www.ncbi.nlm.nih.gov) Entrez Genome Project database contains 257 complete or in progress eukaryotic genome projects. Seventy of these projects are fungal, however many more projects are underway in both public and private laboratories that are not yet accessible. This rapid increase in genomic knowledge has largely driven the emerging discipline of functional genomics, fuelling the development of high-throughput technologies and computational methods for the rapid interpretation and extrapolation of information on a genome-wide scale. Functional genomics aims to functionally annotate every gene within the genome, their interactions with other genes and their involvement in gene regulatory networks, hence, allowing for the study of biological problems at levels of complexity that have never before been possible. Functional genomics and the need for genome-wide expression analysis has been a major driver in the development of DNA, protein and combinatorial chemistry array technologies. DNA microarrays, which are used for simultaneously measuring the level of mRNA gene products from a given biological sample are currently the most advanced of these technologies and will be the focus of this chapter. Although proteins are the ultimate products of genes, measuring mRNA expression levels is a good starting point for functional gene characterization and currently it is a considerably cheaper technology then measuring direct protein levels which utilizes mass spectrometry resources.

In general microarrays are used to measure the concentration of each mRNA from a given sample, providing a snapshot of expression at a single time point or relative to another sample. This is achieved by monitoring the combinatorial interaction of a set of DNA or mRNA fragments with a predetermined library of polynucleotide probes. Before the emergence of this technology, using techniques such as Northern blots or quantitative real-time RT-PCR (qRT-PCR), researchers were only able to measure expression at the mRNA level for a limited number of genes. With the advent of microarrays it is now possible for researchers to uniquely and quantitatively measure the expression of tens to hundreds of thousands of genes at any given time within a given biological sample on a single platform. There is enormous potential for expansion of our scientific knowledge and discovery through the application of microarrays into the investigation of gene-gene interactions and in pharmaceutical and clinical research to enable further understanding of disease and the creation of future diagnostic tools to individualize molecular medicine.

Analysis and handling of the huge amounts of raw gene expression data generated from microarrays has rapidly become one of the major bottlenecks from the utilization of this technology with an increasing reliance on bioinformatic based innovations. Research from the fields of biology, statistics, mathematics, computer science and physics are drawn together to further our understanding of biological processes. Like all experiments the value of the data generated can be greatly affected by the choice of experimental design, and the implementation and analysis phases. With the ultimate goal being to make inferences among biological samples, their genes and associated levels of mRNA expression, many factors must be considered and integrated during each phase. Microarray data must be integrated with nucleotide sequence data, knowledge of protein structure and function, and with phenotypic and clinical data. The overall goal of this chapter is to provide the reader with an overview of current DNA microarray technologies and to introduce issues regarding the experimental design and analysis of microarray data.

This chapter will firstly provide an overview of microarray technology, discussing cDNA and oligonucleotide arrays and their development as well as the steps involved in their assembly and applications within the fields of mycology and biotechnology. Following this issues concerning experimental design and microarray image processing will be discussed. The latter half of this chapter will then talk about cleaning and normalization of raw gene expression data followed by a discussion on methods for statistically analysing the data in order to identify differentially expressed genes. Finally the chapter will conclude with comments about future directions for the usage of microarrays.

For an overview of the computational methods used for the interpretation of microarray data, with a particular focus on unsupervised and supervised methods for clustering microarray data refer to the Tjaden and Cohen chapter in this volume.

2. MICROARRAY PLATFORMS

2.1 Introduction and Generic Features of Microarray Technology

Microarrays are a type of ligand assay based on the same principles as immunoassays, and Northern and Southern Blots. Immunoassay technology first started to appear back in the 1960's (Ekins and Chu 1999). DNA was first hybridized and immobilized onto a solid-phase matrix consisting of plastic and agarose supports back in the 1980's (Polsky-Cynkin et al. 1985). A research team led by Pat Brown and Ron Davis at Stanford University are credited with engineering the first DNA microarray, and in 1995 the same group produced the first modern microarray analysis publication regarding the use of cDNA glass slide microarrays for obtaining gene expression profiles (Schena et al. 1995). Stephen Fodor and his colleagues at Affymetrix are credited with development of the first commercially available short oligonucleotide microarray wafer chip (Lockhart et al. 1996), the GeneChip™, making use of photolithography for the *in situ*-synthesis of nucleotides onto an array (Fodor et al. 1991). Although the first reported use of microarrays was in *Arabidopsis thaliana* (Schena et al. 1995) most of the early microarray research involved utilization of the technology to identify differentially expressed genes in mammalian and yeast fields with the first complete genome, *S. cerevisiae*, being spotted onto an array in 1997 (DeRisi et al. 1997).

Development and automation of microarray technology, particularly within the commercial sector, has been primarily allowed through the application and emergence of advanced technologies such as specialized robotics, fluorescence detection, photolithography, and image processing equipment and software (McLachlan et al. 2004). Whether the microarrays, also known as gene expression arrays, DNA chips, biochips or chips, are commercially- or home-made, cDNA or oligonucleotide based, they all share a number of generic features in regards to their underlying technology such as the probe or 'spot', the target or sample probe and the solid-phase medium of the array platform. The probe or 'spot' typically refers to the single stranded polynucleotide (DNA) that is fixed onto the array. Typically the polynucleotide is of known sequence but may come from a custom unsequenced cDNA library. The term 'target' or 'sample probe' is commonly used to refer to the polynucleotide in the given sample solution which hybridizes to the fixed complementary probe sequence (Kohane et al. 2003). In general there are two basic sources for nucleotide probes on an array; each unique oligonucleotide probe is either individually synthesized base pair by base pair onto the solid array surface (Baldi and Hatfield 2002) or pre-synthesized DNA, cDNA, oligonucleotides or PCR products are directly attached to the array surface. If oligonucleotides are used the probes can be either short, 24-30 bases in length, or long, 60-70 bases in length, while the pre-synthesized PCR and cDNA products are typically hundreds to thousands of base pairs in length (Yauk et al. 2004). The DNA, PCR products or cDNA probes are generally amplified from a vector stored within a bacterial clone or are amplified from open-reading frames or nucleotide fragments of a chromosome (Kohane et al. 2003). Traditionally pre-synthesized probes are attached to the solid array surface using techniques such as robotic spotting or piezoelectricity. Typically the target or sample probe is RNA or cDNA synthesized from the total RNA or mRNA extracted from the biological sample of interest, which for detection purposes is synthesized using fluorescently or biotinylated labeled nucleotides.

Materials commonly used for the solid phase of the array platform include glass or plastic slides similar to a microscope slide, and silica chips. Less commonly used mediums include charged nylon filters, nylon meshes, silicon, nitrocellulose membranes, gels and micro-beads (Kricka and Forina 2001; Baldi and Hatfield 2002). Glass is a good choice of material for the solid phase and it is typically pre-coated with a product such as silicon hydride and poly-lysine or poly-amine to reduce background fluorescence and encourage electrostatic adsorption of the probes onto the slide (Baldi and Hatfield 2002; McLachlan et al. 2004). The solid phase of the array contains a grid with an ordered arrangement of tens to hundreds of thousands of "spots" or "wells" that are each capable of holding a droplet of the probe molecule (Kehoe et al. 1999). If a glass slide is used the spotted or *in situ*-synthesized probe is typically immobilized by air drying and ultraviolet (UV)-irradiation (Cheung et al. 1999). Each individual spot on the grid generally represents an individual gene, thus serving as an experimental assay for the relative levels of expression for that given gene.

Whether the platform is cDNA or oligonucleotide based, the underlying basis of microarray technology is complementary base pairing between probes and targets. This allows for the determination of the relative levels of mRNA expression of target sequences within the sample via measurement of the quantity of labeled target that binds to each immobilized spot of DNA (Colebatch et al. 2002). In a typical two-colour array experiment one sample (e.g. treatment sample) is labeled with a red Cy5 dye and the second sample (e.g. control sample) is labeled with a green Cy3 dye (Figure 1). After co-hybridization and scanning, the colour of the spot indicates the relative abundance of each sample. If a spot is significantly red it indicates that the gene in the treatment is in abundance, if it is green it indicates RNA from that gene in the control population is in abundance, if the spot is yellow it indicates equal levels of binding and hence no-change in gene expression, while a black spot indicates that binding has not occurred. With one-colour arrays, a single sample is hybridized to each chip, and the relative ratios of absolute spot intensities are compared to establish relative abundances.

2.2 Limitations of Microarray Technology

In regards to what microarrays are able to detect and measure there are a number of limitations that must be considered when planning experiments. DNA microarrays measure levels of mRNA only which do not always correlate well with protein levels. In addition cDNA microarrays are unable to detect and quantify the effects of translational or post-translational modifications on the activity of proteins.

Most probe sequences target the 3′ end of mRNA transcripts, which means that they are generally insensitive to different mRNA isoforms, typically being unable to detect the impact of alternative splicing during precursor mRNA processing (Brewster et al. 2004). However this limitation can be overcome by designing probes to specific isoforms.

One must also be aware that whilst a given probe is supposedly on the array, it may not be able to bind its target sequence due to it being a poorly designed probe, or it may in fact cross-hybridize to a different sequence due to homology or contamination of probe source material. Such effects are likely to be platform specific, as different platforms utilize different probe sequences to the same target mRNA.

Microarrays are sensitive platforms, and this sensitivity means that they are often sensitive to nuisance variables such as confluence of cells or date of hybridization. Careful experimental design (what, when and how many samples are hybridized) can ensure one measures true biological signal and good analysis software will ensure the results are reliable and robust against the many noise sources.

Along with the overall cost of microarray experiments some of the more technical limitations include having sufficient quantities of the biological sample of interest to allow extraction of high quality DNA and/or RNA. Such issues will be discussed further in section 6.2.

2.3 Current Microarray Platforms

Microarray technology has, and is continuing to rapidly evolve, resulting in the availability of a number of array platforms with different strengths and weaknesses. The two main types in use are the one-colour systems, in which one hybridizes one sample per microarray, and two-colour systems, in which one competitively hybridizes two labeled samples to a single array.

Generally the spotted cDNA and long oligonucleotide arrays are produced and used with identical or similar methods (Barczak et al. 2003). Spotted array assembly uses a robotic spotter to mechanically pick up small volumes (in the nanoliter or picoliter range) of sequence specific probes onto its pins and then deposit them onto a specific grid address on the glass slide (Figure 1) (Baldi and Hatfield 2002). The size of these spots is dependent on the type of pin and robot that is used. Spot quality is variable and often print run specific. Typically space limitations mean that each clone or oligonucleotide in the library is only printed once on the array but some smaller libraries do permit printing duplicate spots on each slide. If choice is available, well separated duplicate spots are much better than those printed side by side. The main advantage of robotically spotted systems is that they are cheap and highly customizable. Long oligonucleotide libraries tend to be more specific than cDNA libraries. The main disadvantages of cDNA spotted systems are the

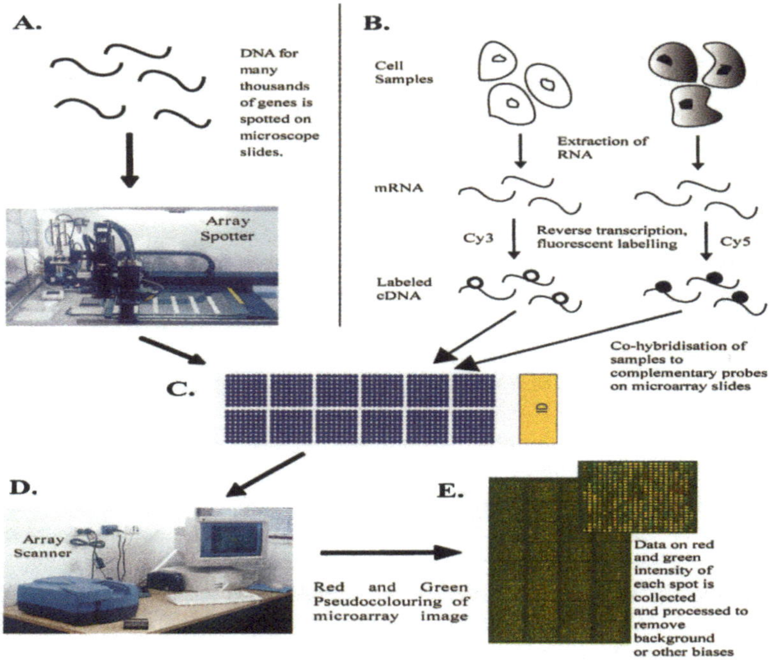

Fig. 1. Overview of the processes involved with the fabrication and usage of two-colour microarrays. A) depicts the fabrication of a two-colour robotically spotted array. B) represents the preparation of target samples, going from RNA extraction from cell samples through to fluorescent labeling and co-hybridization onto the microarray. A microarray slide is depicted in C. Each grid on the microarray is referred to as a subarray. After co-hybridization of fluorescently labeled samples to the array, the microarray is placed into the array scanner as shown in D. The resulting microarray image contains the raw data of the microarray experiment and is pseudocoloured red and green for visualization, and image processing. For one-colour microarrays, the processes from B-E are similar, except that only one sample is hybridized per array.

difficulties associated in maintaining and replicating a large library with potential for contamination and unavoidable variance in cDNA concentrations during these processes. Designing sensitive and specific oligonucleotide probes is not easy and often involves a trial and error approach. However, cost considerations often mean that poorly designed oligonucleotides are retained and used for long periods.

This problem is largely overcome by Agilent Technologies (www.agilent.com, Palo Alto, CA) which have developed a modified ink-jet printer head to *in situ*-synthesize long oligonucleotides (60mers) producing a highly customizable two-colour system with excellent quality spots. The advantages of this system include: the slide quality tends to be higher and more reliable than robotically spotted arrays, and whilst standard layouts are available (e.g. the 44k human genome array), one has complete control over what is printed. The use of 60mer's also increases specificity between the sample probe and target. The disadvantage is that these systems are more expensive, and there is a limit to the number of spots that may be printed on a single slide (less than 50,000).

The Affymetrix (www.affymetrix.com) GeneChip™ utilizes a sophisticated mask based photo-lithographic technique (developed for silicon chips) combined with solid-phase chemistry to *in situ*-synthesize 25mer probes at extremely high density (500,000 features per chip). For GeneChip™ arrays, each gene is typically represented by one probe set. Each probe set typically consists of 11 probe pairs which each cover a different 25 base pair section of the target mRNA (older chips used up to 20 probe pairs). Each probe pair consists of a complementary perfect match (PM) 25 mer oligonucleotide, and a mismatch (MM) 25mer oligonucleotide in which the 13th nucleotide in the sequence is changed to its complement, thereby functioning as a non-specific hybridization control (Brewster et al. 2004). However, the effectiveness of the MM oligonucleotide in this role is questionable as it has been shown that many MM's are also sensitive to the true target signal effectively invalidating such a role (Irizarry et al. 2003b). In contrast to the individually assembled spotted arrays, GeneChip™ arrays are produced as a wafer containing between 40 and 400 individual microarrays that are separated after probe synthesis (Kehoe et al. 1999). The main advantage of Affymetrix chips is they produce high quality and highly reproducible chips suitable for single colour hybridizations. Their main disadvantage is that they are very expensive, and not easily customizable.

Illumina's (www.illumina.com, San Diego, CA) recently developed BeadChip™ technology is a single colour system that uses a fiber optic detection system in conjunction with micro-beads tagged with 50-mer long oligonucleotides. This results in very small feature (spot) size allowing each probe to be represented on average 30 times per chip (Jianbing et al. 2003; Kuhn et al. 2004). The advantages of this system are the high quality, relatively low cost of chips and reagents, and it is highly customizable. The disadvantages are that this is a new system, with poorly developed analysis software, and the high resolution scanner required is expensive.

Other recent and developing technologies include Amersham's (www.amersham.com) CodeLink™ technology which uses a proprietary 3-D aqueous gel matrix slide surface with 30-base oligonucleotide probes. The 3-D gel matrix provides an aqueous environment that holds the probe away from the surface of the slide, allowing for maximum interaction between probe and target. This results in higher probe specificity and array sensitivity. CombiMatrix (www.combimatrix.com, Mukilteo, WA) and Nanogen (www.nanogen.com, San Diego, CA) utilize electrical addressing systems for the manufacture of DNA arrays onto semiconductor chips. It is believed by researchers at CombiMatrix that through using this technology it will be possible to produce a biological array processor with over 1 000 000 sites per square centimeter (Baldi and Hatfield 2002). CombiMatrix technology allows production of highly customized arrays even for very small orders and several fungal genomes are already available. Lynx (www.lynxgene.com, Hayward, CA) have developed a type of "fluid array" which employs a highly advanced tool developed by Sydney Brenner called massively parallel signature sequencing (MPSS) (Brenner et al. 2000). This platform allows for co-hybridization of two sample probes, measuring the absolute mRNA levels for virtually every expressed gene within the given samples (Stolovitzky et al. 2005) but at present it is extremely expensive.

2.4 Further Applications of Microarrays

While gene expression profiling experiments are the most common applications of microarray technology, this technology has been developed for numerous other applications such as chromatin immunoprecipitation (ChIP), Tiling arrays, comparative genomic hybridization (CGH) and single nucleotide polymorphism (SNP) arrays, all of which will be discussed below.

ChIP-on-Chip experiments involve hybridizing two independent samples to the array: one being an immunoprecipitated sample containing all the transcription factor-bound DNA and the other a non-specific sample. This allows for rapid and precise mapping of binding sites for transcription factors and other DNA-binding proteins. ChIP-on-Chip experiments also allow for investigation of the activation state of chromatin, chromatin remodeling and functional studies of histone modification as performed by Bernstein et al. (2005) who coupled ChIP with tiling arrays to investigate histone modification patterns in human and mouse cells. Cawley et al. (2004) also combined ChIP with tiling arrays to map the binding sites for three DNA binding transcription factors for human chromosomes 21 and 22.

Tiling arrays contain evenly spaced probes (overlapping or separated) designed to exhaustively span all non-repetitive intronic and exonic (i.e. non-coding and coding regions) sequence from a given genome. Kapranov et al. 2002 reported use of the first human tiling array platform, designed to interrogate on average every 35 bases of the approximately 35 million base pairs of chromosomes 21 and 22. Application of tiling arrays has revealed that a great deal more genomic sequence is transcribed into RNA than can currently be accounted for using present gene annotation data. Such transcripts have been termed TUFs (transcripts of unknown function) and are referred to as the 'hidden transcriptome'. Tiling arrays also allow for identification of alternate spliced isoforms, trans-splicing (where one gene is spliced to the next gene down-stream providing evidence for multiple alternative splice forms), exon-skipping and evidence for non-coding and antisense RNAs as well. Bertone et al. 2005 constructed a series of high-density oligonucleotide tiling arrays representing the entire human genome to comprehensively identify novel transcribed regions.

Similar to the ChIP-on-Chip experiments, Tenenbaum et al. (2000) hybridized purified endogenous mRNA-binding proteins (mRNP) complexes to cDNA arrays to identify subsets of mRNA contained in endogenous messenger ribonucleoprotein complexes (mRNPs) such as ribosomes that are cell type specific. Arava et al. (2003) performed a similar study, being the first group to perform a complete genome-wide analysis of mRNA translation profiles using *S. cervisiae* by separating free and ribosome bound mRNA on sucrose gradients then hybridizing these to cDNA arrays containing all known and predicted genes of *S. cerevisiae*. Interestingly they found that the number of ribosomes associated with mRNAs i.e. ribosome density, decreased with increasing open reading frame (ORF length). The purpose of their study was to carry out a comprehensive and detailed characterization of mRNA association with ribosomes in yeast cells growing in rich medium to probe the general features of translational behaviour and identify mRNAs that behave distinctively.

CGH (comparative genomic hybridization) arrays are currently the most powerful method for simultaneously detecting and localizing loss or gain of genetic material by allowing for copy number changes within genomes to be assayed through direct hybridization of whole genomic DNA (Mantripragada et al. 2004). By hybridizing entire genomes or specific chromosomal regions of interest, CGH arrays can be used to detect genetic aberrations such as deletions, amplifications, unbalanced translocations and copy number polymorphisms that are often associated with cancer and other complex diseases, and thus mapping of associated breakpoints. CGH can be coupled with tiling arrays to obtain a more complete picture of genome-wide copy number changes. CGH can also be coupled with SNP (single nucleotide polymorphism) arrays.

SNPs are the most frequent form of genetic variation present in the human genome with the SNP Consortium having mapped the presence of over two million SNPs within the human genome (http://www.ncbi.nlm.nih.gov/SNP/). Studies of SNPs offer the possibility to identify disease loci. High-density SNP arrays such as the Affymetrix 10K, 100K or 500K high density SNP mapping arrays provide a high-throughput platform for such studies. SNP arrays have also become valuable tools for loss-of-heterozygosity (LOH) studies and can be coupled with CGH to analyze copy number abnormalities (Zhou et al. 2005). Other applications of SNP arrays include whole genome association, genotyping, genetic linkage analysis, linkage disequilibrium mapping, and genetic epidemiology.

3. APPLICATIONS OF MICROARRAYS WITHIN MYCOLOGY AND BIOTECHNOLOGY

Early microarray experiments focused on the identification of differentially expressed genes in studies involving mainly human and yeast samples. Some of the early yeast microarray experiments involved answering questions regarding the size and diversity of genomes from different yeast strains. Microarray experiments helped identify the discarding of gene encoding DNA fragments from different laboratory yeast strains (Lashkari et al. 1997) and how the gene expression of varying yeast strains changed in response to altered growth conditions (DeRisi et al. 1997). DeRisi et al. (1997) were the first authors to report the use of a microarray containing nearly the whole genome of *S. cerevisiae*. *S. cerevisiae* has been used in a large number of microarray experiments for multiple applications such as investigation of the consequence of loss of gene function (Giaever et al. 2002), the identification of cytoplasmic localized mRNAs (Shepard et al. 2003), functional genomic analysis of a commercial wine strain of *S. cerevisiae* grown under varying nitrogen conditions with high-sugar (Backhus et al. 2001) and for "microarray karyotyping" of several *S. cerevisiae* wine strains to determine the genomic differences between them which may account for some of the observed variations in their fermentation properties (Dunn et al. 2005).

Microarrays can be applied to a wide range of studies including the comparison of disease versus non-diseased tissue, determining the effect of specific gene mutations or gene knockouts within a given cell line or whole organism and also in evolutionary studies. Microarrays are also rapidly becoming a valuable tool for cancer and viral diagnosis and treatment with the American Food and Drug Administration's making the first approval for use of a microarray as a genetic test in 2004. This microarray, called the AmpliChip® Cytochrome P450 Genotyping Test, was manufactured by Roche Molecular Systems, Inc., of Pleasanton California U.S.A. and was cleared for use with the Affymetrix GeneChip™ Microarray Instrumentation System (Affymetrix, Santa Clara, CA), allowing for physicians to individualize drug administration and dosage. In March 2003, a microarray designed to detect a wide range of known viruses and novel viral families was used during a severe acute respiratory syndrome (SARS) outbreak to reveal the presence of a previously uncharacterized coronavirus from a patient sample (Wang et al. 2003). With such applications in mind it is of no surprise that the pharmaceutical companies are increasingly utilizing microarrays throughout the many stages of drug development (Gerhold et al. 2002).

There is also huge potential for using microarrays as diagnostic tools within the field of agriculture. Microarrays have been used for the identification of pathogen's such as individual *Fusarium* fungi species on cereal grain (Nicolaisen et al. 2005), and for studying the complex signaling that exists between plants and their host-pathogen/symbiont relationships with other organisms. Once a genetically modified organism (GMO) has been generated, microarrays can be used to characterize the effect of the gene modification to ensure that it doesn't result in any undesirable phenotypic effects (Brewster et al. 2004). For investigation of fungal-plant interactions, DNA microarrays have been specifically constructed for examination of symbiotic interactions between *Arbuscular mycorrhizal* (AM) fungi and legumes (Franken and Requena 2001), between *Ectomycorrhizal* fungi and eucalyptus trees (Voiblet et al. 2000) and numerous studies examining fungal pathogenic interactions between *Alternaria brassicicola* and *A. thaliana* (Schenk et al. 2000) and *Cochliobolus carbonum* and maize (Baldwin et al. 1999). Microarrays have also been used to investigate the virulence of *Aspergillus fumigatus*, a fungal pathogen of humans (Rementeria et al. 2005) and to investigate how hypoviruses affect fungal development including asexual and sexual sporulation (Allen et al. 2003).

4. EXPERIMENTAL DESIGN OF MICROARRAY EXPERIMENTS

4.1 General Experimental Design of Microarrays

The standard statistical experimental design concepts of control, randomization, and replication apply equally well to microarray experiments. Before embarking on an experiment one must decide what biological questions one seeks to answer. On this basis one can chose a suitable probe library (this will define the chip type), and what samples or pairs of samples, will be hybridized against this probe library. One must also consider the feasibility of randomizing sample collection and date of hybridization to avoid confounding results. It is also important to consider how much replication is possible. Finally, thought should be given to how the data will be analysed. Typically many of these choices are driven by cost, but some forethought can help refine and prioritize the biological aims of the experiment thus maximizing the information gained from an experiment.

Microarray experimental design is largely concerned with choosing the hybridization strategy. However, before looking at this in detail we should first review the basic steps in a microarray experiment. A typical microarray experiment begins with the extraction of mRNA or total RNA from a specific biological sample of interest followed by synthesis of fluorescently labeled cDNA target (Figure 1). It is important that during the extraction of RNA, all traces of genomic DNA are removed in order to keep background levels low as it is a common source of contamination. The isolated mRNA or total RNA is typically reverse transcribed by first-strand cDNA synthesis and labeled for detection during the scanning process. For two-coloured microarrays the most commonly used fluorescent probes are the cyanine dyes Cy3 (green) and Cy5 (red) whilst one-colour arrays typically use biotin/strepatavidin conjugated probes for labeling. The labeled target is then denatured by heating to obtain single stranded polynucleotides from the sample, which upon cooling will hybridize to its complementary probe fixed on the array. In order to promote the specific and complementary binding of the labeled sample to its probe while reducing the level of background noise, it is of critical importance that the hybridization conditions are optimized (Brewster et al. 2004). Following hybridization, the array is washed to remove any unhybridized sample probes, and the amount of target hybridized to each spot is then quantified by scanning and image processing (see section 5). Microarray experiments thus include a large number of technical steps, and the resultant level of background noise can be influenced at many points and can depend upon the skills of the technicians performing extraction, labeling and hybridization. It is important to be aware of how the data is generated, so that it may be appropriately analysed, but we will not discuss this further instead concentrating on issues related to choosing what to hybridize.

As mentioned earlier the most important design step is deciding the aims of the experiment, and prioritizing what comparisons are of most interest. This will guide subsequent choices on platform type, RNA sources, and hybridization strategy. Many of these design choices are inter-related. Two-colour spotted arrays are cheap, but typically more noisy than the one- or two-colour commercial arrays systems (Irizarry et al. 2005). One also desires to perform sufficient biological replication to build confidence in the results. Using a cheaper system allows for the use of more replication or time points in a time course study, however the gain may not be huge due to increased noise of such systems. Sometimes the availability of a specific probe library will make a single platform the obvious choice, and similarly the desire for a specific comparison may make one design standout.

When deciding between whether to use a one- or two-colour system the choice is generally made on the basis of the aims and complexity of the experiment, and the desire to link the experiment to other experiments to be performed at a later date, or in other laboratories. Two-colour microarrays are analogous to matched pairs experiments - by co-hybridizing samples we control for many array and hybridization specific variables. To achieve similar power, one-colour arrays must be technically more stable and reproducible. With Affymetrix, the production methods and robotic control of hybridization ensures that this is generally the case (Irizarry et al. 2005). In general two-colour competitive hybridization is good for small-scale experiments, but as the scale increases, one faces problems in choosing which pairs of samples to co-hybridize, shifting the balance towards one-colour systems. Also if hybridizations are to be conducted over long time scales then one-colour systems may be more appropriate than using a two-colour system with a common reference design (discussed later).

4.2 Replicate Experiments, Reproducibility and Randomization

A typical question concerning researchers contemplating a microarray experiment is "How many replicates are required?". A typical response is "As much as one can afford". Confidence in results is based on their reliability which is derived from performing replicated experiments using different biological samples thus increasing the number of degrees of freedom. Biological replicates must be obtained from different biological samples involving separate RNA extractions for each to ensure an adequate measure of biological variability. The amount of variability between biological replicates will depend upon whether the material is derived from cell-lines, in-bred or out-bred species or strains (which are most variable).

One also has the option of performing technical replication, where technical replicates involve RNA that has been extracted from the same biological sample but has been independently labeled for hybridization. The need for technical replication is related to the level of variability inherent in the microarray platform being used. Dye swap replicates are technical replicates in which the dye labels are reversed e.g. Sample A is labeled with Cy3 in the first slide, and Cy5 in the second, whilst Sample B is labeled with Cy5 in the first slide, and Cy3 in the second. Dye swap replicates can be used in most two colour designs, to reduce any possible effects due to differential dye responses which are not completely removed by normalization procedures (discussed in section 6). One related approach is to perform dye swapping on biological replicates. In the opinion of Glonek and Solomon (2004), if hybridizations are to be replicated then they should be performed as dye-swapped replicates.

It is important for researchers to realize that under no circumstance does technical replication account for biological variability, it purely provides an estimated measure of the level of experimental variability (Kerr 2003) while increasing the likelihood of detecting differentially expressed genes.

Instead of providing adequate biological replicates, some researchers will pool samples with the belief that pooling will provide a means of reducing the biological variability and the number of arrays required for the experiment. However pooling generally doesn't provide a valid basis for adequate statistical analysis of the resulting data set. While pooling does lead to a reduction in observed biological variance it also results in the elimination of all independent biological replication making it impossible to compare individual samples from which the pools were derived (Simon and Dobbin 2003). However, pooling is sometimes necessary when there are insufficient quantities of RNA from an individual sample.

Another important consideration is randomization. For any experiment in which there is a treatment the biological samples should be randomly assigned to the treatment groups. In the opinion of Kerr (2003), if the microarray data is particularly susceptible to technical variation then arrays should also be chosen randomly from the batch of arrays for each planned hybridization to remove any possible systematic variation that may be related to the order in which the arrays were printed.

There have been several comparative cross-platform studies performed which indicate that in general biological variability tends to be higher than technical variability, and that commercial platforms tend to have less technical variability than in-house printed arrays (Yauk et al. 2004; Irizarry et al. 2005).

After determining the overall aim of the microarray experiment, i.e. what biological question is to be answered, one of the first steps involved with the design is selection of the functionally relevant biological sample, whether it is a cell type, tissue or whole organism such as fungi. Treatments and conditions relating to growth and isolation of the samples need to be identified, performed and kept tightly constant across all specimens in order to minimize biological variation arising from environment. Kazan et al. (2001) suggest that in order to gain the best possible comparisons, a separate control accounting for differences imposed by treatments must be used for each treatment in the experiment. Once the biological sample is selected and control and treatment groups obtained then the next stage in the process is generation of the sample which requires the isolation of either total RNA or mRNA. If RNA becomes degraded during the experimental process then it will be unsuitable for labeling by most standard techniques. Some manufacturers do offer alternative protocols when RNA degradation cannot be avoided. To help determine the integrity and quality of the sample the RNA is typically examined by "denaturing" gel electrophoresis. The presence of RNA degradation in Affymetrix arrays can also be detected in-silico using RNA degradation plots and NUSE boxplots which are part of the affy and affyPLM packages of the freely available Bioconductor analysis package (Gentleman et al. 2004).

4.3 Models and Assumptions of Experimental Design

There are several models currently used for the design of microarray experiments. These models cover issues concerning the labeling and allocation of arrays and the order of sample probe hybridization to the arrays. The choice of experimental design is dictated by the biological questions being asked, availability of microarray platforms, suitability of analysis software and constraints related to amounts of RNA and financial considerations. At present the most commonly used models are the reference design, the balanced block design, the loop design, the dual-label or dye-swap design (only applicable to two-colour arrays) and the time course and factorial designs (Figure 2). Each of these will be briefly discussed below.

The reference design (Figure 2B) involves a common reference being used for comparisons with treatment effects across a number of different experiments. To be useful, the reference must contain detectable levels of all genes expressed in samples co-hybridized with it, therefore the reference is often a pooled sample of all experimental conditions. This design often makes the assumption that the effect of dye bias is equal across all comparisons with the reference (Maindonald et al. 2003) as the reference is generally labeled with the same dye on each array. Therefore any gene-specific dye bias not removed by normalization will affect all arrays in a similar fashion (Simon and Dobbin 2003). One reason for using this design is when there is limited availability of RNA from one or all of the samples. The orientation of dye labeling is applied in the same direction so that samples for comparison are always labeled with the same dye and the reference is always labeled with the same dye (Churchill 2002). For spotted arrays, this design uses an aliquot of common reference RNA as one of the samples co-hybridized to each array so that comparisons between the reference and sample of interest can occur on the same spot (Simon and Dobbin 2003). In the opinion of Maindonald et al. (2003) a direct pair-wise comparison for two treatments should be more precise than indirect comparison of all treatments through a reference. Similarly, Churchill (2002) argues that use of a reference sample is unnecessary and results in inefficient experiments as half of the gene-expression measurements are made on the reference sample which is generally of no interest to the researcher. Reference designs can still be appropriate in complex experiments where each sample is involved in several comparisons.

The direct pair-wise comparison (Figure 2A) preferred by Maindonald et al. (2003) is referred to as the balanced block design. The major advantage of this design is that it can limit the number of arrays per experiment thus reducing the overall cost. A disadvantage of this design is that it generally has a higher level of signal-to-noise ratio then the reference design due to the variation of spots between arrays and within arrays. Simon and Dobbin (2003) report that the efficiency of the block design is reduced when there is increased biological sample variation.

When cluster analysis of the resulting data set is the main objective of the experiment then a particular useful experimental model is the loop design (Figure 2C). This model typically involves the co-hybridization of two differing samples onto a single array with the aliquot for each sample being split between two arrays allowing for the arrays to ultimately be used for linking each of the samples together in a loop pattern therefore allowing for all pair-wise comparisons to be made between samples while controlling the size and variability of spots (Simon and Dobbin 2003). In comparison to the reference design, the loop design is more balanced with respect to the dyes as each sample is labeled at least once with each of the dyes used (Kerr and Churchill 2001). A major down-side to this design however is the increased variance due to the requirement for modeling indirect effects relating to the arrays that link two samples of interest (Simon and Dobbin 2003). The loop design is also generally less robust in respect to the occurrence of bad arrays as one or more bad arrays result in breaking of the loop whereas in the reference and block designs they can simple be removed from subsequent data analysis.

Large time-course experiments (Figure 2E-F)are those where samples are collected and measured at many different time points usually in response to a treatment and comparisons are then made between the time points (i.e. between 0, 6 and 12 hrs). A crucial factor for the validity of time-course designs is the actual times used and the overall number of time points. If poorly designed these experiments will become costly in terms of equipment and consumables and it is generally impossible to perform pair-wise comparisons on all samples.

The previous designs are examples of single factor designs in which we study different levels of a specific factor. Often we will be interested in investigating several factors at once, such as comparing several different treatments over time. Such designs are known as factorial designs (Figure 2D), and allow investigating each effect plus the presence of interactions between factors (e.g. which genes would show changes in slope if different treatments were plotted over time). In conjunction with analysis utilizing linear models, such as methods implemented in Limma (see section 6.6) factorial designs allow highly effective identification of genes that for example respond to stimulation differently in test and control groups. Further

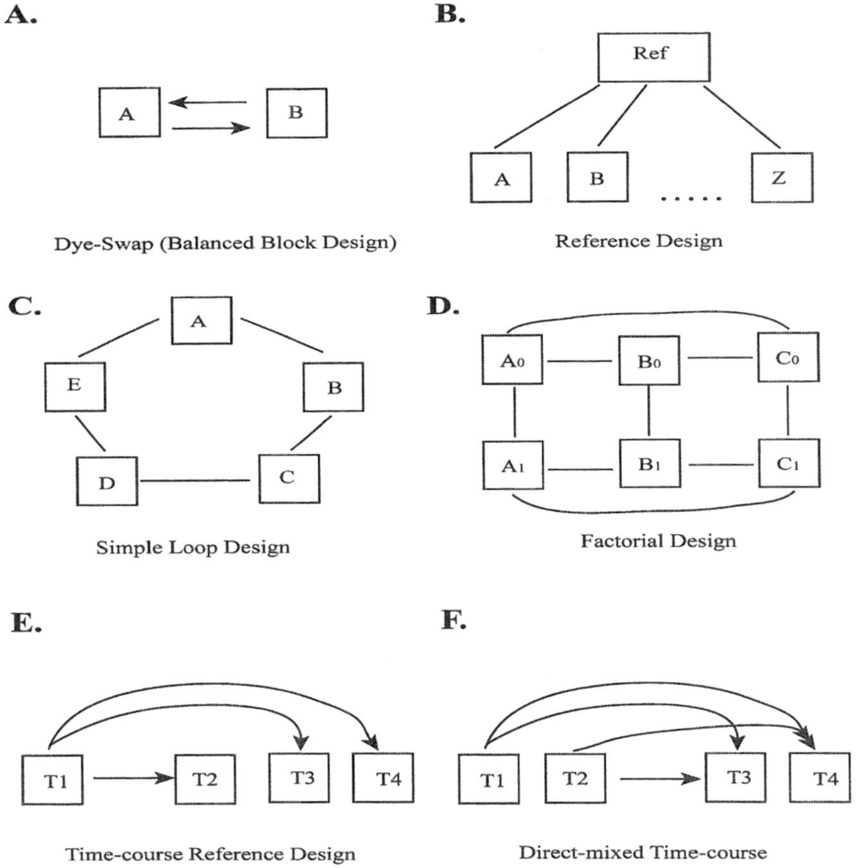

Fig. 2. Graphical representation of experimental designs. Boxes represent hybridized arrays and lines or arrows represent comparisons between samples. For two-coloured arrays, by convention arrow tails represent the Cy3 (green) labeled sample while arrow heads represent the Cy5 (red) labeled sample. (A) allows for direct pair-wise comparison of all genes between two biological samples. (B) allows for indirect comparisons to be made between samples and a common reference. (C) represents a loop design. (D) displays a factorial design where three samples taken at 0 hours are being compared with three samples taken at 1 hour. Such designs can become much more complex with increasing contrasts of interest. When this occurs it is more useful to use a reference design. (E) and (F) represent varying types of time-course designs. All designs can involve the use of dye-swaps.

considerations on how to optimally choose hybridizations when comparing several factors such as different cell lines and treatments are discussed by Glonek and Solomon (2004).

5. SCANNING OF MICROARRAYS AND IMAGE PROCESSING

After hybridization the gene expression data must be extracted from the microarray and analysed. This requires scanning of the array to measure the fluorescent signal for each spot on the array followed by analysis of the resulting image to extract the foreground and background intensity values that are used in subsequent analysis. One typically has little control over scanning equipment hence microarray scanners will not be discussed here. An area where choice is more readily available is the image processing software. Image analysis and the resulting acquisition of data is an important aspect of microarray experiments and can potentially have a large impact on subsequent data analysis. The commercial platforms (e.g. Affymetrix and Agilent) tend to have their own customized image analysis software packages, which leave little choice to the end user. However, in-house spotted arrays are typically more variable, requiring more care in the choice of image analysis software to ensure unnecessary variation is not introduced. Due to the wide variety of microarray platforms and various forms of labeling, no single microarray scanning device or image analysis software is suitable for all purposes. In this section we will only address the scanning and processing of images from two- and one-colour microarrays that emit a fluorescent signal.

Currently, there is a wide range of both commercial and freeware image analysis software available. Some of these are designed specifically for glass slide arrays while others can be used in conjunction with a variety of array platforms such as the nylon filter arrays. Regardless of platform, all image analysis software is generally designed to perform three fundamental processes: Addressing or gridding, segmentation, and intensity extraction or data acquisition. Addressing or gridding is the process of locating each spot on the slide and assigning it to a coordinate by taking advantage of the rigid layout of the spots. Segmentation is the method used to differentiate and classify the foreground pixels for each spot and background pixels. Information extraction or data acquisition is the process of calculating the foreground and background intensities for each spot based on the assignment of pixels during segmentation.

During the image processing stage a number of problems can arise which can result from insufficient labeling and concentration of the sample probes or too little of the probe on the solid-phase for binding of the sample probe as well as insufficient exposure time (Cheung et al. 1999). Other problems specifically relate to the presence of poor-quality spots which can have a drastic impact on the data set if not reduced (McLachlan et al. 2004). Spots are classified as poor quality when they have variable diameters and contours, a background signal that is higher then the foreground signal or the presence of spatial artifacts (McLachlan et al. 2004). A good image analysis program should therefore have the capability of collecting quality measures for each spot that can be used to flag unreliable spots or arrays. However, in general flagged spots should be down-weighted in analysis, rather than completely eliminated (Ritchie 2004).

5.1 Scanning Microarray Images

All microarrays that have been hybridized with a fluorescently labeled target use an optical system to scan slides to produce a digital record. This record contains the fluorescence intensity for every pixel at each grid location on the array. The intensity is proportional to the number of sample probes hybridized to the spotted probe (Cheung et al. 1999). Commonly used scanners are typically based on a confocal laser microscopy system where a separate laser is used as a source of excitation light for each fluorescent dye, and a photomultiplier (PMT) tube is used as the detector. In brief the fluorescent dyes become 'excited' by the laser light, absorbing its energy, and resulting in the emission of photons (Cy3 dyes produce a band from 510-550nm, whilst Cy5 dyes produce a band in the 630-660nm range). A detector such as the PMT is scanned across the surface and measures the intensity at each point (Baldi and Hatfield 2002; Yang et al. 2002a). Depending on the scanner used a number of settings such as the power of the excitation laser and the voltage of the PMT can be varied to improve the sensitivity of image acquisition so that low signals can be detected.

Images generated from microarrays contain the raw data of the experiment. Typically a 16-bit grey scale image is produced for each fluorescence frequency. For two-colour systems these can be combined into a single falsely coloured red-yellow-green image.

5.2 Addressing or Gridding

The scanned microarray image is imported into an image analysis program, where the first stage of analysis is to locate each spot on the slide by addressing or gridding of the array. The majority of image analysis software systems now provide reasonable automatic or semi-automatic gridding procedures with slight variations occurring between each. Aside from the comment that the results of automated gridding should be visually scanned to ensure the accuracy of the process, addressing and gridding will not be discussed here in detail; however this has been reviewed recently by Smyth et al. (2003).

5.3 Segmentation (Identifying Fore and Background pixels)

Segmentation of a microarray image is the process of dividing the image into different regions based on certain properties. For spotted arrays it involves the classification of pixels as being foreground or background (Yang et al. 2001) In regards to two-colour arrays, there is presently a number of differing segmentation methods used for production of the spot mask. It has been shown that the choice of segmentation method can introduce variability into the resulting microarray data, hence care must be taken when selecting the method for use (Ahmed et al. 2004; Ritchie 2004).

The four main groups of segmentation methods used for two-colour microarray images are: fixed circle segmentation; adaptive circle segmentation; adaptive shape segmentation; and histogram segmentation (Yang et al. 2001). Given that spots on many in house printed arrays are often irregularly shaped, fixed or adaptive circle systems should be avoided (Yang et al. 2001). Histogram segmentation uses a thresholding system for classifying pixels as either foreground or background (Chen et al. 1997), but often results in the over- and under-estimation of foreground and background intensities (Smyth et al. 2003). Adaptive shape segmentation methods such as the watershed (Beucher and Meyer 1993) and seeded region growing (Adams and Bischof 1994) implemented within the Spot software make no assumptions in regards to the spots circularity and size and hence are suitable for use with both commercial and non-commercially produced arrays.

5.4 Intensity Extraction or Data Acquisition

Determining spot intensity requires computing the average pixel value of the foreground pixels of a spot. Background intensity needs to then be approximated using a suitable method and subtracted from the spot intensity to provide a foreground fluorescent intensity. For two-colour arrays foreground and background intensity needs to be calculated for both the Cy3 (green) and Cy5 (red) channels. Correction for background intensity is necessary as it is likely that not all of the measured spot intensity comes from the fluorescent label of the hybridized sample. Signals can also result from factors such as non-specific hybridization, from fabrication artifacts on the glass caused by chemicals, dust and spatial variation. If the method for background estimation is poorly chosen, correction of foreground can result in negative intensities and hence missing values when log intensities are computed, typically resulting in the loss of low-intensity data (Smyth et al. 2003). Poor choice of background correctors can also introduce extra noise (variability) making detection of true differential expression more difficult (Yang et al. 2002a; Ritchie 2004). For both one-colour and two-colour arrays the reported intensity level (spot, background, PM or MM) is a summary of fluorescence measurements detected in a series of pixels.

As for segmentation, there a number of differing methods implemented in the software for calculation of background intensity. Generally however, these methods can be classified on the basis of whether they use a constant or global value, a local background estimate or a morphological background estimated by applying a non-linear filter to a local window around the spot (Yang et al. 2002a). Local background methods only consider the intensities of small regions surrounding the spot mask while global background generally considers the intensity of background for the whole array. In the experience of Smyth et al. (2003) local background methods result in over estimation of the background while global methods can result in an under-estimation of the background. Morphological opening tends to give a less variable background estimate which is not upwardly biased by the presence of bright pixels (Yang et al. 2002a; Ritchie 2004). Ritchie (2004) has performed an in-depth study of different background estimators and advises that if the purpose of the experiment is to select differentially expressed genes then the use of a morphological opening background estimator is recommended. Morphological opening has been available in the Spot (http://www.cmis.csiro.au/index.htm) software package for several years, and more recently has been included in GenePix (http://www.axon.com) and ImaGene's (http://www..biodiscovery.com) offerings. If one only has the choice of a local background corrector (e.g. an older version of GenePix implementing median scale normalization) or no background correction,

Ritchie (2004) advises that more reliable (less variable) results are obtained with no background correction.

After background correction, data analysis software for the two colour-arrays calculates the log-differential expression ratio being $M = \log_2 R/G$ for each spot and the log-intensity of the spot will being $A = 1/2\log_2 RG$, which is a measure of the overall brightness of the spot.

6. ANALYSIS OF MICROARRAY DATA

Microarray data analysis techniques have rapidly evolved, currently there are a wide variety of methods available to identify differentially expressed genes and infer functional information. These include analysing single genes to investigate how the behaviour of each gene changes between a control and a treatment or multiple gene analysis where clusters of genes are analyzed to determine common functionality, pattern-identification, gene-gene interactions and gene regulatory networks. Cluster analysis of microarray data is covered in the proceeding Tjaden and Cohen chapter, and therefore won't be discussed further. Overall success in identifying differentially expressed genes during the microarray data analysis stage is heavily dependent on the suitability of the chosen experimental design, which also governs what type of analysis to use and this was discussed previously in section 4.3.

Regardless of platform, the raw microarray data acquired from an image needs to be processed in order to remove poor quality spots and then normalized to correct for systematic variation before further downstream analysis. Downstream analysis involves identifying genes that are differentially expressed. This requires the selection and calculation of a suitable statistic to be used for ranking of the genes, followed by selection of an appropriate cut-off point for differential expression whereby genes having a rank value above the cut-off are considered to be differentially expressed and those having a value below are considered not to be. Within this process it is often common to calculate the false-discovery rate, in order to determine the number of expected false-positives and false-negatives that are to be included or excluded from the final list of differentially expressed genes. More precisely false-positives are genes having no differential expression that appear within the final list of differentially expressed genes while false-negatives are genes having true differential expression that are excluded from the final list.

Before normalization and further analysis the raw microarray data usually undergoes a logarithmic transformation to the base 2. Transformation of the raw data can help minimize some of the systematic variation by eliminating measurements for poor quality spots, and may facilitate in the identification of differentially expressed genes. In the case of two-colour arrays, the logarithmic transformation converts the intensity ratios into differences between the two channels at each spot, making up-regulated and down-regulated values of the same scale comparable as the non-transformed ratios tend to treat up- and down-regulated genes differently. There are a number of variations to the standard logarithmic transformation such as shift transformations (Newton et al. 2001; Kerr et al. 2002); curve fitting transformations (Yang et al. 2002b) and variance stabilizing transformation methods (Rocke and Durbin 2001; Cui et al. 2003).

The final stage of microarray data analysis is biological interpretation of the results to determine their functional relevance, and confirmation of observed differential expression by other experimental means. Biological interpretation of the data requires the utilization of additional bioinformatics techniques for the correlation of expression data with other types of data such as genomic, proteomic or metabolomic data.

6.1 Sources of Experimental and Biological Variation

Poor quality or noise within the microarray data arises from the many sources of variation throughout the experimental process. If not removed, the noise will ultimately affect the observed levels of differential expression. Throughout the preceding sections of this chapter some of the sources of experimental variation have already been mentioned however they will be re-iterated here.

Variation arising during the early stages of the experimental process may occur due to the use of poor quality RNA, RNA degradation, and the presence of genomic DNA within the RNA sample. A number of variations can also arise during both the labeling and hybridization stages. If the conditions of hybridization, such as temperature and duration, are not optimized and kept constant, further variations will arise during the stages of scanning and image processing and there will be a higher occurrence of non-specific hybridization of the samples to the probes. Non-specific hybridization and the presence of foreign artifacts on the arrays such as dust, clothing fibers, skin and scratching of the slides is also a significant problem for both of the common array platforms. During the labeling process non-controllable sequence bias can occur, resulting from some of the fluorescent dye labels showing preferential binding to some nucleotides over others. Another variation specific to spotted arrays relates to the lengths of the probes. As the probe length can vary from a hundred to a thousand bases in length, there is a higher likelihood of the probes cross-linking which results in a decreased number of available probe molecules for the binding of sample probes. Also, the actual process of robotically spotting the probes onto the array can introduce a great deal of systematic variation due to inconsistencies occurring with location, size and shape of spots on the individual arrays and between the arrays. These variations typically result from there being slight differences between the print-tips on the robotic arrayer, or from using arrays that were generated at different times. One of the most common forms of bias affecting spotted microarrays is that of dye-bias also known as red-green bias. Dye-bias arises due to there being differences between the labeling efficiencies and scanning properties for the two fluorescent dyes, Cy3 and Cy5. For any array platform, there is often a major problem with saturation of the probes, occurring when the probe intensities reach the maximum level of intensity acquired by the scanner. Saturation can result in loss of information regarding differential expression by masking highly expressed genes. Saturation effects can be minimized during the scanning process by adjusting the settings of the scanner, however, this may result in the exclusion of low expressed genes hence, normalization of the data may be more desirable and will be discussed below.

6.2 Normalization of Microarray Data

One would like to remove or minimize any non-biological variation present. This process is generically referred to as normalization. Normalization is platform specific, with different approaches required for one- and two-colour systems.

A wide range of normalization algorithms are available such as local versus global and linear versus nonlinear normalization. Selection of a suitable algorithm requires some assessment to be made in regards to what type and degree of systematic variation is present and whether normalization is required within-arrays, between-arrays or both. It is important that the normalization process does not remove or reduce any variation arising from biological differences between RNA samples or the printed probes. While normalization is generally considered necessary, over normalizing the data can introduce biases that are more detrimental to the identification of differentially expressed genes than a small difference in scale. Selection of a suitable normalization method can be aided by viewing exploratory plots such as M vs. A plots (Figure 3) to investigate if there is any obvious curvature deviating from the horizontal line at zero, or boxplots of each array to determine the difference in spread of log-ratios for each array, or of print-tip groups for each individual array in the case of spotted arrays. The majority of normalization methods aim to scale individual intensities so that the mean or median intensities are balanced within and between arrays, allowing for meaningful comparisons to be made.

Many of the normalization methods make the assumption that the majority of genes are not differentially expressed, hence the average, or geometric mean, of the ratio is one and that the average, or arithmetic mean, of the log ratio is zero i.e. it is assumed that the expression level for the average gene does not change during the experimental conditions. Normalizing all or the majority of genes present on the array generally provides the most reliable and stable estimation of spatial and intensity dependent trends present within the data (Smyth et al. 2003). However, at times it is more useful to normalize a subset of genes back to a set of control genes or a set of housekeeping genes present on the array.

6.2.1 Normalization of gene chip (single colour) arrays

One-colour arrays such as the Affymetrix GeneChip™ allow for hybridization of only a single sample to each individual array. Appropriate normalization is therefore vital, as different arrays need to be compared against each other in order to determine a meaningful estimate of the level of differential expression in a given gene. In the following discussion we will also consider probe set summarization methods, as some normalization methods are applied to probeset summaries, whilst others are applied at the level of individual probes.

For several years Affymetrix have supplied their Microarray Analysis Suite version 5 (MAS5) for probeset summarization and array normalization. Probe set summarization is performed by using a Tukey biweight estimator to robustly obtain the average difference between the PM and modified MM signals for each probe pair in the probe set. If PM > MM, then the modified value is just the MM, but if PM < MM, the MM is modified to ensure the difference is always positive. This approach is used as the MM is supposed to measure non-specific signal binding to the PM (i.e.

background signal), and thus should always be less than or equal to MM, however in practice this is often not the case.

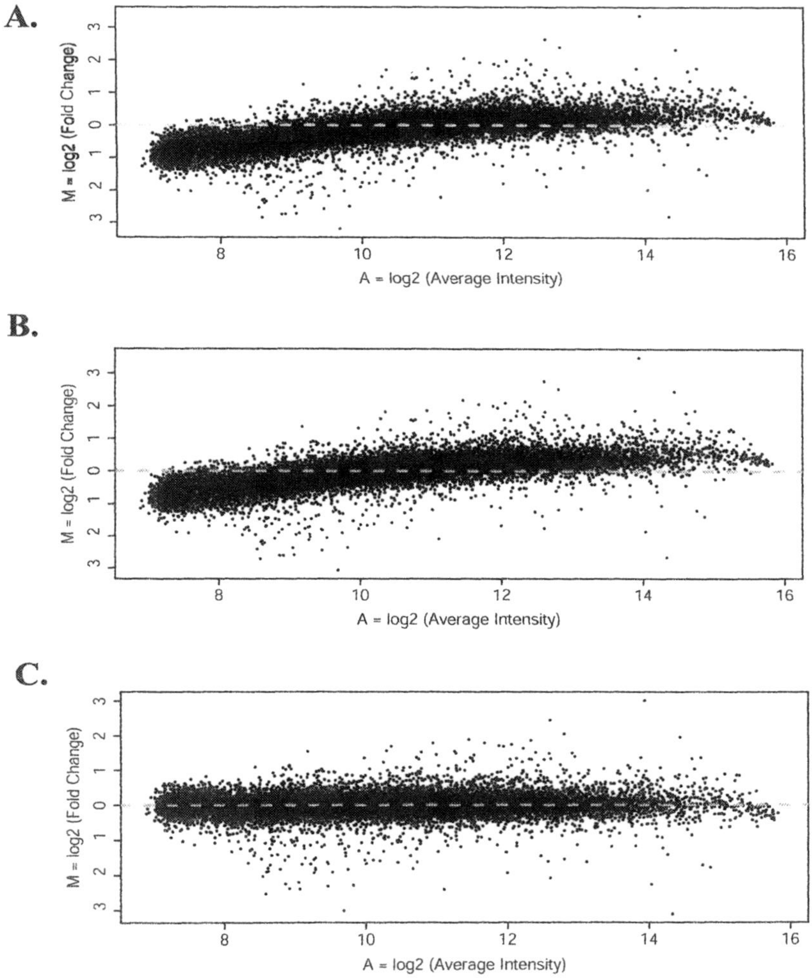

Fig. 3. Exploratory M vs. A plots are used to view the effect of normalization on two-colour microarray data. MA plots are a rotated plot of the deviation from red to green allowing easy visual identification of intensity dependent effects. A) displays a raw MA plot, demonstrating the need for normalization. B) displays the plot after global normalization as performed by the GenePix software. This normalization is generally not recommended as it simply shifts the median M values to zero and doesn't remove intensity dependent effects which are extremely common. C) displays the plot after performing the recommended print-tip intensity dependent loess normalization. This normalization approach involves fitting individual loess lines through each of the print tip groups on the array to bring the mean M in all print tip groups to zero.

Li and Wong (2001) noted that variation between probes in a probe set was often substantially (5 times or more) than the variation in values for a given probe across arrays. This massive within probe set variance thus advocates treatment of probe specific effects when trying to summarize probesets and Li and Wong (2001) advocated a multiplicative model in their dChip software. Irizarry et al. (2003a,b) used a series of experiments spiking in RNA of known concentrations to study probe sets effects and further noticed that many MM's showed concentration dependent effects (that is they were often sensitive to true signal in addition to any non specific background signal) and concluded that a more appropriate approach was a additive model, which they used in their highly successful RMA approach. RMA performs a background correction and normalization step on individual probes, before obtaining a probeset summary. The probeset summary comes from an additive model where the observed intensity in a probe is modeled as the sum of the true probeset expression value, a probe affinity term, and an error term. RMA then uses the estimate probeset expression value as its measure. Before considering the RMA approach we will examine the normalization system used in MAS5, which is based upon the summarized average difference values for each probe set.

The normalization approach used by Affymetrix in MAS5 is to scale intensities so that each array has the same average value. A reference array is defined (typically a control sample), and ratio of intensity in the reference to a given array for each probeset is obtained. The scaling factor which is applied to this array is the trimmed mean of these ratios. This method is obviously highly dependent upon the choice of a good reference array and does not perform well if there are non-linear relationships between arrays. To rectify the problem of non-linearity Schadt et al. (2001) and Li and Wong (2001) both propose normalization methods that make use of non-linear smooth curves, fitting a non-linear regression of the baseline array values onto the experimental array values. However these approaches also depend upon the choice of a suitable reference array.

Whilst developing the RMA method, Bolstad et al. (2003) considered both existing, and several new approaches to normalization. In particular they considered complete data methods, in which all available data (rather than pairs of arrays) are used to determine the normalization. Firstly they utilize a background correction method in which they estimate the observed PM signal as being due to true signal plus a noise signal (due to non-specific hybridization and optical noise). The observed distribution of all PM values on the $\log_2$ scale has a log normal form (normal + exponential decay) from which appropriate background values may be estimated.

Once data is background corrected the quantile normalization method analyses intensity distributions by performing pairwise comparisons of quantile-quantile plots for multiple arrays. Assuming that there is an underlying common distribution of intensities across arrays, the method then aims to give each array the same intensity distribution by taking the mean quantile and substituting it as the value of the data item in the original dataset (Bolstad et al. 2003). Bolstad et al. (2003) found that quantile normalization was able to reduce the variation of a probe set measure across multiple arrays to a greater degree than the Affymetrix scaling method and the non-linear method by Schadt et al. (2001). Specifically they found that performance of the quantile method was most favorable in terms of speed as well as bias and variability measures and thus it is the recommended normalization method for high-density oligonucleotide arrays.

The RMA method for normalization and probe set summary provides an approximate 5 fold reduction in variance compared to the MAS5 approach giving a massive increase in sensitivity allowing the detection of true differential expression. However, a slight bias tradeoff is made, in that RMA compresses fold change estimates by 10-20% compared to MAS5. This fold change compression has more recently been addressed in a updated version of RMA known as GCRMA (Wu and Irizarry 2004). This approach uses sequence specific models for background estimation, resulting in similar estimates of true signal with MAS5, whilst retaining the low variance. Thus for probe set summarization and array normalization we would highly advise using RMA or GCRMA.

Finally we should note that Affymetrix have recently updated their analysis algorithms, dropping their MAS5 approach and utilizing the PLIER algorithm. Whilst few details are available, the PLIER algorithm is a model based approach broadly similar to RMA, and thus represents a substantial improvement over MAS5.

6.2.2 Normalization of two-colour spotted arrays

A major source of variation affecting the analysis of two-colour microarray data is that arising from dye bias. Normalization methods therefore need to minimize this bias by balancing the fluorescence intensities of the green (Cy3) and red (Cy5) dyes, as dye bias and other variations can result in a shift in the average ratio of the Cy3 and Cy5 channels, thus the intensities may also need to be rescaled. To reveal the extent of dye bias it is useful to view M vs. A plots for each array (Figure 3). Intensity-dependent normalization may not be the only type of normalization required. Yang et al. (2002b) address three main forms of normalization being: within slide normalization; paired-slide normalization for dye-swap experiments and between-slide normalization.

Within-slide normalization methods, i.e. those that are applied to a single array, can be carried out by performing a form of location and intensity dependent normalization for each individual slide and one of the several forms of global normalization. As with the one-colour arrays, global normalization aims to correct the log-ratio values by subtracting a constant value, typically estimated from the mean or median M-values of a subset of genes whose expression is expected to remain constant. For two-colour arrays, global normalization makes the assumption that the red and green intensities can be related by a constant factor, with the aim being to shift the log-ratios to zero (Figure 3B). Sadly, despite there being evidence of spatial or intensity dependent dye biases in most experiments, global normalization methods are still generally the most widely used despite their inability to correct such types of variation (Yang et al. 2002b).

In general a more appropriate technique is the robust intensity-dependent Loess normalization method (Figure 3C) (Yang et al. 2002b). This method assumes the deviation from an M value of 0 varies in an intensity dependent way (i.e. over the range of A values observed. This is most obvious in MA plots where one observes curvature in the raw data (Figure 3A). At each value of A, a robust locally weighted regression line is obtained to robustly locate the central M value of the points. This value is then subtracted from all points at this value of A, thus shifting the central cluster of points back to the zero line (Figure 3C). Outliers (which in this case are differentially expressed genes) do not influence this calculation. Intensity-dependent Loess normalization may be applied globally over an array, or individually to each print tip group on an array.

This latter case may be necessary due to slight variations within the print-tips, the robotic spotter tip length or opening may vary during the array assembly process leading to spatial variation across the slide. Print tip intensity dependent loess also performs a de-facto spatial normalization as well, although this may occasionally perform poorly if there is strong intensity gradient within the print tip group (perhaps due to a local hybridization artifact). It is also possible to use spot quality weights in these methods, so poor quality spots do not influence the normalization procedure. In general it is recommended that print-tip intensity dependent loess is used as the default normalization method for two-colour microarrays. Occasionally more specialized forms of normalization are required such as 2D spatial normalization (Cui et al. 2003).

When dealing with replicate experiments, the relative gene expression levels may have different spreads in their log-ratios due to differences in experimental conditions. If significant, an adjustment of scale in which M-values from a series of arrays are scaled so that each array has the same median absolute deviation will be required to balance out the relative expression levels between experiments and hence between arrays (Yang et al. 2002b). After within-slide normalization, all normalized log-ratios will be centered around zero, regardless of the normalization method i.e. whether lowess or non-linear, however it is useful to examine boxplots displaying the spread of log-ratios for individual arrays to determine if scaling is required between arrays (Smyth et al. 2003). Failing to perform scale-normalization could lead to one or more slides having undue weight when averaging log-ratios across experiments to an average of log-ratios across slides. One common method of scale normalization is to divide each intensity by the total of the intensities on the slide, so that all slides then have the same total intensity. Another lowess normalization performed in a similar manner to that for GeneChip involves fitting a robust regression line through the M vs. A plot instead of a lowess curve.

An alternative form of normalization for two-colour arrays is the single-channel normalization method proposed by Yang and Thorne (2003) which allows for meaningful information to be individually obtained from the Cy3 and Cy5 channels of two-colour microarrays. This method removes systematic intensity bias that is not due to real gene expression separately from each channel both within and between arrays. Ultimately, single channel analysis allows for comparisons of absolute intensities between separate arrays for which no direct comparisons have been made. The cost is that single channel data from two-colour systems is considerably more noisy, requiring roughly four times the number of arrays that would be needed had a direct comparison been made.

6.3 Identifying Differentially Expressed Genes

A common aim of microarrays is to reliably identify differentially expressed genes between two conditions. This can be quantified by a *t*-test, which is simply a measure of the mean difference between conditions (mean M value), divided by the standard error in this difference (standard error in M = standard deviation/square root of n). One obtains significant *t* values if the absolute value of the ratio is large, as this implies that the observed mean difference is much larger than the variance in its measurement.

Genes identified as being differentially expressed are those that display a significant change in their expression between two samples of interest. Identification of differentially expressed genes within the normalized data set requires two steps; firstly selection and calculation of a suitable statistic for ranking the genes in order of evidence for differential expression from strongest to weakest and secondly selection of a suitable cut-off value for the ranking statistic where any gene having a value falling above the cut-off is considered to be differentially expressed. Although relatively simple in principle, in reality identifying differentially expressed genes is actually quite a complex problem due to the measured intensity values being affected by numerous sources of fluctuation and noise (Draghici et al., 2003).

A common, however flawed method, for ranking the genes is to simply consider the average fold change, or M values for each gene. Use of M values as the ranking statistic is a poor choice as it ignores any variability between replicates and there is no means by which to calculate the level of confidence you can have in regards to the supposed differential expression. Using simple fold change cut-offs can lead to an increased number of false-positives and false-negatives (Cui and Churchill, 2003).

Commonly used statistical methods that can be used to rank genes from replicated data are the Student's *t*-test and its many variations, one-way analysis of variance (ANOVA), empirical Bayes analysis and the Wilcoxon (or Mann-Whitney) test. A simulated comparative study by Troyanskaya et al. (2002) found that both the *t*-test and the Wilcoxon test resulted in a low number of false-positives while successfully identifying a large number of the differentially expressed genes. The Student's *t*-test, or simply *t*-test, is one of the simplest and most common methods that can be used to compare two conditions provided that true biological replication has been used in the experiment. In general, methods based on calculating the *t*-statistic are able to identify differentially expressed genes by examining the difference between the means, relative to the spread, or variance of the data by determining the ratio of the difference between two means and measuring the variability, or error variance, between the two data sets for each gene (Cui and Churchill, 2003). The ordinary *t*-statistic however is still not ideal in the context of microarrays as it is sensitive to genes with unusually low variance, resulting in an excessive number of false-positives in the list of differentially expressed genes. Genes identified as having a small estimated sample, or error variance, may still have a good chance of giving a large *t*-statistic even when they are not differentially expressed (Smyth et al. 2003). Smyth (2004) has developed a empirical Bayes based moderated *t*-test which produces reduced false-positive rates compared to the standard *t*, and more ad-hoc moderated *t*-tests such as that used in SAM. This moderated *t*-test was based on the B statistic developed by Lonnstedt and Speed

(2002). The B statistic is essentially a calculated log posterior odds ratio of differential expression versus non-differential expression that takes into account gene-specific variances while combining the information across many genes. Smyth (2004) extended the hierarchical model of Lonnstedt and Speed (2002), resetting the statistic in the context of general linear models with arbitrary coefficients and contrasts of interest. The hybrid classical/Bayes approach is proposed by Smyth (2004) in terms of moderated *t*-statistics, where the posterior odds of differential expression are shown to depend on the data through the moderated *t*-statistics. This approach can be further generalized to a moderated F-statistic, allowing for tests to be conducting that simultaneously involve two or more contrasts (Smyth 2004). Motivated by both of these preceding methods, Tai and Speed (2004) propose a one-sample multivariate empirical Bayes statistic (the *MB*-statistic) to rank genes from replicated microarray time course experiments.

ANOVA models, such as the classical F test, are basically generalization of the *t*-test that are more suitable for use when two or more conditions are to be compared and can be roughly divided into fixed, random and mixed effects models. The fixed-effects and mixed ANOVA models are generally a more powerful method to use when there are several sources of variation in the data and when consideration needs to be given to multiple factors (Cui and Churchill, 2003). Basically, ANOVA models make multiple estimations of variance in order to determine the overall level of variability within multi-factorial experiments by comparing the variation among replicated samples within and between conditions to determine differential expression. A novel method proposed by Draghici et al. (2003) makes use of a log-linear statistical model and an ANOVA approach to model the noise characteristic of multi-channel arrays. This model is then used to identify differentially expressed genes for a given confidence level (Draghici et al. 2003).

6.4 Determining Significance, False Positive and False Negatives

Once the genes have been suitably ranked the next step in the process is selection of a suitable cut-off value for the differential expression. At the same time the significance or confidence that can be given to the observed differential expression needs to be determined while giving someone allowance and control in regards to the amount of multiple testing needed to conduct a test for each gene such as controlling the family-wise error rate (FWER) or the false discovery rate (FDR) (Smyth, 2004). A simple, however informal, graphical method for assigning significance is to display the genes by their ranking statistic in a normal or *t*-distribution plot then selecting the genes whose points deviate markedly from the grouped bulk of genes. Depending on the user, manual selection of a cut-off value can typically result in either an over- or under-estimation in regards to differential expression.

False-positives and negatives can be respectively classed as being either a Type I or Type II error (Cui and Churchill, 2003). Calculation of both help determine the confidence that one can have in the results of their data, and in general both types of errors need to be balanced when selecting a cut-off value for differential expression. The problem of multiple testing is that it can increase the number of false-positives and false-negatives within the final list of differentially expressed genes. One of the most stringent approaches to the multiple testing problem is to control the FWER which determines the probability of accumulating one or more false-positives errors within the final list of differentially expressed genes, thus increasing confidence that the final list is free from such errors. The simplest procedure for controlling the FWER is the Bonferroni correction (Cui and Churchill, 2003) while Dudoit et al. (2000) propose a more rigorous method for controlling the FWER making use of a re-sampling method which computes a step-down adjusted p-value for each gene. A less stringent and possibly more powerful method for addressing the multiple testing problem is to control the FDR, which determines the expected proportion of false-positives within the list of differentially expressed genes. In contrast to methods that determine significance levels, the FDR is typically computed after the list of differentially expressed genes has been generated therefore providing a post-data method of confidence. Due to its low stringency, the FDR provides an increased number of genes identified as being truly differentially expressed then that of the FWER.

Finally we should note that p-values from microarray experiments are at best approximate. P-values should be seen as an evidence based ranking system. Genes with small p-values have strong evidence, whilst those with large values have weak evidence. The range of p-values gives us a measure of the relative confidence between those at the top of the list, and those further down. Experience has shown that approaches such as moderated t-tests, and FDR adjustments are on the right track, but that exact p-values should be treated with some caution. For this reason, the setting of p-value cut-offs is an arbitrary process, and one should perform exploratory plots before deciding on appropriate cut-off values.

6.5 Verification of Differential Expression

The final stage in the analysis of microarray data requires bioinformatics analysis of the final list of differentially expressed genes, to either characterize the non-annotated gene or to determine the functions and pathways that each gene is involved with. Results of the bioinformatics analysis will aid selecting the genes of most interest from within this final list. Selected genes will then need to have their differential expression confirmed by some of the more traditional or alternative methods for measuring gene expression such as northern blots, qRT-PCR, in situ-hybridization, ribonuclease protection assays, and serial analysis of gene expression (SAGE). Further biological studies may involve altering gene function with targeted mutations, antisense technology or protein inhibition. Ultimately the goal is to come to a conclusion to the biological question that was trying to be answered by the microarray experiment.

6.6 Introduction to Software

Currently there is a diverse range of public and commercially available software for the analysis of microarray data (Table 1). The most commonly used commercial analysis software systems are GeneSpring and GeneSite, neither will be discussed further. Many of these, particularly the commercial software, are represented by a Graphical User Interface (GUI), making analysis simpler and more accessible to a wide range of people by providing a predefined set of operations for the analysis of

Table 1. List of commercially and freely available analysis software for microarrays

Tool Name	Description	URL http://
GeneSpring	t-test, various clustering	www.sigenetics.com
GeneSight	t-test, various clustering, non-linear normalization	www.biodiscovery.com
R	environment for statistical computing	www.r-project.org
Bioconductor	set of microarray analysis tools	www.bioconductor.org
affy,	background correction, normalization, and probeset summarization for Affymetrix GeneChip™ arrays using robust multichip analysis (RMA)	www.bioconductor.org/packages
affylmGUI	Graphical User Interface for limma and affy packages	www.bioconductor.org/packages
affyPLM,	Quality control (NUSE, RLE) for affymetrix GeneChip™.	www.bioconductor.org/packages
GCRMA	background correction, normalization, and probeset summarization for Affymetrix GeneChip™ arrays using Gene Chip Robust Multichip Analysis (GCRMA)	www.bioconductor.org/packages
limma	Empirical Bayes analysis, normalization, analysis of two-colour and single colour arrays	www.bioconductor.org/packages
limmaGUI	Graphical User Interface for limma package	www.bioconductor.org/packages
marray	Diagnostic plots, reading data, normalization of two-colour arrays	www.bioconductor.org/packages
TM4	various normalization, t-test, Mann-Whitney test, clustering, dissimilarity measures and graphical options	www.tigr.org/software
BASE	LIMS and data analysis system for microarray experiments	base.thep.lu.se
Microarray Analysis Suite (MAS) version 5.0	t-test, Mann-Whitney test, various normalization	www.affymetrix.com
MAANOVA	ANOVA programs for microarray data. Diagnostics, normalization, ANOVA, Clustering (Matlab, R and Java)	www.jax.org/research/churchill/software/anova
Cyber-T	Differential gene expression, t-test or t-test with Bayesian framework	http://visitor.ics.uci.edu/genex/cybert/index.shtml
dChip	Differential gene expression, t-test, model-based analysis for GeneChips™	www.dchip.org
MAExplorer	Differential gene expression, scatter plots, k-means clustering dendrograms	www-lecb.ncifcrf.gov/MAExplorer

microarray data. Other programs are command-line driven, such as the set of microarray analysis tools provided by the Bioconductor project (Gentleman et al. 2004) that are designed for use with the R (Ihaka and Gentleman 1996) statistical computing environment. The Bioconductor project is an international initiative for the collaborative creation of extensible open development software for computational biology and bioinformatics. It contains many peer reviewed and award winning algorithms such as RMA which is part of the Affy package (Gautier et al. 2004), robust normalization for two colour microarrays as advocated by Yang and Speed (2002), and in the Marray package (Yang and Dudoit 2005) and linear modeling and empirical Bayes adjusted *t* and F-statistics (Smyth 2003) in the Limma package (Smyth et al. 2005). While some of the analysis software is platform specific, others can readily accept microarray data generated from one- and two-colour microarray platforms. R provides an extensive environment for detailed bioinformatics data mining of microarray datasets such as clustering, principal components analysis, chromosomal clustering, Gene Ontology clustering and over-representation analysis.

Affymetrix provides its own analytical software the Affymetrix Microarray Suite (Table 1), for the analysis of its GeneChip™ data; however a number of publicly available tools have been developed for the storage, management and analysis of Affymetrix probe level data, such as the Affy package of Bioconductor. Affy provides a number of algorithms for background correction such as the robust multichip analysis (RMA) method of Irizarry et al. (2003a,b), which performs background correction, normalization and probe set summarization as well as providing an implementation of Affymetrix's MAS 5.0 algorithm (Gautier et al. 2004). A number of the implemented methods are designed to produce a range of diagnostic plots for the data e.g., 2-D spatial images, boxplots and histograms.

Both the Marray and Limma R packages provide functions for the analysis of two-colour spotted microarray data, providing functions to produce diagnostic plots of spot statistics such as boxplots, scatter-plots, and spatial colour images. Specifically, Limma allows for the use of the empirical Bayes linear modeling approach described by Smyth (2004) for the analysis of designed experiments and identification of differentially expressed genes. Limma also permits appropriate treatment of two levels of replication such as duplicate spots printed on slides and replicated slides or alternatively, a mix of technical and biological replication.

Quality control of Affymetrix chips is provided in the Affy and affyPLM packages. These include RNA degradation plots, NUSE (normalized unscaled standard error) and RLE (Relative Log Error) box plots that are particularly useful for identifying poor quality arrays (Gautier et al. 2004; Bolstad 2005). The linear model functions of the Limma package and those for identifying differentially expressed genes are applicable to all microarrays platforms including Affymetrix GeneChips™ and other single-channel microarray experiments. The Marray package provides some alternative functions for reading and normalizing spotted microarray data, providing flexible location and scale normalization routines for log-ratios. These two packages have a reasonable level of overlap however the Limma package is based on a more general separation between within-array and between-array normalization (Smyth et al. 2005). Wettenhall and Smyth (2004) have also generated a graphical user interface model of the Limma package, limmaGUI (Wettenhall and Smyth 2004). A sister package to limmaGUI has been developed, affylmGUI (http://bioinf.wehi.edu.au/affylmGUI/R/library/affylmGUI), providing a GUI for analysis of Affymetrix microarray data. These GUI's provide a simple point-and-click interface to many of the commonly-used Limma and Affy functions, and provide automated construction of appropriate design and contrast matrices for analysis with Limma so users have to only specify the contrasts they wish to compare.

Some other publicly available alternatives to the open-source software of Bioconductor and R are the TM4 (Saeed et al. 2003) microarray analysis suite of Java-based tools and the BioArray Software Environment (BASE) (Saal et al. 2002) which provides a Web-based approach, using standard browsers to interact with a central microarray database and appropriate data analysis tools. The major advantage of the BASE approach is that it removes the need for users of the system to have to ensure that their software is current and the calculations underlying any analysis are passed on to more powerful central servers, keeping the user's desktop computers free.

When selecting the most suitable software to use for analysis of microarray data, the end choice will ultimately depend on the bioinformatics savvy and statistical knowledge of the user as well as the availability of money if considering the use of commercial software.

7. CONCLUSION

Since the advent of high-throughput microarray technology there has been a focus on improving this technology and developing suitable experimental designs and statistical algorithms for image processing, data cleaning, and identifying differentially expressed genes. In parallel, there has also been a focus on the development of user friendly software for the implementation of these algorithms and techniques for image processing and analysis phases of microarray experiments. This review has summarized the current status of microarray technology and issues concerning the experimental design, image processing and statistical analysis of microarray experiments while the proceeding chapter by Tjaden and Cohen will go into more detail regarding the statistical algorithms used for clustering microarrays.

Future directions of microarrays will follow in a similar manner with their being a strong focus on increasing the high-throughput capability of microarrays, improving protocols, improving the analysis of the vast quantities of raw gene expression data and the generation of user-friendly analysis software while reducing the overall cost of performing a microarray experiment. Although microarrays have been used for fungal research, virtually from as soon as the technology emerged, there is still huge potential for increasing our mycology knowledge base by utilizing microarray technology and then applying this knowledge within the pharmaceutical and agricultural industries via biotechnology based research.

Out of the 70 fungal genome projects currently accessible from NCBI (http://www.ncbi.nlm.nih.gov), 9 of these contain completed sequences of fungal genomes, 37 of the projects are in the process of assembling fungal genome sequence fragments and 29 of them are still in the process of sequencing. On top of this there are still many more fungal genome projects that are underway in both public and private laboratories that have not yet been made accessible. The sequence information from all of these projects can be used to develop fungal species and strain specific whole genome microarrays allowing for high-throughput gene expression studies on a genome-wide scale. Generation of whole-genome arrays will allow for a rapid elucidation of the cellular mechanisms that allow fungi to exist either alone or in a host-pathogen or host-symbiont relationship. Such studies may also lead to the identification of novel pathways by which fungi affect organisms such as plants or humans that can then be targeted by the pharmaceutical or agricultural industries for the development of drugs or fungicides for eradication of invasive, disease causing fungal pathogens. In regards to the symbiotic relationship of fungi with plants, whole-genome expression analysis studies will help in the elucidation, and possibly identification of cellular and novel mechanisms by which fungi are needed for growth of the host plant. Overall, utilization of microarrays within mycology will provide insight into the function of varying cellular functions of fungi at the gene level.

REFERENCES

Adams R and Bischof L (1994) Seeded region growing. IEEE Transactions on Information Technology in Biomedicine 16: 1651-1656.

Ahmed AA, Vias M, Iyer NG, Calsda C and Brenton JD (2004) Microarray segmentation methods significantly influence data precision. Nucleic Acids Research 32: e50.

Allen TD, Dawe AL and Nuss DL (2003) Use of cDNA microarrays to monitor transcriptional responses of the chestnut blight fungus *Cryphonectria parasitica* to infection by virulence-attenuating hypoviruses. Eukaryotic Cell 2: 1253-1265.

Arava Y, Wang Y, Storey JD, Liu CL, Brown PO and Herschlag D (2003) Genome-wide analysis of mRNA translation profiles in *Saccharomyces cerevisiae*. Proceedings of the National Academy of Sciences of the United States of America 100: 3889-3894.

Backhus LE, DeRisi J and Bisson LF (2001) Functional genomic analysis of a commercial wine strain of *Saccharomyces cerevisiae* under differing nitrogen conditions. FEMS Yeast Research 1: 111-125.

Baldi P and Hatfield W (2002) DNA Microarrays and Gene Expression From Experiments to Data Analysis and Modeling, Cambridge University Press, Cambridge, United Kingdom.

Baldwin D, Crane V and Rice D (1999) A comparison of gel-based, nylon filter and microarray techniques to detect differential RNA expression in plants. Current opinion in Plant Biology 2: 96-103.

Barczak A, Rodriques MW, Hanspers K, Koth LL, Tai YC, Bolstad BM, Speed TP and Erle DJ (2003) Spotted long oligonucleotide arrays for human gene expression arrays. Genome Research 13: 1775-1785.

Bernstein BE, Kamal M, Lindblad-Toh K, Bekiranov S, Bailey DK, Huebert DJ, McMahon S, Karlsson EK, Kulbokas EJ, 3rd, Gingeras TR, Schreiber SL and Lander ES (2005) Genomic maps and comparative analysis of histone modifications in human and mouse. Cell 120: 169-181.

Bertone P, Gerstein M and Snyder M (2005) Applications of DNA tiling arrays to experimental genome annotation and regulatory pathway discovery. Chromosome Research 13: 259-274.

Beucher S and Meyer F (1993) The morphological approach to segmentation: the watershed transformation. Mathematical morphology in image processing. Optical Engineering 34: 433-481.

Bolstad B (2005) affyPLM: Fitting Probe Level Models www.bioconductor.org/ repository/devel/ vignette/ AffyExtensions.pdf,

Bolstad BM, Irizarry RA, Astrand M and Speed TP (2003) A comparison of normalization methods for high density oligonucleotide array data based on variance and bias. Bioinformatics 19: 185-193.

Brenner S, Johnson M, Bridgham J, Golda G, Lloyd DH, Johnson D, Luo S, McCurdy S, Foy M, Ewan M, Roth R, George D, Eletr S, Albrecht G, Vermaas E, Williams SR, Moon K, Burcham T, Pallas M, DuBridge RB, Kirchner J, Fearon K, Mao J and Corcoran K (2000) Gene expression analysis by massively parallel signature sequencing (MPSS) on microbead arrays. Nature Biotechnology 18: 630-634.

Brewster JL, Beason KB, Eckdahl TT and Evans IM (2004) The Microarray Revolution. Biochemistry and Molecular Biology Education 32: 217-227.

Cawley S, Bekiranov S, Ng HH, Kapranov P, Sekinger EA, Kampa D, Piccolboni A, Sementchenko V, Cheng J, Williams AJ, Wheeler R, Wong B, Drenkow J, Yamanaka M, Patel S, Brubaker S, Tammana H, Helt G, Struhl K and Gingeras TR (2004) Unbiased mapping of transcription factor binding sites along human chromosomes 21 and 22 points to widespread regulation of noncoding RNAs. Cell 116: 499-509.

Chen Y, Dougherty ER and Bittner ML (1997) Ratio based decisions and the quantitative analysis of cDNA microarray images. Journal of Biomedical Optics 2: 364-374.
Cheung VG, Morley M, Aguilar F, Massimi A, Kucherlapati R and Childs G (1999) Making and reading microarrays. Nature Genetics 21: 15-19.
Churchill GA (2002) Fundamentals of experimental design for cDNA microarrays. Nature Genetics 32 Suppl: 490-495.
Colebatch G, Trevaskis B and Udvardi M (2002) Functional genomics: tools of the trade. New Phytologist 153: 27-36.
Cui X, Kerr MK and Churchill GA (2003) Transformations for cDNA Microarray Data. Statistical Applications in Genetics and Molecular Biology 2: 1-20.
DeRisi JL, Iyer VR and Brown PO (1997) Exploring the metabolic and genetic control of gene expression on a genomic scale. Science 278: 680-686.
Draghici S, Kulaeva O, Hoff B, Petrov A, Shams S and Tainsky MA (2003) Noise sampling method: an ANOVA approach allowing robust selection of differentially regulated genes measured by DNA microarrays. Bioinformatics 19: 1348-1359.
Dudoit S, Yang YH, Callow MJ and Speed TP (2000) Statistical methods for identifying differentially expressed genes in replicated cDNA microarray experiments. Department of Statistics, UC Berkeley, CA, pp. Technical Report 578.
Dunn B, Levine RP and Sherlock G (2005) Microarray karyotyping of commercial wine yeast strains reveals shared, as well as unique, genomic signatures. BMC Genomics 6.
Ekins R and Chu FW (1999) Microarrays: their origins and applications. comment. Trends in Biotechnology 17: 217-218.
Fodor SP, Read JL, Pirrung MC, Stryer L, Lu AT and Solas D (1991) Light-directed, spatially addressable parallel chemical synthesis. Science 251: 767-773.
Franken P and Requena N (2001) Analysis of gene expression in *arbuscular mycorrhizas*: new approaches and challenges. New Phytologist 150: 517-523.
Gautier L, Cope L, Bolstad BM and Irizarry RA (2004) affy - analysis of Affymetrix GeneChip data at the probe level. Bioinformatics 20: 307-315.
Gentleman RC, Carey VJ, Bates DM, Bolstad B, Dettling M, Dudoit S, Ellis B, Gautier L, Ge Y, Gentry J, Hornik K, Hothorn T, Huber W, Iacus S, Irizarry R, Leisch F, Li C, Maechler M, Rossini AJ, Sawitzki G, Smith C, Smyth G, Tierney L, Yang JYH and Zhang J (2004) Bioconductor: open software development for computational biology and bioinformatics. Genome Biology 5: R80.
Gerhold DL, Jensen RV and Gullans SR (2002) Better therapeutics through microarrays. Nature Genetics 32 Suppl: 547-551.
Giaever G, Chu AM, Ni L, Connelly C, Riles L, Veronneau S, Dow S, Lucau-Danila A, Anderson K, Andre B, Arkin AP, Astromoff A, El-Bakkoury M, Bangham R, Benito R, Brachat S, Campanaro S, Curtiss M, Davis K, Deutschbauer A, Entian KD, Flaherty P, Foury F, Garfinkel DJ, Gerstein M, Gotte D, Guldener U, Hegemann JH, Hempel S, Herman Z, Jaramillo DF, Kelly DE, Kelly SL, Kotter P, LaBonte D, Lamb DC, Lan N, Liang H, Liao H, Liu L, Luo C, Lussier M, Mao R, Menard P, Ooi SL, Revuelta JL, Roberts CJ, Rose M, Ross-Macdonald P, Scherens B, Schimmack G, Shafer B, Shoemaker DD, Sookhai-Mahadeo S, Storms RK, Strathern JN, Valle G, Voet M, Volckaert G, Wang CY, Ward TR, Wilhelmy J, Winzeler EA, Yang Y, Yen G, Youngman E, Yu K, Bussey H, Boeke JD, Snyder M, Philippsen P, Davis RW and Johnston M (2002) Functional profiling of the *Saccharomyces cerevisiae* genome. Nature 418: 387-391.
Glonek GF and Solomon PJ (2004) Factorial and time course designs for cDNA microarray experiments. Biostatistics 5: 89-111.
Ihaka R and Gentleman R (1996) R: Language for data analysis and graphics. Journal of Computational Graphics 5: 299-314.
Irizarry RA, Bolstad BM, Collin F, Cope LM, Hobbs B and Speed TP (2003a) Summaries of Affymetrix GeneChip probe level data. Nucleic Acids Research 31: e15.
Irizarry RA, Hobbs B, Collin F, Beazer-Barclay YD, Antonellis KJ, Scherf U and Speed TP (2003b) Exploration, normalization, and summaries of high density oligonucleotide array probe level data. Biostatistics 4: 249-264.
Irizarry RA, Warren D, Spencer F, Kim IF, Biswal S, Frank BC, Gabrielson E, Garcia JG, Geoghegan J, Germino G, Griffin C, Hilmer SC, Hoffman E, Jedlicka AE, Kawasaki E, Martinez-Murillo F, Morsberger L, Lee H, Petersen D, Quackenbush J, Scott A, Wilson M, Yang Y, Ye SQ and Yu W

(2005) Multiple-laboratory comparison of microarray platforms.[see comment]. Nature Methods 2: 345-350.

Jianbing F, Diping C, Chanfeng Z, Lixin Z and Wenyi F (2003) High-density fiber optic array technology and its applications in functional genomic studies. Chinese Science Bulletin 48: 1903-1905.

Kapranov P, Cawley SE, Drenkow J, Bekiranov S, Strausberg RL, Fodor SP and Gingeras TR (2002) Large-scale transcriptional activity in chromosomes 21 and 22. Science 296: 916-919.

Kazan K, Schenk PM, Wilson I and Manners JM (2001) DNA microarrays: new tools in the analysis of plant defence responses. Molecular Plant Pathology 2: 177-185.

Kehoe DM, Villand P and Somerville S (1999) DNA microarrays for studies of higher plants and other photosynthetic organisms. Trends in Plant Science 4: 38-41.

Kerr MK (2003) Design considerations for efficient and effective microarray studies. Biometrics 59: 822-828.

Kerr MK, Afshari CA, Bennett L, Bushel B, Martinez J, Walker NJ and Churchill GA (2002) Statistical analysis of a gene expression microarray experiment with replication. Statistica Sinica 12: 203-217.

Kerr MK and Churchill GA (2001) Experimental design for gene expression microarrays. Biostatistics 2: 183-201.

Kohane IS, Kho AT and Butte AJ (2003) Microarrays for an Integrative Genomics, The MIT Press Massachusetts Institute of Technology Cambridge, Massachusetts London, England.

Kricka LJ and Forina P (2001) Microarray technology and applications. Clinical Chemistry 47: 1479-1482.

Kuhn K, Baker SC, Chudin E, Lieu M, Oeser S, Bennett H, Rigault P, Barker D, McDaniel TK and Chee MS (2004) A novel, high-performance random array platform for quantitative gene expression profiling. Genome Research 14: 2347-2356.

Lashkari DA, DeRisi JL, McCusker JH, Namath AF, Gentile NC, Hwang SY, Brown PO and Davis RW (1997) Yeast microarrays for genome wide parallel genetic and gene expression analysis. Proceedings of the National Academy of Sciences of the United States of America 94: 13057-13062.

Li C and Wong WH (2001) Model-based analysis of oligonucleotide arrays: Expression index computation and outlier detection. Proceedings of the National Academy of Sciences of the United States of America 98: 31-36.

Lockhart DJ, Dong H, Byrne MC, Follettie MT, Gallo MV, Chee MS, Mittmann M, Wang C, Kobayashi M, Horton H and Brown EL (1996) Expression monitoring by hybridization to high-density oligonucleotide arrays. see comment. Nature Biotechnology 14: 1675-1680.

Lonnstedt I and Speed T (2002) Replicated microarray data. Statistica Sinica 12: 31-46.

Maindonald JH, Pittelkow YE and Wilson SR (2003) Some considerations for the design of microarray experiments. Science and Statistics 40: 367-390.

Mantripragada KK, Buckley PG, Stahl TD and Dumanski JP (2004) Genomic microarrays in the spotlight. Trends in Genetics 20: 87-94.

McLachlan GJ, Do K and Ambroise C (2004) Analyzing Microarray Gene Expression Data, John Wiley & Sons, Hoboken, New Jersey.

Newton M, Kendziorski C, Richmond C and Blattner F (2001) On differential variability of expression ratios: improving statistical inference about gene expression changes from microarray data. Journal of Computational Biology 8: 37-52.

Nicolaisen M, Justesen AF, Thrane U, Skouboe P and Holmstrom K (2005) An oligonucleotide microarray for the identification and differentiation of trichothecene producing and non-producing *Fusarium* species occurring on cereal grain. Journal of Microbiological Methods 62: 57-69.

Polsky-Cynkin R, Parsons GH, Allerdt L, Landes G, Davis G and Rashtchian A (1985) Use of DNA immobilized on plastic and agarose supports to detect DNA by sandwich hybridization. Clinical Chemistry 31: 1438-1443.

Rementeria A, Lopez-Molina N, Ludwig A, Vivanco AB, Bikandi J, Ponton J and Garaizar J (2005) Genes and molecules involved in *Aspergillus fumigatus* virulence. Revista Iberoamericana de Micologia 22: 1-23.

Ritchie ME (2004) Quantitative quality control and background correction for two-colour microarray data, Department of Medical Biology, The Walter and Eliza Hall Institute of Medical Research, University of Melbourne.

Rocke DM and Durbin B (2001) A model for measurement error for gene expression arrays. Journal of Computational Biology 8: 557-569.

Saal LH, Troein C, Vallon-Christersson J, Gruvberger S, Borg A and Petersen C (2002) BioArray Software Environment (BASE): a platform for comprehensive management and analysis of microarray data. Genome Biology 3: software0003.0001-0003.0006.

Saeed AI, Sharov J, White J, Li J, Liang W, Bhagabati N, Braisted J, Klapa M, Currier T, Thiagarajan M, Sturn A, Snuffin M, Rezantsev A, Popov D, Ryltsov A, Kostukovich E, Borisovsky I, Liu Z, Vinsavich A, Trush V and Quackenbush J (2003) TM4: A Free, Open-Source System for Microarray Data Management and Analysis. Biotechniques 34: 374-378.

Schadt EE, Li C, Ellis B and Wong WH (2001) Feature Extraction and Normalization Algorithms for High-Density Oligonucleotide Gene Expression Array Data. Journal of Cellular Biochemistry Supplement 37: 120-125.

Schena M, Shalon D, Davis RW and Brown PO (1995) Quantitative monitoring of gene expression patterns with a complementary DNA microarray. see comment. Science 270: 467-470.

Schenk PM, Kazan K, Wilson I, Anderson JP, Richmond T, Somerville SC and Manners JM (2000) Coordinated plant defense responses in *Arabidopsis* revealed by microarray analysis. Proceedings of the National Academy of Sciences of the United States of America 97: 11655-11660.

Shepard KA, Gerber AP, Jambhekar A, Takizawa PA, Brown PO, Herschlag D, DeRisi JL and Vale RD (2003) Widespread cytoplasmic mRNA transport in yeast: identification of 22 bud-localized transcripts using DNA microarray analysis. Proceedings of the National Academy of Sciences of the United States of America 100: 11429-11434.

Simon RM and Dobbin K (2003) Experimental design of DNA microarray experiments. Biotechniques Suppl: 16-21.

Smyth GK (2004) Linear models and empirical Bayes for assessing differential expression in microarray experiments. Statistical Applications in Genetics and Molecular Biology 3: Article 1.

Smyth G, Thorne N and Wettenhall J (2005) limma: Linear Models for Microarray Data User's Guide, http://bioinf.wehi.edu.au/limma/usersguide.pdf,

Smyth GK, Yang YH and Speed T (2003) Statistical issues in cDNA microarray data analysis. Methods in Molecular Biology 224: 111-136.

Stolovitzky GA, Kundaje A, Held GA, Duggar KH, Haudenschild CD, Zhou D, Vasicek TJ, Smith KD, Aderem A and Roach JC (2005) Statistical analysis of MPSS measurements: Application to the study of LPS-activated macrophage gene expression. Proceedings of the National Academy of Sciences of the United States of America 102: 1402-1407.

Tenenbaum SA, Carson CC, Lager PJ and Keene JD (2000) Identifying mRNA subsets in messenger ribonucleoprotein complexes by using cDNA arrays. Proceedings of the National Academy of Sciences of the United States of America 97: 14085-14090.

Tai YC and Speed T (2004) A multivariate empirical Bayes statistic for replicated microarray time course data. Department of Statistics, University of California, Berkeley, (submitted), pp. Technical Report #667.

Troyanskaya OG, Garber ME, Brown PO, Botstein D and Altman RB (2002) Nonparametric methods for identifying differentially expressed genes in microarray data. Bioinformatics 18: 1454-1461.

Voiblet C, Duplessis S, Encelot N and Martin F (2000) Identification of symbiosis-regulated genes in *Eucalyptus globulus-Pisolithus tinctorius ectomycorrhiza* by differential hybridization of arrayed cDNAs. The Plant Journal 25: 181-191.

Wang D, Urisman A, Liu Y, Springer M, Ksiazek TG, Erdman DD, Mardis ER, Hickenbotham M, Magrini V, Eldred J, Latreille JP, Wilson RK, Ganem D and DeRisi JL (2003) Viral Discovery and Sequence Recovery Using DNA Microarrays. PLoS Biology 1: 257-260.

Wettenhall JM and Smyth GK (2004) limmaGUI: A graphical user interface for linear modeling of microarray data. Bioinformatics 20: 3705-3706.

Wu Z and Irizarry RA (2004) Preprocessing of oligonucleotide array data. Nature Biotechnology 22: 656-658.

Yang YH, Buckley MJ, Dudoit S and Speed TP (2002a) Comparison of methods for image analysis on cDNA Microarray Data. Journal of Computational & Graphical Statistics 11: 108-136.

Yang YH, Buckley MJ and Speed TP (2001) Analysis of cDNA microarray images. Briefings in Bioinformatics 2: 341-349.

Yang YH and Dudoit S (2005) Bioconductor's marray package: Plotting component, www.bioconductor.org/repository/devel/vignette/marrayPlots.pdf,

Yang YH, Dudoit S, Luu P, Lin DM, Peng V, Ngai J and Speed TP (2002b) Normalization for cDNA microarray data: a robust composite method addressing single and multiple slide systematic variation. Nucleic Acids Research 30: e15.

Yang YH and Speed T (2002) Design Issues for cDNA Microarray Experiments. Nature Reviews 3: 579-588.

Yang YH and Thorne NP (2003) Normalization for Two-color cDNA Microarray Data.

Yauk CL, Berndt ML, Williams A and Douglas GR (2004) Comprehensive comparison of six microarray technologies. Nucleic Acids Research 32: 1-7.

Zhou X, Rao NP, Cole SW, Mok SC, Chen Z and Wong DT (2005) Progress in concurrent analysis of loss of heterozygosity and comparative genomic hybridization utilizing high density single nucleotide polymorphism arrays. Cancer Genetics and Cytogenetics 159: 53-57.

ELSEVIER

Applied Mycology and Biotechnology
An International Series
Volume 6. Bioinformatics

2

Methods for Protein Homology Modelling

Melissa R. Pitman and R. Ian Menz
School of Biological Sciences, Flinders University, South Australia.
(ian.menz@flinders.edu.au)

Homology modelling has become a useful tool for the prediction of protein structure when only sequence data are available. Structural information is often more valuable than sequence alone for determining protein function. Homology modelling is potentially a very useful tool for the mycologist, as the number of fungal gene sequences available has exploded in recent years, whilst the number of experimentally determined fungal protein structures remains low. Programs available for homology modelling utilise different approaches and methods to produce the final model. Within each step of the homology modelling process, many factors affect the quality of the model produced, and appropriate selection of the program can significantly improve the quality of the model. This review discusses the advantages and limitations of the currently available methods and programs and provides a starting point for novices wishing to create a structural model. We have taken a practical approach as we hope to enable any scientist to utilise homology modelling as a tool for the analysis of their protein, or genome, of interest.

1. INTRODUCTION

Over the last decade, the number of gene sequences available has increased exponentially, as genomes of organisms from all kingdoms have been sequenced, including close to 70 fungal and over 100 animal species, including humans. To deal with these advancements, there has been an explosion in the research and development of software to organise and analyse the genome sequence databases. However, a full understanding of the importance of this genomic information cannot be gained until the functions of all the gene products are determined.

The function of a protein is primarily dictated by its three dimensional structure, but methods for determining the three dimensional structure of a protein are time-consuming and expensive. The process of structure determination commonly includes development of a protein expression system, protein purification,

Corresponding author: R. Ian Menz

crystallisation and finally structure determination, where each successive step may take years to accomplish. For this reason, although the number of protein sequences available has increased exponentially, the number of experimentally derived protein structures lags far behind. For example, although there are more than 27,000 protein sequences in the NCBI database for *Neurospora crassa*, the first filamentous fungal genome to be sequenced, two years after completion of the genome sequence, the Protein Data Bank (PDB) structural database contains only nine *N. crassa* protein structures.

Over several decades there has been extensive research into *in silico* (computer) methods for structure determination. The ultimate aim of this approach is the development of a method for determining the 3D structure of a protein from the sequence alone. One strategy, known as homology modelling, utilises the redundancy of protein structure by using homologous proteins, or structurally related proteins belonging to the same family, to predict the structure of an unknown protein. Although there are many millions of proteins, the number of unique structural folds is two to three orders of magnitude lower (Xu 2003). The assumption is that all members of a protein family are related by divergent evolution from a common ancestor and must therefore share the same basic fold. Thus if a protein belongs to a family in which the structures of several proteins have been determined empirically, an atomic model can be built by comparison with those structures. The structural genomics initiatives aim to characterise most protein sequences by an efficient combination of targeted high-throughput experimental structure determination and prediction (Baker et al. 2003), suggesting that homology modelling will become an increasingly important tool for biologists.

Applications for protein structures produced by homology modelling include identification of regions of importance within a protein for further experimental studies such as mutation analysis. Furthermore, if homology modelling is combined with other computational methods such as ligand docking, the models produced can be used to screen proteins for potential interaction with substrates, inhibitors or co-factors, hence aiding in functional analysis. Such methods have been essential in pharmacology and functional genomics applications. One of the advantages of computational methods for structure prediction is that whole genomes can be analysed. For example, in a large-scale protein structure modelling project based on the *Saccharomyces cerevisiae* genome, 1,071 protein sequences were modelled using 236 proteins of known structure (Sanchez and Sali 1998).

The following section outlines the general steps involved in homology modelling whilst the third section focuses on the practical aspects of protein homology modelling. The final section includes considerations for modelling fungal proteins.

2. HOMOLOGY MODELLING

Modelling programs fall into two major categories: user-based, and fully automated. In the user-based, semi-automated programs the user is required to take a hands-on approach utilising software to run the process locally, while the fully automated "blackbox" systems use remote software for model production *via* a server. Semi-automated modelling requires more user input and so our discussion of the modelling steps will be focussed on the user-based approach. The fully automated modelling servers use a similar overall approach, and will be further discussed in section 3.

2.1. General Steps in Homology Modelling

There are four major steps in protein homology modelling (Figure 1). The first step is to identify protein structure(s) to act as template(s). Secondly, the sequence of the protein of known structure is aligned with the protein to be modelled (the target sequence). Thirdly, the alignment is used to guide how the target sequence is overlayed on the 3D-coordinates of the template structures to generate the initial model. Finally, the model is optimised using structural, stereochemical and energy calculation techniques. Often, this process is repeated until a suitable model is obtained. The main difference between the various modelling methods is how the 3D model is calculated from the alignment.

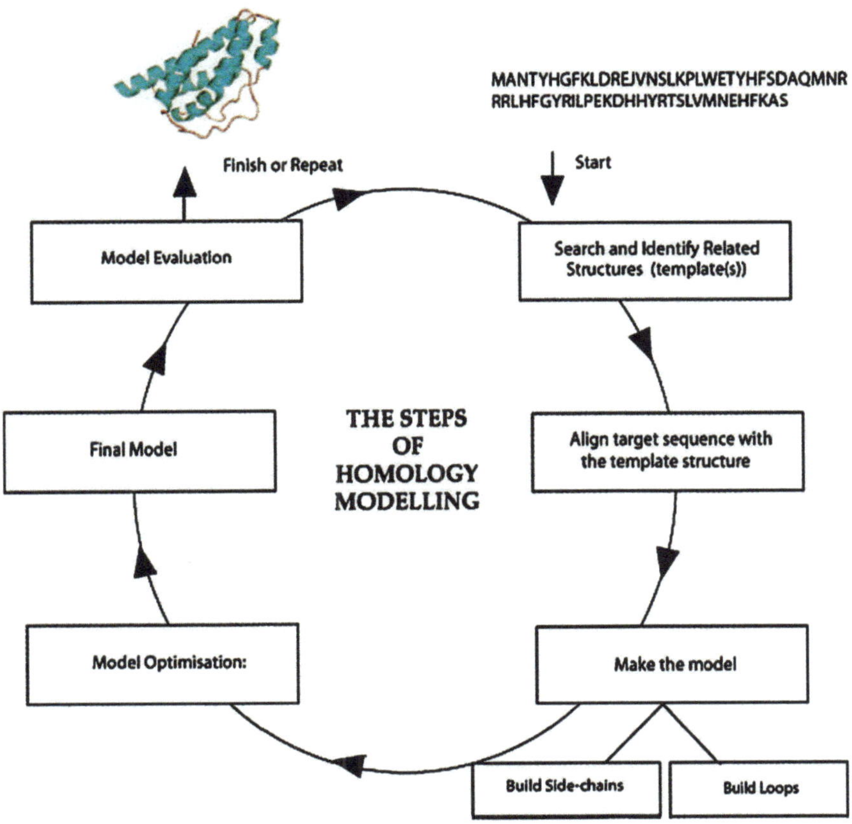

Fig. 1. The steps involved in homology modelling

2.2. Identification of Template Structures

Homology modelling requires at least one sequence of known structure with significant amino acid sequence similarity to the target sequence (Peitsch 2002). In order to find suitable templates, the target sequence is used to search a protein structure database for homologous proteins. As a general rule for homology modelling, the minimum percentage of amino acid sequence identity required between the target and template is 30% (Rost 1999). Below 25% sequence identity it is difficult to assume common ancestry and hence homology by sequence alone (Chung and Subbiah 1996). In most cases, the higher the sequence identity, the more accurate the model and use of more than one template structure in the modelling process can often improve accuracy. It has been well established that the majority of errors in models arise from errors in the initial alignment of the target and template sequences, making the alignment the most important step in the overall process. If structural homologs are known, for example from structural classification databases such as SCOP, CATH or FSSP, then the homologs can be retrieved directly from the PDB. Alternatively, if only the target protein sequence is known, then proteins with homology and whose structure have been determined, can be identified by performing a BLAST search using the interface provided on the NCBI website.

Functionally important similarities between proteins are not always evident from comparison of the raw sequences and may only be recognisable by comparison of the three-dimensional structures. Consequently, many proteins of known structure that could potentially share structural similarity with the target sequence are overlooked as template structures because they share little sequence homology with the target sequence. To address this problem, profile methods have been developed, which identify patterns of conservation from alignment of related sequences and use these patterns to find proteins with more distant similarity (Altschul and Koonin 1998). Profile-based methods may prove to be beneficial in increasing the accuracy of detection of homologs and have been employed in the program PSI-BLAST (Altschul et al. 1997).

The process of finding template structures can also be difficult if the target protein has a unique function or is a membrane protein. Although membrane proteins represent 30-40% of the proteins expressed by a cell, they are grossly under-represented in the protein structure database, making up only 2% of the protein structures determined. As the number of known membrane protein structures increases due to structural genomics efforts the number of potential templates are likely to improve.

2.3. Alignment of the Template and Target Sequences

The alignment of template and target sequences is the most important step in the modelling process, as the accuracy of the final model is heavily influenced by this step. If the level of sequence identity is low (~30%), it can be beneficial to align the target sequence with protein sequences of other family members, even if their structures are not available, in order to ensure regions of functional or structural importance are aligned correctly with the template sequence. An example to illustrate the importance of using a multiple sequence alignment is shown below (Fig. 2).

In some cases, the modelling program is able to produce a multiple sequence alignment from the sequences used as input, however in cases of low sequence identity it may be preferable to use other alignment methods (programs) that allow for manipulation of parameters, such as gap penalties, to ensure that errors are avoided. If a multiple sequence alignment is used and includes members of the protein family it may be useful to utilise any experimental information to assess the quality of the sequence alignment or manually alter the alignment. Alignment programs such as CLUSTALX (Thompson et al. 1997) and PileUp (Edelman et al. 1994) can be used to produce a multiple sequence alignment.

In the alignment of JMJMJMJM and BWBWBWBW there are three possibilities:

```
J M J M J M J M
| | | | | | | |
B W B W B W B W
```

or

```
J M J M J M J M
  | | | | | | |
  B W B W B W B W
```

or

```
  J M J M J M J M
  | | | | | | |
B W B W B W B W
```

If you add another sequence with some homology, the alignment becomes more accurate.

```
J M J M J M J M
  | | | | | | |
  M B M B M B M B
  | | | | | | |
B W B W B W B W
```

Therefore, in regions of low sequence homology i.e. loops it may be beneficial to include other sequences from the protein family to improve the accuracy of the alignment.

Fig. 2. Explanation of a pathological alignment problem. The original sequences are hard to align unless a third homologous sequence is included. Adapted from (Bourne and Weissig 2003).

2.4. Model Production

There are three overall approaches to homology modelling, fragment-based assembly, segment-matching methods and satisfaction of spatial restraints, each of which is similarly accurate if used optimally (Fiser and Sali 2001). Specific examples of modelling programs that utilise the different approaches will be discussed in section 3.2. Separate procedures are required to model loops and side-chains.

2.4.1. Fragment based methods

This method, also known as rigid body assembly, is the first method developed for homology modelling and is still widely used. Fragment based methods use the alignment of template and target sequence to identify structurally conserved regions (SCRs). SCRs tend to be structural elements such as alpha helices or beta strands and typically include regions of functional importance such as the active site of an enzyme. The regions between SCRs, which tend to have lower sequence similarity, are assigned as variable regions (VRs) and generally comprise the loop structures. Once the SCRs have been assigned to the template sequence, the SCR coordinates are copied onto the corresponding residues in the target structure. Using more than one template structure to construct the framework has been shown to increase the accuracy of the model produced (Srinivasan and Blundell 1993; Sali 1995). The benefit of this approach is that the regions of structural conservation have good geometry and require minimal optimisation.

2.4.2. Segment matching methods

Segment matching methods are based on the observation (Unger et al. 1989) that most hexa-peptide segments of protein structure can be clustered into about 100 classes (Marti-Renom et al. 2000). Such methods assemble short segments from template structures to construct the model (Sali 1995). From the template-target sequence alignment the template coordinates for conserved segments are copied onto the target. To connect the gaps, the program splits the target structure into a set of short segments and searches the database for segments that match the framework of the target structure. The matching is based on three criteria: sequence similarity, conformational similarity, and compatibility with the target structure using van der Waal's interactions (Wallner and Elofsson 2005). In some programs such as SegMod, the backbone and side-chains are constructed simultaneously using this approach. As this method implements a database search of segments, insertions and deletions in the target structure can also be modelled (Marti-Renom et al. 2000). Some side-chain and loop modelling can be seen as segment matching because an analogous method is employed.

2.4.3. Satisfaction of spatial restraints

Restraint based homology modelling methods generally treat the model as a whole instead of breaking it into specific regions, as is the case with the other approaches. The template structures are used to produce geometric and biochemical restraints, such as limits on distances between pairs of Cα atoms and ranges of backbone and side-chain dihedral angles. The homology-derived restraints are usually supplemented by stereochemical restraints on bond lengths, bond angles, dihedral angles, and nonbonded atom-atom contacts obtained from a molecular mechanics force field (Marti-Renom et al. 2000). The positions of the atoms within the model are manipulated to generate a model that best fits the restraints.

2.4.4. Loop and side-chain modelling

The procedures used to produce the final model depend on which modelling method was used to generate the backbone structure. If the modelling program is based on fragment-based methods, then the polypeptide backbone for the SCRs is built as previously described, but the loops and potentially the side-chains have to be modelled by another mechanism. In the spatial restraints method, the loops are generally included in the restraints built from the template, but the side-chains are added to the backbone by a separate mechanism. However, if the loops are poorly conserved, they can be modelled separately using a loop modelling method.

2.4.4.1. Loop modelling

Although some loops are functionally active and thus are relatively highly conserved, most loops have no function other than to connect secondary structural elements such as helices and sheets and are generally regions of low sequence conservation. Consequently, corresponding loops in related proteins may adopt significantly different conformations. Therefore, loop modelling can be seen as a mini protein-folding problem, where the conformation of the loop has to be calculated mainly from the sequence information (Fiser et al. 2000). However, since short segments of sequence usually do not provide sufficient information to determine structure, the regions surrounding the loop, the core stem regions that span the loop and the structure that surrounds the loop, must all be considered in the loop modelling process. Loop modelling methods generally fall into two basic groups: database search methods and *ab initio* methods.

Database search methods identify a segment of main-chain that fits the two stem regions flanking a loop, but are not part of it (Fiser et al. 2000). The loop database contains the sequence and structure of loops determined from all known protein structures. The database is searched to find many different alternative segments that fit the stem residues and the selected segments are then sorted according to geometric criteria or sequence similarity between the template and target loop sequences. The selected segments are then superimposed and annealed on the stem regions. After this procedure, the predicted loop structures require optimisation to improve the overall conformation.

Database methods are considered more accurate than *ab initio* methods but as the loop length increases, so does the number of geometrically possible conformations, and the efficiency of the database search is reduced. So, only for loops of seven residues or less are most of the conceivable conformations present in the database of known protein structures (Fidelis et al. 1994). Fortunately, when families of homologous proteins are analysed, insertions longer than eight residues are rare (Pascarella and Argos 1992; Benner et al. 1993; Flores et al. 1993; Sali 1995). As the number of known structures increases, the number of known loop structures will increase and hence the accuracy of database loop modelling methods will improve.

In *ab initio* methods the structure of the loop is predicted based on a conformational search of the space to be filled. This prediction process is guided by a scoring or energy function for the suitability of the loop produced. There are many different methods available, which differ in the search algorithms, energy functions (to score the results of the searches), and optimisation algorithms used. An extensive list of these search algorithms and optimisation techniques has been published previously (Sali 1995; Contreras-Moreira et al. 2002) and specific examples will not be dis+cussed in this chapter. Generally, *ab initio* methods are efficient at modelling smaller loop regions but for larger loops, substantial numbers of loop configurations need to be generated to fully sample the conformational space, thus limiting the efficiency of the method.

2.4.4.2. Side-chain modelling

The general approach for the modelling program is to place the target side-chains as similarly as possible to the corresponding template side-chains, but in many cases this is not feasible due to amino acid differences between the target and the template. In these cases, libraries of possible side-chain conformations or 'rotamers' are used to find a likely conformation for the side-chain. This approach is based on the general observation that the most frequently observed rotamers tend to be the most energetically favoured. The rotamer databases are usually in the form of side-chain torsional angles for preferred conformations of a particular side-chain (Al-Lazikani et al. 2001). When the side-chain to be modelled is much larger than the template structure, there is a high possibility of steric conflicts (or clashes) which need to be addressed during model optimisation.

For each side-chain to be modelled, the possible rotamers must be assembled, sorted and selected, based on particular criteria. A number of approaches have been applied for rotamer search procedures, all of which yield similar results (Xiang and Honig 2001). The main differences are in how the initial conformation is selected and in the criteria used to select the conformations. The accuracy of side-chain modelling depends on the rotamer library used, the choice of force-field used to optimise the conformation, combinatorial complexities, the quality of the protein backbone and bond angle and length parameters (Xiang and Honig 2001). Because greater constraints are imposed on side-chains in buried regions of the protein, these are predicted with more accuracy than those that lie on the surface (Chakravarty et al. 2005). For accurate modelling of exposed residues, it is necessary to simulate a force field to mimic constraints such as solvent effects.

2.5. Model Refinement

Model refinement involves idealisation of bond geometry and removal of unfavourable non-bonded contacts (Peitsch 2002). Energy minimisation packages such as CHARMM, AMBER or GROMOS are usually incorporated into the modelling programs to facilitate model optimisation. Energy minimisation methods have a small radius of convergence; the atoms are only moved within a small area to find the local energy minimum. This is mainly used to remove steric clashes, such as clashes between side-chains, and ensures sensible covalent geometry is maintained around each atom (Contreras-Moreira et al. 2002). In comparison another energy minimisation technique, molecular dynamics, allows a larger deviation of the atom from its original position in order to find the global energy minimum. Molecular dynamics (or conformational sampling) is used for structural optimisation by overcoming energy barriers separating local energy minima (Leach 1999).

2.5.1. Energy minimisation

The landscape of a protein molecule possesses an enormous number of energy minima, but the goal of energy minimization is to find only the local energy minimum around a particular conformation. The energy at this local minimum may be much higher than the energy of the global minimum but the benefit is that only moderate changes are made in the position of the atom. This process can be used to relieve strain in models where loops and side-chains were placed in poor conformations during the model building process. Every minimisation cycle has the potential to rectify significant stereochemistry errors in the model by adjusting short distances between atoms, but the cost may be the introduction of many less significant errors, moving the structure away from the original model after many cycles. Thus, current modelling programs either restrain the atom positions during the process and/or apply only a few hundred steps of energy minimisation (Bourne and Weissig 2003).

2.5.2. Molecular dynamics

Molecular dynamics simulates the natural motion of the molecular system. The energy provided in a molecular dynamics procedure allows the atoms to move and even collide into neighbouring atoms. This is a form of conformational searching since if enough thermal energy is provided, the molecule will be able to cross the energy barriers that separate local minima on the conformational potential energy surface for that molecule (Leach 1999). Simulated annealing is a type of molecular dynamics experiment which is popular when optimising protein models. In this process you simulate a higher temperature, which allows the state of the system to alter, and then lower the simulated temperature to bring the system back to a more stable state, sampling a large conformational space. The cycle is repeated several times so that multiple conformations can be obtained and later analysed. Molecular dynamics simulations on a 10-100nsec time scale perform well with an explicit representation of the protein and solvent environment (Fan and Mark 2004). However, too many cycles of molecular dynamics will shift the model away from the original target and hence potentially degrade the quality of the model.

2.6. Model Evaluation

In evaluating the model there are many different aspects to consider; the residue placement, the interaction of neighbouring residues and the atoms within the residues. One of the main considerations is the stereochemical properties of the model, which includes analysis of properties such as bond lengths, correct chirality, correct ring structure and other geometric properties. Physical properties must also be assessed such as favourable packing within the model and non-clashing non-bonded atoms (no "bad contacts"). The model also needs to have reasonable amino acid geometry which can be assessed by a Ramachandran Plot. General protein properties need to be assessed, for example does the model contain multiple unusual side-chain conformations, buried charges, or residues that are overly strained in their environment. While many of these types of faults may have been resolved to a degree during the optimisation process, errors can still remain. Model evaluation programs analyse these properties and are designed to highlight regions that need further optimisation, often by manual adjustment. There are two main types of model evaluation program those which assess stereochemical properties and those which assess spatial properties. Finally, the model must be able to support all the existing biochemical data that has been elucidated for the target protein. This functional analysis can only be achieved by manual inspection of the model.

2.6.1. Evaluating stereochemical properties

The main basic requirement for a protein model is correct stereochemistry. Validation programs check for anomalies, such as phi/psi angle combinations that are placed in disallowed regions, steric collisions, and unfavourable bond lengths and angles. Programs such as PROCHECK (Laskowski 1993) and WHATCHECK (Hooft et al. 1996) analyse these stereochemical features of the residues in the model and give an evaluation of the overall quality of a model or structure. Analysis of bond geometry by looking at Ramachandran plots is important in order to highlight unrealistic conformations within the model. Certain conformations of phi and psi angles are forbidden in protein structures because they result in steric hindrance, or clashes between atoms. A good model will generally have 90% of its residues in the allowable regions of a Ramachandran plot (Laskowski 1993).

2.6.2. Evaluating spatial properties

Spatial features such as formation of a hydrophobic core, residue and solvent accessibilities, packing and spatial distribution of charged groups, can also be used to evaluate the model (Marti-Renom et al. 2000). Programs that assess these types of parameters include PROSAII (Sippl 1993), ANOLEA (Melo et al. 1997) and VERIFY3D (Eisenberg et al. 1997). These programs evaluate the environment of each residue in a model with respect to the expected environment as found in high-resolution X-ray structures. Verify 3D analyses the 3D-1D profile of a protein structure, which involves the statistical preferences for the following criteria: the area of the residue that is buried, the fraction of side-chain area that is covered by polar atoms (oxygen and nitrogen) and the local secondary structure (Eisenberg et al. 1997). PROSAII relies on empirical energy potentials derived from the pair wise interactions observed in well defined protein structures (Sippl 1993). The main limitation of this method is that it relies on energy calculations and the contributions of individual residues to the overall free energy of folding vary considerably, even when normalised by the number of atoms or interactions made (Marti-Renom et al. 2000).

2.6.3. Manual inspection

The validation process includes manual inspection of the protein model to ensure that the model supports any experimental data. This often entails superimposing the model with the template structures for comparison. Software such as the

SUPERPOSE module of the CCP4 (Collaborative Computational Project 1994) suite of crystallography programs, and Swiss-PDB Viewer perform structural alignments of the model with other similar structures, such as the templates. Commercial homology modelling programs often include their own model evaluation software i.e. ProTable in SYBYL (Clark et al. 1989). The quality of the superposition process is generally measured by a root mean square deviation (RMSD) value, which is the sum of the squared distance between each corresponding Cα atom position in the two structures following superposition. The core Cα atoms of protein models which share 35-50% sequence identity with their templates, will generally deviate by 1.0-1.5 Å from their experimental counter parts (Chothia and Lesk 1986; Peitsch 2002). Manual inspection and manipulation of the model can be performed using molecular graphics software such as O (Jones et al. 1999), Swiss-PDB Viewer (Guex and Peitsch 1997) and Pymol (DeLano 2002). Manual manipulation and visualisation are one of the most important steps to determine the accuracy of the model and to check if the model matches observed experimental data. This process may include altering side-chain rotamers to match a template structure or employing docking programs such as AUTODOCK (Morris et al. 1998), ICM-Dock (Abagyan et al. 1997) or GOLD (Verdonk et al. 2003) to dock known substrates into the active site or known protein-binding molecules to the surface of the model.

2.7. Limitations of Homology Modelling

There have been major advancements in modelling programs in the last decade, however, there are still many areas where homology modelling could be improved. The main contributor to errors in homology modelling is the underlying complexity of proteins; "there is a fine balance of competing interactions between the solvent and the protein as well as alternate packing arrangements of side-chains that cannot be easily captured in simplified representations" (Fan and Mark 2004). Although X-ray crystal structures are seen as the ideal, it should not be forgotten that these can also contain errors. Protein structures are flexible and can exhibit different conformations depending on their environment. To add further uncertainty, the template structure used may contain errors, which are subsequently incorporated into the resulting model. This can mainly be avoided by using structures with higher resolutions or by using more than one template.

One of the major limitations of homology modelling is that the integrity of the model is almost completely reliant on the sequence alignment and therefore, the level of sequence identity between the template and target structures. All modelling programs or methods, will generate erroneous results if the sequence alignment is incorrect. The alignment problem further extends to the loop modelling and side-chain modelling methods as these processes are strongly influenced by the backbone of the model. If the level of sequence identity is high the side-chains are generally well placed in the protein core but are subject to variations at the surface. At the solvent interface (internal and external) there tends to be fewer restraints than in the tightly packed protein core. Unless solvent restraints are simulated during the modelling process the interface regions tend to be less tightly packed and fill a greater volume than what would occur in the actual structure (Contreras-Moreira et al. 2002).

Side-chain modelling programs generally assume that backbone structure is fixed. Hence, the process focuses solely on optimising the side-chain rotamer conformation. However this is unrealistic, as in a protein the backbone would be flexible and could shift to accommodate a larger side-chain if the template and target have differing side-chains. Allowing some backbone flexibility during side-chain modelling procedures would result in a more realistic model, however, ideally the side-chain and backbone should be optimised simultaneously (Vasquez 1996). As yet optimisation procedures are not ideal and molecular dynamics and energy minimisation often move the structure away from the original model or template, potentially introducing further errors to the model. There has been substantial progress in this area, but refinement, still remains one of the bottlenecks of homology modelling (Moult 2005).

Errors in model evaluation can come from the parameters used. Root mean square deviation is a poor indicator of quality when only parts of the model are well predicted. This is because the poorly modelled regions produce such large RMSDs that it is impossible to know if the model contains well-modelled regions at all. One solution to this problem is to score only well-modelled regions when comparing the model and template structures. The ideal modelling evaluation tool would be fully automated and produce one simple numerical measure representing the quality of the model which would be used as a standard measurement within the modelling field (Siew et al. 2000). MaxSub is an evaluation program which has many of these qualities, however, a standard overall measurement remains elusive in the field (Siew et al. 2000).

Developers recognise that there is a need for further improvement of structure prediction methods and the bi-annual Critical Assessment of Protein Structures (CASP) provides them with a way of measuring improvement. In CASP trials, sequences of proteins, for which the structure has been determined, but not released are used to predict the three-dimensional structure of the protein. Upon completion, the predictions are then compared to the actual structure, highlighting areas of improvement in the modelling procedure or areas that require further work. The CASP trials have been running for a decade and have been a catalyst for the steady advancement of the field. At the recent CASP6 there was evidence of improved refinement and side-chain modelling, albeit only in small structures, however this is a promising sign of the improvements to come (Moult 2005).

3. PRACTICAL HOMOLOGY MODELLING

In the sections above we have discussed the procedures involved in homology modelling. In this section we will discuss points that need to be considered in order to begin the modelling process. One of the major decisions to be made is the type of homology modelling package to choose. Depending on the preference and the experience of the modeller, a choice must be made as to whether a manual or fully automated approach will be taken. Each has advantages and disadvantages; the main difference being control of the process.

3.1. Automated Homology Modelling

Although there are a number of downloadable homology modelling programs, the future of homology modelling as a tool for all biologists lies in the fully automated methods. Automated homology modelling programs are run via web-based servers. These servers run the process remotely and the resulting model is emailed back in the form of a pdb file. This process is easy and requires you to know little or nothing about the modelling process. In cases where structures for homologs with high levels of sequence identity (>50%) are available this may be an adequate approach, however if only low identity homologs are available, this approach is likely to be problematic.

Results from the CASP1 experiment held in 1994 suggested that fully automated homology modelling procedures were less accurate than those using manual intervention (Mosimann et al. 1995; Bates et al. 1997). It was suggested that manual intervention at sequence alignment, choice of parents, loop selection and conserved residue interactions improved the outcome (Bates et al. 1997). Since then fully automated approaches have increased in popularity and subsequently there is a separate assessment experiment developed for fully automated programs; Critical Assessment of Fully Automated Procedures or CAFASP. In the last CAFASP3, which was run simultaneously to the CASP5, the top 5-10 modelling servers were able to produce relatively accurate models for all the targets (Fischer et al. 2003). Apart from independent homology modelling servers there are also meta-servers which utilise the results of a number of independent structure prediction servers to produce the final model. Surprisingly, it was found that the performance of the best meta-server predictors was roughly 30% higher than the best independent server (Fischer et al. 2003). This result represents a major advance for fully automated programs.

There are several advantages to using fully automated programs. Many of these relate to convenience. Web-based servers have fewer software issues; there is no need to download, install or maintain the homology modelling programs, which means that it does not matter what platform your computer runs on i.e. unix or windows. One of the issues with semi-automated approaches is that the databases in the programs need to be updated regularly; however web-based servers are generally linked to the appropriate databases and are always up-to-date. In many cases the programs are maintained by the developer, which means that new methods or improvements are available as soon as they are implemented.

The main disadvantage to using a fully automated approach is the lack of control over the process. In sections 2.3 and 2.7 the importance of the sequence alignment was highlighted. However, with most fully automated programs manual inspection or manipulation of the alignment can not be performed. In the case of homologs with low (~30%) sequence identity this could be detrimental and result in a poor model. Due to the obvious need for manual intervention, some of the servers now allow user intervention in the model building process. For example, SWISS-MODEL (Guex and Peitsch 1997) allows a choice of templates and 3D-JIGSAW (Bates et al. 2001) allows for both template selection and manual adjustments of the query to template alignments (Contreras-Moreira et al. 2002). However, in some cases automated programs only allow you to use a PDB code as input for your template selection. This can be detrimental if you prefer to use only a particular protein subunit from the structure file or if you need to modify the structure file in some way.

Careful selection of the appropriate automated program may result in a more accurate model. Some of the programs are not well known and may not be as accurate as others. It is worthwhile determining which modelling and refinement methods a particular program or server uses. Programs that have performed well in the CAFASP experiments are a good choice for modelling as this experiment allows comparison of accuracy. However, some programs used in the experiments are not yet available to the public. Table 1 below lists a selection of the available automated modelling servers. Automated programs allow homology modelling to be available to a wider audience, including non-experts. However, caution and expertise will always be required for critical evaluation and analysis of the results (Forster 2002).

Table 1. Automated Homology Modelling Programs

Name	Type	Description	Web Address	Reference
3D-Jigsaw	FB	Allows some user interaction	http://www.bmm.icnet.uk/servers/3djigsaw/	(Bates et al. 2001)
ROBETTA	FB	Meta-server	http://robetta.bakerlab.org/	(Kim et al. 2004)
Swiss-Model	FB	Allows you to choose and use multiple templates	http://swissmodel.expasy.org/	(Guex and Peitsch 1997)
WHAT IF	FB	Allows the user to perform template selection and alignment	http://swift.cmbi.kun.nl/WIWWWI/	(Vriend 1990)
CPH-Models	SM	Uses profile methods for searching templates and SEGMOD for modelling	http://www.cbs.dtu.dk/services/CPHmodels/	(Lund et al. 2002)
EsyPred3D	SR	Uses MODELLER for model production	http://www.fundp.ac.be/urbm/bioinfo/esypred/	(Lambert et al. 2002)

Method: FB= Fragment Based, SR=Spatial Restraints, SM= Segment Matching

3.2. Manual Modelling Programs

When deciding which modelling program to use there are several factors to consider. One aspect to consider, is the platform on which the modelling program will run. Nearly all modelling programs have been designed to run on a unix/linux or Silicon Graphics platform, however, steadily Windows and Mac versions of the modelling, visualisation and evaluation programs are becoming available. Another important consideration is cost. Fortunately, many of the modelling programs that form the basis of commercial homology modelling programs are also available in a free academic version e.g. MODELLER. However, there are benefits in having the commercial version, many of them being extra features and comprehensible graphical user interfaces. Table 2 contains examples of semi-automated homology modelling programs and their different features.

Table 2. Homology modelling programs and their methods

Name	Method	Avail.	Platform	Description	Web Address	Source
COMPOSER/SYBYL	FB	C	L	Available only in the commercial SYBYL package.	www-cryst.bioc.cam.ac.uk, www.tripos.com	Tripos, St Louis
NEST	FB	F	SGI/L	Also available as a web automated prediction server	http://honiglab.cpmc.columbia.edu/programs/nest.html	(Petrey et al. 2003)
ICM	SR	C	All	A free-ware structure .browser version can be downloaded without modelling or docking features.	www.molsoft.com	(Abagyan et al. 1994)
InsightII	SR	C	SGI/L	Uses MODELLER for homology modelling within a user interface	http://www.accelrys.com/products/insight/index.html	(Sali and Blundell 1993)
MODELLER	SR	F	All	Is able to be scaled up for genome modelling	http://salilab.org/modeller/	(Sali and Blundell 1993)
LOOK	SM	C	L	Uses Segmod and ENCAD for modelling	http://www.bioinformatics.ucla.edu/genemine/	(Levitt 1992)
Swiss-Model	FB	F	All	Part of the DeepView (SwissPDBViewer) program. Uses ProModII for modelling.	http://www.expasy.org/spdbv/	(Guex and Peitsch 1997)

Method: FB= Fragment Based, SR=Spatial Restraints, SM= Segment Matching; Availability: C= Commercial, F = Freeware; Platform: SGI= Silicon Graphics Workstation, L=Linux, All= Linux, Unix, Mac, SGI and Windows

The advantage in using a semi-automated modelling program compared to a fully automated program is once again, control. Depending on your level of knowledge you can have some input into the process. With many programs you can participate in template selection, alignment and refinement processes. As your level of expertise increases, so does your ability to have a greater user input and in turn, a significant effect on the resulting model. For example, with spatial restraint based modelling you can participate in the model production by supplementing the homology-derived restraints with restraints derived from a number of sources such as site-directed mutagenesis and NMR experiments (Marti-Renom et al. 2000). This type of user-input can greatly improve the accuracy of the resulting model.

In order to help highlight the differences between the types of programs and the issues that need to be considered when choosing a modelling program the following section analyses the differences between three programs that use different modelling approaches and refinement methods. The programs are: COMPOSER which uses a fragment-based method, SegMod which uses a segment matching approach, and MODELLER which uses the satisfaction of spatial restraints method.

3.2.1. A fragment-based example: COMPOSER

COMPOSER is a module in the commercial molecular modelling software package SYBYL (Tripos, St. Louis). In COMPOSER each of the steps of homology modelling is represented in the graphical user interface. In the first module, FIND HOMOLOGS the input sequence is used to search the internal structure database, originally taken from the PDB, in order to select homologous structures. The user is able to control the level of sequence identity by assigning a threshold value. Once the search for homologs is complete the user can select which ones will be used in the analysis. The template and target sequences are then aligned to find structurally conserved regions. The alignment of the SCRs and the target sequence can be manually manipulated if required. Alternatively, an alignment file can be directly used as input to the program, giving the user control over the alignment method used. In the model building process, the backbone coordinates of the template are copied to the model. If more than one template is used, the SCR from the template with the highest identity is used. The side-chains are added to the SCRs by a rule-based procedure, using a rotamer database. The variable regions, or loops, are then modelled from the template if there is enough similarity, or from a protein loop database. The side-chains are then built for the VRs by the same method as above. COMPOSER does not contain a refinement procedure although other modules in the SYBYL package can be used.

The advantages of this program are that it allows the user to manipulate the alignment generated or accepts an alignment produced by other software as input. These two features aid in the production of a more accurate alignment and hence increases the likelihood of producing an accurate model. However, one major disadvantage of COMPOSER is the lack of an internal refinement module and therefore, you also require a separate refinement program. The other drawback of this software is that the homolog searching, loop building and side-chain building procedures all require local databases which need to be updated on a regular basis.

3.2.2. A segment-matching approach example: SEGMOD

SEGMOD (Levitt 1992) is a module in the freeware package GeneMine3.5. This package contains other modules that facilitate homolog selection and alignments. The target sequence used as input is divided into short segments. These segments are then used to search structure databases to find matching structural fragments. These are then fitted onto the framework of the template sequence. This process is repeated and ten independent models are built. These models are then averaged to produce the final model. SEGMOD can also use coordinates from multiple structures or from selected regions of one or more structures. This is good for multi-domain proteins, each with homology to other structures. SEGMOD is able to model up to 120 residues for which no template structure exists, i.e. loop segments. If there are insertions and deletions in the middle of the sequence the program will find the best possible structural solution based on known examples representing the way nature has handled similar situations. The program also finds the best way to model both the backbone and side-chains using its own database of structural segments whereas traditional homology modelling programs treat these problems separately. The program uses ENCAD, a molecular dynamics simulation program, for energy minimization refinement where you can choose to use 250 or 500 rounds of energy minimization. The program can easily model multiple polypeptide chains. It also produces some evaluation data in the output, i.e. conformational strain before and after refinement.

The advantage of this program is that it produces several models and then averages them which may be useful for increasing the accuracy of the resulting model. It also allows you to easily model multi-subunit proteins. The program also contains its own built-in refinement module which is convenient.

3.2.3. A spatial restraints approach example: MODELLER

MODELLER is available as a freeware stand-alone package or as part of the commercial software packages, INSIGHTII (Sali and Blundell 1993) and QUANTA (Oldfield and Hubbard 1994). As the freeware version is more widely available to users we will describe this version. The user is responsible for producing an alignment which is used as input to the program. The program builds models based on restraints: homology-derived restraints which are extracted from the alignment of the template and target; stereochemical restraints, which include bond lengths and bond angles, which are obtained by the CHARMM molecular mechanics force field, and dihedral angles and non-bonded atomic distances, which are obtained from a representative set of all known protein structures; and lastly and also optionally, any restraints that can be added by the user i.e. cross-linking or predicted secondary structure. The model produced best satisfies the restraints that have been determined. Loops are modelled by using an optimisation-based approach which does not utilise a database. The loops made are optimised by molecular dynamics using simulated annealing. The program also has the option of an automated alignment and modelling routine, however this is not recommended unless the sequence identity between the target and template is greater than 50%. Like SEGMOD the program allows the user to easily model multimeric proteins.

MODELLER differs from SEGMOD as it uses a different force field (i.e. CHARMM vs ENCAD) and MODELLER uses simulated annealing.

Despite these differences, SEGMOD and MODELLER were found to be in the top three programs tested in a comparative experiment of homology modelling programs (Wallner and Elofsson 2005). COMPOSER was not tested in this experiment however, NEST (Petrey et al. 2003), which uses fragment-based methods ranked equally with SEGMOD and MODELLER. This experiment also revealed some weaknesses in the different programs. MODELLER, the spatial restraints program, was found to have convergence problems i.e. producing models with extended structures and sub-optimal side-chains, while the three fragment-based programs in the experiment produced models with poor stereochemistry in some cases. The segment-based program SEGMOD generated models with bad backbone conformation for some targets (Wallner and Elofsson 2005). Many of these problems were only observed with low sequence identity targets suggesting that at low sequence identity modelling is challenging for most programs (Wallner and Elofsson 2005). In general, fragment-based methods tend to have problems dealing with gaps in the sequence, which suggests that when using a non-optimal alignment the choice of modelling program is important (Wallner and Elofsson 2005).

4. CONSIDERATIONS FOR MODELLING FUNGAL PROTEINS

Fungal genomes are important targets for both genomic and structural genomic projects. This is primarily due to the use of yeast and filamentous fungi as comparative systems for eukaryotic genetics and proteome function. There is also an interest in fungal pathogens due to their impact on human health and agriculture (Birren et al. 2003). The objective of the fungal genomics projects is to sequence and identify all the genes and hence, gene products for a particular organism. There has been an explosion in the number of fungal genome projects, many of which are summarised in Table 3. As a result, the number of fungal protein sequences will increase, producing more targets for both structural genomics projects and individual homology modellers.

The structural genomics projects aim to use these protein sequences and select a number of representative proteins for experimental protein structure determination. These structures can then be used as templates to predict the structures of homologous proteins. These efforts increase the value of the genomic data and aid in determining the functions of the proteins. Such analysis is beneficial for increasing the understanding of fungal proteomes and will aid in finding potential targets that could be utilised in developing diagnostics or therapies for fungal pathogens.

Structural genomics efforts for fungal genomes are only in the early stages and the number of experimentally derived protein structures of fungal proteins remains low. This is highlighted in Figure 3 which displays the proportion of known protein structures for each of the genomics projects. Notably, a substantial proportion of the protein structures are from *Saccharomyces* however this is expected as it was the first genome sequenced and was completed almost a decade ago.

There are a number of structural genomics groups working on target proteins for a wide-range of organisms, which include some fungal species. Three major

Table 3. Summary of genome sequencing groups and target species. Many species are distributed between more than one sequencing centre. This is not an exhaustive list.

Institution/Group	Species
Broad Institute	*Ajellomyces capsulatus*
	Aspergillus nidulans
	Batrachochytrium dendrobatidis
	Candida [tropicalis]
	Chaetomium globosum
	Clavispora lusitaniae
	Coccidioides immitis
	Coprinopsis cinerea
	Cryptococcus neoformans
	Kluyveromyces waltii
	Lodderomyces elongisporus
	Neurospora crassa
	Phaeosphaeria nodorum
	Pichia quilliermondii
	Podospora anserina
	Rhizopus oryzae
	Saccharomyces [paradoxus, bayanus, mikatae]
	Schizosaccharomyces [octosporus, japonicus]
	Ustilago maydis
DOE Joint Genome Institute	*Phakopsora [meibomiae, pachyrhizi]*
Genolevures	*Candida [glabrate, tropicalis*
	Debaryomyces hansenii
	Kluyveromyces [marxianus, thermotolerens, lactis]
	Pichia [angusta, farinose]
	Saccharomyces [cerevisiae, uvarum, kluyveri, exiguus, servazzii]
	Yarrowia lipolytica
	Zycosaccharomyces rouxii
Genome Sciences Centre	*Filobasidiella neoformans*
Genoscope	*Encephalitozoon cuniculi*
	Podospora anserina
International Gibberella Zeae Genomics Consortium	*Gibberella zeae*
International Rice BLAST Genome Consortium	*Magnaporthe grisea*
Marine Biological Laboratory	*Antonospora locustae*
Microbia	*Aspergillus terreus*
Qiagen	*Pichia angusta*
S.pombe European Sequencing Consortium	*Schizosaccharomyces pombe*
Sanger	*Candida albicans*
	Saccharomyces cerevisiae
Stanford University	*Candida albicans*
	Cryptococcus
	Saccharomyces cerevisiae
The Institute for Genomic Research	*Aspergillus [fumigatus, flavus]*
	Coccidioides posadasii
	Cryptococcus neoformans
Washington University	*Ajellomyces [capsulatus, dermatitidis]*
	Saccharomyces [kudriavzevii, bayanus, castellii, kluyveri]
Zoologisches Institut der Univ. Basel, Switzerland	*Eremothecium gossypii*

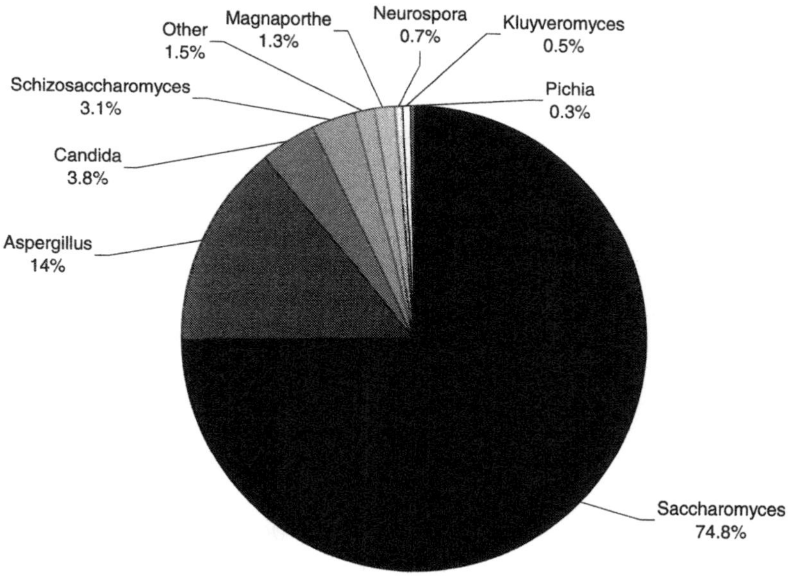

Fig. 3. Approximate proportion of protein structures for Fungal Genera. The total number of structures for all genera = 1721. Genera with no known protein structures were not included and genera with less than five known structures were grouped as 'Other'.

groups are working on *Saccharomyces cerevisiae* structural targets; Structural Proteomics in Europe (SPINE-EU), NorthEast Structural Genomics Consortium and the Joint Center for Structural Genomics, USA. Combined there are 713 overall fungal protein targets, 223 of these have been successfully expressed and purified, whilst the structure of only 14 have been determined and submitted to the PDB (http://www.rcsb.org/pdb/). The South Paris Yeast Structural Genomics Project is only in preliminary development and will focus solely on *Saccharomyces cerevisiae*. At this time there does not appear to be any other structural genomics projects focused solely on fungal proteins, however this is likely to change in the future as more fungal genomes become available and the field of fungal structural genomics expands.

5. CONCLUSION

Protein homology modelling is becoming an increasingly important tool for discovering the functional significance of genomic data. There are a variety of different software tools available ranging from fully automated protein modelling servers to software packages that allow, or require a great deal of user input. In general the greater the amount of user intervention the greater the accuracy of the model generated. These packages all use a variety of different methods or approaches but when used optimally all the methods have comparable accuracies. Regardless of how the homology model is determined the quality or accuracy of the model is primarily dependent on the particular sequence being modelled and the level of homology with the template structure.

The current methods are capable of producing three dimensional protein models with sufficient accuracy to investigate the molecular role of specific amino acids and how these influence parameters such as substrate and inhibitor specificity. Hence, they are an extremely useful commodity for understanding the function of a protein in the absence of experimental structural data. However, there are still many known limitations to homology modelling and the development and improvement of the tools is ongoing, and predominantly driven by structure prediction experiments such as CASP and CAFASP. Therefore, the potential and significance of homology modelling will continue to grow in the future.

REFERENCES

Abagyan RA, Totrov MM and Kuznetsov DA (1997) ICM: a new method for protein modelling and design: applications to docking and structure prediction from the distorted native conformation. J Comp Chem 15: 488-506.

Al-Lazikani B, Jung J, Xiang Z and Honig B (2001) Protein structure prediction. Curr Opin Chem Biol 5: 51-56.

Altschul SF and Koonin EV (1998) Iterated profile searches with PSI-BLAST-a tool for discovery in protein databases. TiBS 23: 444-7.

Altschul SF, Madden TL, Schaffer AA, Zhang J, Zhang Z., Miller W and Lipman, DJ (1997) Gapped BLAST and PSI-BLAST: a new generation of protein database search programs. Nucleic Acids Res 25: 3389-402.

Baker, EN, Arcus, VL and Lott, JS. (2003) Protein structure prediction and analysis as a tool for functional genomics. Appl Bioinformatics 2: S3-10.

Bates PA, Jackson RM and Sternberg MJ (1997) Model building by Comparison: A Combination of Expert Knowledge and Computer Automation. Proteins: Struct Func and Gen 29 (Suppl 1): 59-67.

Bates PA, Kelley LA, MacCallum RM and Sternberg MJE (2001) Enhancement of Protein Modelling by Human Intervention in Applying the Automatic Programs 3D-JIGSAW and 3D-PSSM. Proteins: Struct Func and Gen 45 (Suppl 5): 39-46.

Benner SA, Cohen MA and Gonnet GH (1993) Empirical and structural models for insertions and deletions in the divergent evolution of proteins. J Mol Biol 229: 1065-82.

Birren B, Fink G and Lander E (2003) A White Paper for Fungal Comparative Genomics. Whitehead Institute Centre for Genome Research, Cambridge, MA, USA.

Bourne PE and Weissig H (2003) Structural Bioinformatics, Wiley-Liss, Inc., Hoboken, New Jersey, USA.

Chakravarty S, Wang L and Sanchez R (2005) Accuracy of structure-derived properties in simple comparative models of protein structures. Nucleic Acids Res 33: 244-259.

Chothia C and Lesk AM (1986) The relation between the divergence of sequence and structure in proteins. EMBO J 5: 823-826.

Chung SY and Subbiah S (1996) A structural explanation for the twilight zone of protein sequence homology. Structure 4: 1123-7.

Clark M, Cramer III, RD and Van Opdenbosch N (1989) Validation of the general purpose tripos 5.2 force field. J Comput Chem 10: 982-1012.

Collaborative Computational Project, N. (1994) The CCP4 Suite: Programs for Protein Crystallography. Acta Crystallograph Sect D 50: 760-763.

Contreras-Moreira B, Fitzjohn PW and Bates PA (2002) Comparative modelling: an essential methodology for protein structure prediction in the post-genomic era. Appl Bioinformatics 1: 177-90.

DeLano WL (2002) The PyMOL Molecular Graphics System. DeLano Scientific, San Carlos, CA, USA.

Edelman I, Olsen S and Devereux J (1994) Program Manual for the Wisconsin Package, Versions 8, 9, & 10. Genetics Computer Group, Accelrys, a subsidary of Pharmacopeia Inc. USA

Eisenberg D, Luthy R and Bowie J (1997) VERIFY3D: assessment of protein models with three-dimensional profiles. Meth Enzymol 277: 396-404.

Fan H and Mark AE (2004) Refinement of homology-based protein structures by molecular dynamics simulation techniques. Protein Sci 13: 211-220.

Fidelis K, Stern PS, Bacon D and Moult J (1994) Comparison of systematic search and database methods for constructing segments of protein structure. Protein Eng 7: 953-60.

Fischer D, Rychlewski L, Dunbrack RL, Jr., Ortiz AR and Elofsson A (2003) CAFASP3: the third critical assessment of fully automated structure prediction methods. Proteins: Struct Func and Gen 53 (Suppl 6): 503-16.

Fiser A, Do, RK and Sali A (2000) Modeling of loops in protein structures. Protein Sci 9: 1753-73.

Fiser A and Sali A (2001) Comparative protein structure modelling with MODELLER: A practical approach. The Rockefeller University, New York.
Flores TP, Orengo CA, Moss DS and Thornton JM (1993) Comparison of conformational characteristics in structurally similar protein pairs. Protein Sci 2: 1811-26.
Forster M (2002) Molecular modelling in structural biology. Micron 33: 365-384.
Guex N and Peitsch MC (1997) SWISS-MODEL and the Swiss-PdbViewer: An environment for comparative protein modeling. Electrophoresis 18: 2714-2723.
Hooft,RWW, Vriend G, Sander, C and Abola EE (1996) Errors in protein structures. Nature 381: 272-272.
Jones TA, Zou JY and Kjeldegaard C (1999) Improved Methods for binding protein models in electron density maps and the location of errors in these models. Acta Crystallograph Sect A 47: 110-119.
Kim DE, Chivian D and Baker D (2004) Protein structure prediction and analysis using the Robetta server. Nucleic Acids Res 32: W526-31.
Lambert, C., Leonard, N., De Bolle, X. and Depiereux, E. (2002) ESyPred3D: Prediction of proteins 3D structures. Bioinformatics 18: 1250-1256.
Laskowski, R.A. (1993) PROCHECK: a program to check the stereochemical quality of protein structures. J Appl Cryst 26: 283-291.
Leach AR (1999) Molecular Modelling: Principles and Applications, Pearson Education.
Levitt M (1992) Accurate modeling of protein conformation by automatic segment matching. J Mol Biol 226: 507-533.
Lund O, Nielsen M, Lundegaard C and Worning P (2002) CPHmodels 2.0: X3M a Computer Program to Extract 3D Models., In CASP5 conference A102, California.
Marti-Renom MA, Stuart AC, Fiser A, Sanchez R., Melo F and Sali A (2000) Comparative Protein Structure Modeling of Genes and Genomes. Annu Rev Biophys Biomol Struct 29: 291-325.
Melo F, Devos D, Depiereux E and Feytmans E (1997) ANOLEA: a www server to assess protein structures. Proc Int Conf Intell Syst Mol Biol 97: 110-113.
Morris, G.M., Goodsell, D.S., Halliday, R.S., Huey, R., Hart, W.E., Belew, R.K. and Olson, A.J. (1998) Automated Docking Using a Lamarckian Genetic Algorithm and and Empirical Binding Free Energy Function. J Comp Chem 19: 1639-1662.
Mosimann S, Meleshko R. and James M.N.G. (1995) A critical assessment of comparative modeling of tertiary structures of proteins. Proteins 23: 327-336.
Moult J (2005) A decade of CASP: progress, bottlenecks and prognosis in protein structure prediction. Curr Opin Struct Biol 15: 285-289.
Oldfield, T.J. and Hubbard, R.E. (1994) Analysis of Ca Geometry in Protein Structures. Proteins: Struct Func and Gen 18: 324-337.
Pascarella S and Argos P. (1992) Analysis of insertions/deletions in protein structures. J Mol Biol 224: 461-71.
Peitsch M.C. (2002) About the use of protein models. Bioinformatics 18: 934-8.
Petrey D, Xiang X, Tang CL, Xie L, Gimpelev M, Mitors T, Soto CS, Goldsmith-Fischman S, Kernytsky, A., Schlessinger A, Koh IYY, Alexov E and Honig B (2003) Using Multiple Structure Alignments, Fast Model Building, and Energetic Analysis in Fold Recognition and Homology Modeling. Proteins: Struct Func and Gen 53: 430-5.
Rost B (1999) Twilight zone of protein sequence alignments. Protein Eng 12: 85-94.
Sali A (1995) Modeling mutations and homologous proteins. Curr Opin Biotechnol 6: 437-51.
Sali A and Blundell, T.L. (1993) Comparative protein modelling by satisfaction of spatial restraints. J Mol Biol 234: 779-815.
Sanchez R and Sali A (1998) Large-scale protein structure modeling of the *Saccharomyces cerevisiae* genome. Proc Natl Acad Sci USA 95: 13597-13602.
Siew, N., Elofsson, A., Rychlewski, L. and Fischer, D. (2000) MaxSub: an automated measure for the assessment of protein structure prediction quality. Bioinformatics 16: 776-785.
Sippl, M.J. (1993) Recognition of Errors in Three-Dimensional Structures of Proteins. Proteins 17: 355-362.
Srinivasan N and Blundell TL (1993) An evaluation of the performance of an automated procedure for comparative modelling of protein tertiary structure. Protein Eng 6: 501-12.
Thompson JD, Gibson TJ, Plewniak F, Jeanmougin F and Higgins DG (1997) The ClustalX windows interface: flexible strategies for multiple sequence alignment aided by quality analysis tools. Nucleic Acids Res 24: 4876-4882.
Unger R, Harel D, Wherland S and Sussman JL (1989) A 3D-building blocks approach to analyzing and predicting structure of proteins. Proteins 5: 355-73.
Vasquez, M. (1996) Modeling side-chain conformation. Curr Opin Struct Biol 6: 217-221.
Verdonk, M.L., Cole, J.C., Hartshorn, M.J., Murray, C.W. and Taylor, R.D. (2003) Improved Protein-Ligand Docking Using GOLD. Proteins 52: 609-623.
Vriend, G. (1990) WHAT IF: A molecular modeling and drug design program. J Mol Graph 8: 52-56.

Wallner, B. and Elofsson, A. (2005) All are not equal: A benchmark of different homology modeling programs. Protein Sci 14: 1315-1327.
Xiang, Z. and Honig, B. (2001) Extending the accuracy limits of prediction for side-chain conformations. J Mol Biol 311: 41-430.
Xu, J. (2004) Protein Structure Prediction by Linear Programming. PhD dissertation, Univeristy of Waterloo, Waterloo ON, Canada.

Applied Mycology
and Biotechnology
An International Series
Volume 6. Bioinformatics

Phylogenetic Network Construction Approaches

Vladimir Makarenkov, Dmytro Kevorkov and Pierre Legendre
Département d'informatique, Université du Québec à Montréal, C.P. 8888, succ. Centre-Ville, Montréal (Québec) Canada H3C 3P8 (makarenkov.vladimir@uqam.ca); Département d'informatique, Université du Québec à Montréal, C.P. 8888, succ. Centre-Ville, Montréal (Québec) Canada H3C 3P8. (kevorkov@lacim.uqam.ca); Département de Sciences Biologiques, Université de Montréal, C.P. 6128, succ. Centre-ville, Montréal, (Québec) Canada H3C 3J7 (pierre.legendre@umontreal.ca).

This chapter presents a review of the mathematical techniques available to construct phylogenies and to represent reticulate evolution. Phylogenies can be estimated using distance-based, maximum parsimony, or maximum likelihood methods. Bayesian methods have recently become available to construct phylogenies. Reticulate evolution includes horizontal gene transfer between taxa, hybridization events, and homoplasy. Genetic recombination also creates reticulate evolution within lineages. Several methods are now available to construct reticulated networks of various kinds. Twelve such methods and the accompanying software are described in this review chapter.

1. INTRODUCTION

Evolution of species has long been assumed to be a branching process that could only be represented by a tree topology. In a tree topology, a species is linked to its closest ancestor; other interspecies relationships cannot be taken into account. Well-known evolutionary mechanisms such as hybridization or horizontal gene transfer can only be represented appropriately using a network model.

Patterns of reticulate evolution have been found in a variety of evolutionary contexts, giving rise to a number of recent studies. In bacterial evolution, lateral gene transfer (i.e. horizontal gene transfer) is the mechanism allowing bacteria to exchange genes across species (Sonea and Panisset 1976 1981; Doolittle 1999; Sonea and Mathieu 2000; Sneath 2000). In plant evolution, allopolyploidy leads to the

Corresponding author: Vladimir Makarenkov

appearance of new species encompassing the chromosome complements of the two parent species. Reticulate patterns are also present in micro-evolution within species in sexually-reproducing eukaryotes (Smouse 2000). Examples of molecular data sets containing regions with reticulate histories can be found in Fitch et al. (1990). (multigene families), Robertson, Hahn and Sharp (1995). (virus strains), and Guttman and Dykhuizen (1994). (bacterial genes). For example, the phylogeny of 24 inbred strains of mice obtained by Atchley and Fitch (1991, 1993). included several strains with hybrid origins. Hatta et al. (1999). conducted a molecular phylogenetic analysis providing strong evidence that reef-building corals have evolved in repeated rounds of species separation and fusion, a process leading to a reticulate evolutionary history. Odorico and Miller (1997). discovered patterns of variation due to reticulate evolution in the ribosomal internal transcribed spacers and 5.8s rDNA among five species of *Acropora* corals. The reticulate origin of some root knot nematodes of the genus *Meloidogyne*, which are widespread agricultural pests, was discussed by Hugall, Stanton and Moritz (1999). Cheung et al. (1999). established clear evidence that the evolution of class-I alcohol dehydrogenase genes in catarrhine primates has been reticulate. Phylogenetic analyses of two archaeal genes in *Thermotoga maritima* revealed multiple transfers between archaea and bacteria (Nesbø et al. 2001). The latter analyses confirmed the hypothesis that lateral gene transfer (LGT) events have occurred between bacteria and archaea.

According to McDade (1995). analytical tools enabling one to generate reticulate topologies that accurately depict hybrid history represent a wide-open field for research. When traditional cladistic/phylogenetic methods are applied in such cases, they may produce confusing results since they are constrained to generate only tree-like patterns. Homoplasy is another source of confusion in the reconstruction of phylogenetic trees; it can be represented by supplementary branches added to phylogenetic trees (Makarenkov and Legendre 2000). In their review on reticulate evolution, Posada and Crandall (2001). considered several definitions of net-like evolution, accompanied by proposals of how the involved biological procedures should be represented mathematically. Nakhleh et al. (2003). reported a suite of useful techniques for studying the topological accuracy of methods for reconstructing phylogenetic networks. Linder et al. (2003, 2004). have recently provided an overview of the methods and software meant to depict reticulation events in different evolutionary contexts.

The present article is organized as follows: section 2 recalls the main approaches used to infer phylogenetic trees from sequence and distance data; section 3 describes different evolutionary contexts where patterns of reticulate evolution can occur; section 4 presents a number of algorithms and software for reconstructing evolutionary networks; we conclude with an extensive list of references.

2. PHYLOGENETIC TREE RECONSTRUCTION METHODS

A classical way to illustrate phylogenetic relationships among species is to model them using a phylogenetic tree (i.e. a phylogeny or an additive tree). In this section we discuss the main approaches for inferring phylogenetic trees. For a comprehensive discussion of the methods for inferring phylogenies readers are referred to Swofford et al. (1996). Li (1997). and Felsenstein (2003).

There exist two main approaches for inferring phylogenies. The first one, called the *phenetic approach,* makes no reference to any historical relationship. It operates by measuring distances between species and reconstructs the tree using a hierarchical clustering procedure. The second one, called the *cladistic approach,* considers possible pathways of evolution, inferring the features of the ancestor at each node and choosing an optimal tree according to some model of evolutionary change. The phenetic approach is based on similarity whereas the cladistic approach is based on genealogy. Four basic types of methods for building phylogenies will be presented in detail in this section: distance-based methods (which belong to the phenetic approach), maximum parsimony, maximum likelihood, and Bayesian methods (which belong to the cladistic approach). The two most comprehensive software packages, widely used by the community of computational biologists, are PHYLIP (PHYLogeny Inference Package), a set of freeware programs developed by Felsenstein (2004). and PAUP (Phylogenetic Analysis Using Parsimony) developed by Swofford (1998). Both PAUP and PHYLIP contain the most popular distance-based, maximum likelihood and maximum parsimony methods. They also provide visualization tools as well as bootstrap and jackknife tree validation support. In addition, the user manuals available for both packages are recognized as essential guides, serving as a comprehensive introduction to phylogenetic analysis for beginners as well as important sources of references for experts in the field.

2.1. Distance-Based Methods

Distance-based methods estimate pairwise distances prior to computing a branch-weighted phylogenetic tree. If the pairwise distances are sufficiently close to the number of evolutionary events between pairs of taxa, these methods reconstruct a correct tree (Kim and Warnow 1999). This assumption is true for many models of biomolecular sequence evolution, in which case distance-based methods give sufficiently accurate results (Li 1997). The main advantage of distance-based methods is their small time complexity that makes them applicable to the analysis of large data sets.

If the rate of evolution is constant over the entire tree and the "molecular clock" hypothesis holds, corrections to the pairwise distances required during inference of the phylogenetic tree may be small. However, the "molecular clock" assumption is usually inappropriate for distantly related sequences and the reconstruction of a correct phylogenetic tree becomes problematic under this hypothesis. If the molecular clock assumption does not hold, the observed differences among sequences do not accurately reflect the evolutionary distances. In that case, multiple substitutions at the same site obscure the true distances and make sequences seem artificially closer to each other then they really are. Correction of the pairwise distances that accounts for multiple substitutions at the same site should be used in such cases. There are many Markov models for modeling sequence evolution; each of them implies a specific way to estimate and correct pairwise distances. Furthermore, these corrections have substantial variance when the distances are large. Among the most popular sequence-distance transformation models we have the Hamming, Jukes Cantor (Jukes and Cantor 1969). Kimura 2-parameter (Kimura 1981). and LogDet (Steel 1994). distances. When the goal is to infer relationships with high divergence between sequences, it can be difficult to obtain reliable values for the distance matrix; as consequence, the distance-based algorithms have little chance of succeeding. More detailed description of some distance-based methods is presented below:

UPGMA: The UPGMA [Unweighted Pair-Group Method using Arithmetic averages (Rohlf 1963).] method was originally proposed for taxonomic purposes. It could be used for phylogeny inferring as well, but one has to assume that the rate of nucleotide or amino acid substitution is the same for all evolutionary lineages. UPGMA always produces an ultrametric tree (i.e. a dendrogram). In practice, this method recovers the correct tree with reasonably high probability when the "molecular clock" hypothesis applies and the evolutionary distance is large for all pairs of sequences. This method can be useful to biologists interested in constructing species trees.

At present, however, many investigators use relatively short DNA sequences for which the "molecular clock" hypothesis is often not valid. Therefore, one should be cautious about UPGMA trees. This method produces a rooted tree because of the assumption of a constant rate of evolution, though it is possible to remove the root if necessary. We illustrate the application of the UPGMA procedure using a set of four species characterized by the sequences TAGG, TACG, AAGC, and AGCC. Using the number of differences as an estimate of the dissimilarity among species, we obtain the distance matrix shown in Table 1.

Table 1. Distance matrix for the four sequences TAGG, TACG, AAGC, and AGCC

	TAGG	TACG	AAGC	AGCC
TAGG	0	1	2	4
TACG		0	3	3
AAGC			0	2
AGCC				0

The smallest distance in Table 1 is 1 (between the sequences TAGG and TACG). Consequently, the first cluster to be formed is {TAGG, TACG} and the phylogeny will contain the tree fragment shown in Fig. 1.

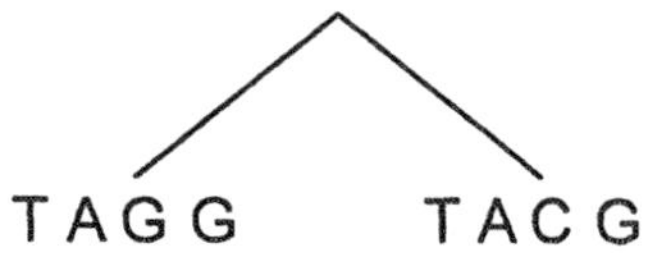

Fig. 1. The first cluster {TAGG, TACG} created by the UPGMA algorithm.

The combined node {TAGG, TACG}, formed by the nodes TAGG and TACG, replaces them in the initial distance matrix to obtain the reduced distance matrix shown in Table 2.

Table 2. Reduced distance matrix

	{TAGG, TACG}	AAGC	AGCC
{TAGG, TACG}	0	½ (2+3) = 2.5	½ (4+3) = 3.5
AAGC		0	2
AGCC			0

The next cluster with the closest nodes (distance = 2) is {AAGC, AGCC}. These two sequences have two differences in the homologous sites. The final cluster fusion links clusters {TAGG, TACG} and {AAGC, AGCC} (Fig. 2).

Fig. 2. Phylogenetic tree obtained by UPGMA for the set of sequences in Table 1.

Neighbor-joining (NJ): Neighbor-joining (Saitou and Nei 1987; Studier and Keppler 1988). is arguably the most popular among the distance-based methods. For some time, the success of NJ was inexplicable for computational biologists, due to the lack of approximation bounds. One of the first bounds was found by Atteson (1999). who showed that this method would be able to return the true phylogeny given that the observed distance is sufficiently close to the true evolutionary distance. Compared to UPGMA, NJ is designed to correct the unequal rates of evolution in different branches of the tree. NJ has a low $O(n^3)$ time complexity, where n is the number of species, and like other distance methods performs well when the divergence between sequences is low. In its first step, NJ considers a bush tree with n leaves and n branches. The tree is gradually transformed into a binary phylogenetic tree with the same n leaves and $2n$-3 branches by merging at each iteration a pair of branches corresponding to the shortest possible tree. Computationally, the tree generation by NJ is similar to UPGMA. When two nodes are linked, their common ancestral node is added to the reduced matrix and the terminal nodes with their respective branches are removed from it. Contrary to UPGMA, neighbor-joining does not produce a dendrogram (ultrametric distance) but an additive tree (additive distance).

Bio Neighbor-joining (BioNJ): The BioNJ (Gascuel 1997a). method is an improved version of the neighbor-joining method of Saitou and Nei (1987). The branch length estimation and distance matrix reduction formulae in NJ provide low variance estimators (Gascuel 1997a). In the paper describing BioNJ, Gascuel (1997a). showed how to improve the accuracy of NJ by incorporating minimum variance optimization in the NJ reduction formula. BioNJ follows an agglomerative scheme similar to that of NJ. It works iteratively, picking a pair of taxa, creating a new node which represents the cluster of these taxa, and reducing the distance matrix by replacing the two taxa by this node. BioNJ uses a simple, first-order model of the variances and covariances of evolutionary distance estimates. This model is well adapted when the estimates are obtained from aligned sequences. At each step it permits the selection, from the class of admissible reductions, of the reduction that minimizes the variance of the new distance matrix. In this way, BioNJ obtains better estimates to choose the pair of taxa to be agglomerated during the next steps. Like NJ, the BioNJ method has a time complexity of $O(n^3)$ for n species. This makes it applicable to the analysis of large data sets. The performances of the two methods are similar when the substitution rates are low, or when they are the same in various lineages. When the substitution rates are high and varying among lineages, BioNJ outperforms NJ in terms of topological accuracy (Gascuel 1997a).

Among other popular distance-based methods, let us mention ADDTREE by Sattath and Tversky (1977). Unweighted Neighbor-Joining (UNJ) by Gascuel (1997b). the Method of Weighted least-squares (MW) by Makarenkov and Leclerc (1999). and FITCH by Felsenstein (1997).

Recommended software: PHYLIP (Felsenstein), PAUP (Swofford), MEGA (Kumar, Tamura and Nei), DAMBE (Xia), T-REX (Makarenkov), and BIONJ (Gascuel).

2.2. Maximum Parsimony

In contrast to the distance-based methods, parsimony infers phylogenetic trees by evaluating the possible mutations between sequences. In general terms, the aim of parsimony methods is to find the phylogenetic tree with minimum total length. That is the tree with the smallest number of evolutionary changes explaining the observed data. For instance, the phylogenetic tree with minimum total length for the sequences CAAG, CCAG, GCAT, and GCTT is presented in Fig. 3.

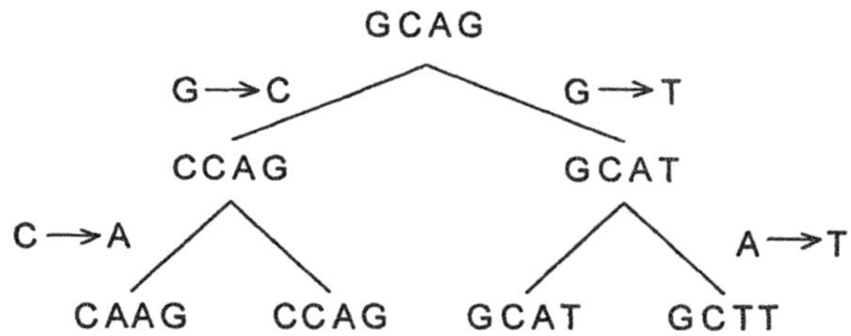

Fig. 3. The phylogenetic tree with minimum total length for the sequences CAAG, CCAG, GCAT, and GCTT.

There are several variations of parsimony. The two simplest and most widely used variations are the Fitch (Fitch 1971). and Wagner (Farris 1970). parsimonies. The Fitch parsimony uses no constraints at all, whereas the Wagner parsimony uses a minimum of constraints on permissible character-state changes. The Wagner method assumes that characters are measured on an interval scale; thus, this method is appropriate for binary, ordered multistate and continuous characters. The Fitch method allows unordered multistate characters (e.g. in nucleotide or protein sequences). Wagner parsimony assumes that any transformation from one character state to another implies a transformation through any intervening states, as defined by the ordering relationship. The Fitch parsimony allows any state to transform directly into any other state. Both methods permit free reversibility. It means that the change of a character state in either direction is assumed to be equally probable, and character states may transform from one state to another and back again. A consequence of reversibility is that a tree may be rooted at any point with no change in tree length.

The Dollo (Farris 1977). and Camin-Sokal (Camin and Sokal 1965). parsimonies are less common. Dollo parsimony does not allow free reversibility. Each character state can appear only once in a tree. If the distribution of character states is not entirely accounted for by the tree, it must be explained by extra reversals (losses). This has been proposed as a way to analyze restriction site data, where the probability of a loss is much higher than that of a gain. Camin-Sokal was the first parsimony method described in the literature. In that method, the tree is rooted and the root contains all ancestral states. Evolution is assumed to be irreversible; only multiple gains are allowed.

Often, more than one tree with minimum total length may be found by maximum parsimony methods. In order to guarantee to find the best possible tree, an exhaustive evaluation of all possible tree topologies has to be carried out. Parsimony will correctly reconstruct a phylogenetic tree if the number of sequence changes per sequence position is small. In the case of a large number of changes, the proportion of homoplastic changes increases. This can cause errors during tree reconstruction, especially during the analysis of long unbranched lineages, or if the tree contains a mixture of short and long branches. Parsimony methods accurately reconstruct phylogenetic trees in which multiple changes at the same site rarely occur alongside a single branch (Hillis 1996; Kim 1996). Maximum parsimony methods are usually much slower than distance-based procedures.

Recommended software: PHYLIP (Felsenstein), PAUP (Swofford), MEGA (Kumar, Tamura and Nei), and NONA (Goloboff).

2.3. Maximum Likelihood

The maximum likelihood approach for inferring phylogenies from sequence data was introduced by Felsenstein (1981). The Felsenstein (1981). method does not impose any constraint on the constancy of evolutionary rate among lineages. It assigns quantitative probabilities to mutational events, rather than merely counting them. This method compares possible phylogenetic trees on the basis of their ability to predict the observed data. The tree that has the highest probability of producing the observed sequences is preferred. Similarly to maximum parsimony, maximum likelihood reconstructs ancestors at all nodes of each considered tree, but it also assigns branch lengths based on the probabilities of mutations. For each possible tree topology, the assumed substitution rates are varied to find the parameters that give the highest likelihood of producing the observed sequences.

From many points of view, maximum likelihood seems to be an appealing way to estimate phylogenies (Whelan et al. 2001). All possible mutational pathways that are compatible with the data are considered. Likelihood functions are known to be a consistent and powerful basis for statistical inference (Edwards 1972). This method represents well the evolutionary relationships among sequences. It takes into account various parameters of the evolutionary process, such as the relative probabilities of transitions versus transversions, or the degree to which the rate of evolution differs across sites. The biologist does not need to know the correct values of these parameters; they are estimated in the tree evaluation process.

The main obstacle to the widespread use of maximum likelihood is computational time. Algorithms that find the maximum likelihood score must search through a multidimensional space of parameters. This makes the solution of large-scale problems (>100 sequences) extremely time consuming. Maximum likelihood estimation may be subject to systematic errors. This happens if the model of evolution used to evaluate the likelihood of given trees does not reflect the actual evolutionary processes.

Felsenstein has developed one of the first maximum likelihood programs, DNAML (DNA Maximum Likelihood program), which is included in the PHYLIP package. The program has been used extensively and has proved of great utility in phylogenetic analyses. Computer simulations have shown that the method is highly efficient in estimating true phylogenies under various situations involving violation of evolutionary rate constancy among lineages (see for instance, Hasegawa and Yano 1984; Hasegawa et al. 1991). An improved version of the DNAML program is based on the algorithm by Felsenstein and Churchill (1996). Several models of base substitution are available in DNAML; for example, a model allowing the expected frequencies of the four bases to be unequal and one allowing the expected frequencies of transitions and transversions to be different. DNAML has also several ways of allowing different rates of evolution to occur at different sites. Another program available in the PHYLIP package, DNAMLK (DNA Maximum Likelihood program with molecular clock), implements the maximum likelihood method for DNA sequences under the constraint that the derived phylogenies must be consistent with a molecular clock hypothesis.

Recommended software: PHYLIP (Felsenstein), PAUP (Swofford), MEGA (Kumar, Tamura and Nei), NONA (Goloboff), and PHYML (one of the fastest ML methods by Guindon and Gascuel).

2.4. Bayesian Phylogenetics

The Bayesian approach is relatively new in phylogenetics (Huelsenbeck and Ronquist 2001; Larget and Simon 1999; Li et al. 2000; Rannala and Yang 1996; Yang and Rannala 1997). This method is closely related to maximum likelihood. The optimal hypothesis is the one that maximizes the posterior probability. The posterior probability for a hypothesis is proportional to the likelihood multiplied by the prior probability of that hypothesis. Prior probabilities of different hypotheses depend on the scientist's assumptions concerning the possible phylogenetic relationships in the data. In many cases, researchers have no information about prior probability distributions. One way of solving this is to specify a uniform prior, in which every possible value of a parameter is given the same probability *a priori*. Compared to maximum likelihood, the advantages of Bayesian methods are higher computational speed and a possibility to incorporate in them complex models of sequence evolution.

Complex parameter-rich models are a problem for maximum likelihood. When the ratio of data points to parameters is low, the estimation of parameters in maximum likelihood can be unreliable. In Bayesian analysis, the final result does not depend on one specific value, but considers all possible parameter values. Even if there are enough data to estimate many parameters, the hill-climbing algorithms that are used to find the maximum likelihood point can be slow or unreliable as the number of parameters increases (particularly if there are complex interactions among some of the parameters). This is not the case for Bayesian methods, because they rely on an algorithm that does not attempt to find the highest point in the space of all parameters.

The best-known Bayesian phylogenetic software programs are MRBAYES written by Huelsenbeck (Huelsenbeck and Ronquist 2001). and BAMBE written by Larget and Simon (1999). MRBAYES uses nucleic acid sequences, protein sequences, and morphological characters to derive phylogenies. It assumes a prior distribution of tree topologies and uses Markov Chain Monte Carlo (MCMC) methods to search the tree space and to infer the posterior distribution of topologies. The BAMBE package infers phylogenetic trees from DNA sequence data. The program uses a prior distribution of trees and implements an arrangement algorithm described in the paper by Mau et al. (1997). The resulting posterior distribution can be used to characterize the uncertainty about not only the tree, but the parameters of the substitution model as well.

Recommended software: MRBAYES (Huelsenbeck) and BAMBE (Larget and Simon).

3. EXISTING MECHANISMS OF RETICULATE EVOLUTION

Classically, the evolution of species has been depicted using phylogenetic trees. An example of such a tree, taken from a famous and controversial paper by Doolittle (1999). is shown in Fig. 4. This way of representing evolution has been questioned by recent developments in molecular phylogenetics. As pointed out by Doolittle (1999). molecular phylogeneticists will have failed to find the true tree of life, not because their methods are inadequate or because they have chosen the wrong genes, but because the history of life cannot be properly represented as a tree. Indeed, the mechanisms of horizontal gene transfer, hybridization, homoplasie, and homologous recombination necessitate the use of network models to illustrate them. Fig. 5 shows an example of a horizontal gene transfer network involving species from the kingdoms of *Bacteria*, *Eukarya*, and *Archaea*.

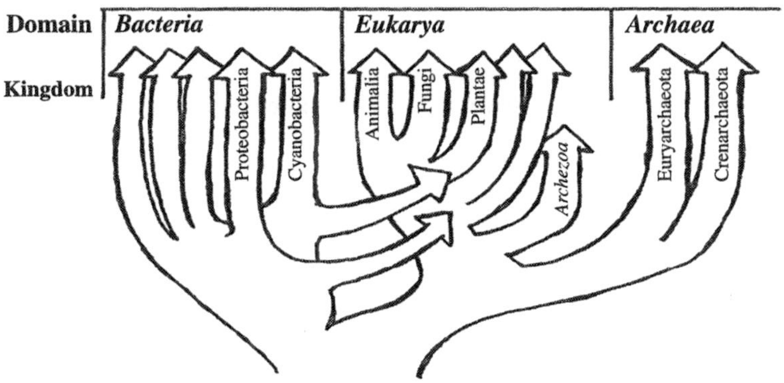

Fig. 4. An example of a phylogenetic tree with a strict hierarchical classification (from Doolittle[1] 1999).

[1] Reprinted with permission from Doolittle WF (1999). Phylogenetic classification and the universal tree. Science 284:2124-2128. Copyright 1999 AAAS.

The fact that most archaeal and bacterial genomes contain genes from multiple sources is challenging for molecular biologists. Following Sonea and Panisset (1976, 1981, Sonea and Mathieu 2000). who showed that horizontal gene transfer (HGT) was a common evolutionary mechanism among bacteria, Doolittle (1999). emphasized the importance of HGT in the evolution of bacteria and higher groups of organisms.

Another reticulate process, hybridization, is prevailing in plants and some groups of animals. In plant evolution, hybridization is critically important as a source of novel gene combinations and as a mechanism of speciation. For instance, in plant breeding desirable traits can be moved from one cultivated or even wild species into another cultivated species (Walter *et al.* 1999). According to one estimate (Stace 1984). there are about 70 000 naturally occurring interspecies plant hybrids in the world.

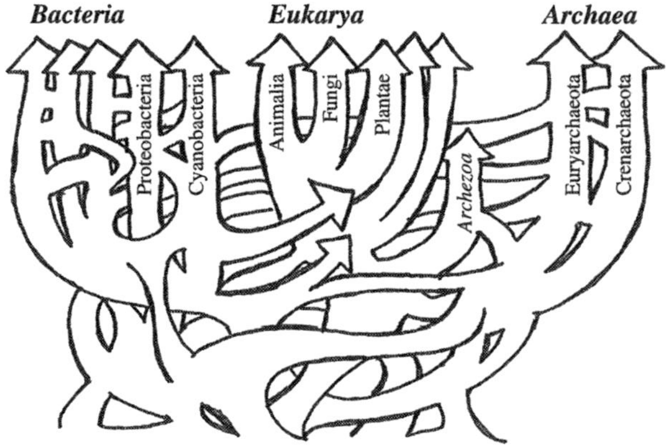

Fig. 5. A reticulated tree, or species network, which might more appropriately represent life's history (from Doolittle[1] 1999, Fig. 3).

Reticulate evolution shows the lack of independence between lineages. When a reticulation event occurs, two or more independent evolutionary lineages interact at some level of biological organization. In this section, we discuss the most important mechanisms of reticulate evolution which led to the development of the computational methods and software tools that will be described in the next section.

3.1. Horizontal Gene Transfer (HGT)

Horizontal gene transfer is a direct transfer of genetic material from one lineage to another. A HGT between the ancestors of Species 3 and 4 took place in the scenario shown in Fig. 6. Because only a few genes, and sometimes only a part of a gene, are transferred from one organism to another, two evolutionary scenarios (Fig. 7) can take place after a HGT event occurred. The first one, presented in Fig. 7a, is appropriate for the genes acquired through the horizontal transfer shown in Fig. 6, whereas the second one, shown in Fig. 7b, is plausible for all the other genes inherited from the direct species ancestors.

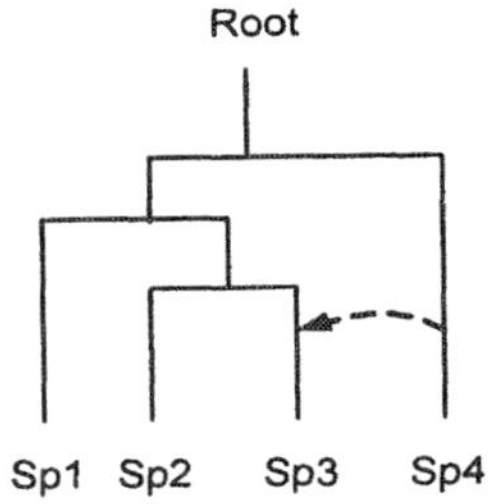

Fig. 6. Horizontal gene transfer.

Horizontal gene transfer is common among bacteria. *Bacteria* and *Archaea* developed the ability to adapt to new environments using the acquisition of new genes through horizontal transfer rather than by the alteration of gene functions through numerous point mutations. Because they are unable to reproduce sexually, bacterial organisms have adopted several mechanisms to exchange genetic materials. The major mechanisms of HGT are the following:

- *Transformation* – This process is most common in bacteria that are naturally transformable. Bacteria take up naked DNA fragments from the environment. This is a common mode of horizontal gene transfer; it can mediate the exchange of any part of a chromosome. Typically, only short DNA fragments are exchanged in this way.
- *Conjugation* – This type of DNA transfer is mediated by conjugal plasmids or conjugal transposons. Even though conjugation requires cell-to-cell contact, it can occur between distantly related bacteria or even between bacteria and eukaryotes. Long fragments of DNA can be transferred by conjugation.
- *Transduction* – This is the transfer of DNA by phage. It requires that the donor and recipient share cell surface receptors for phage binding. It is typically limited to closely related bacteria. The length of DNA transferred by transduction is limited by the size of the phage head.

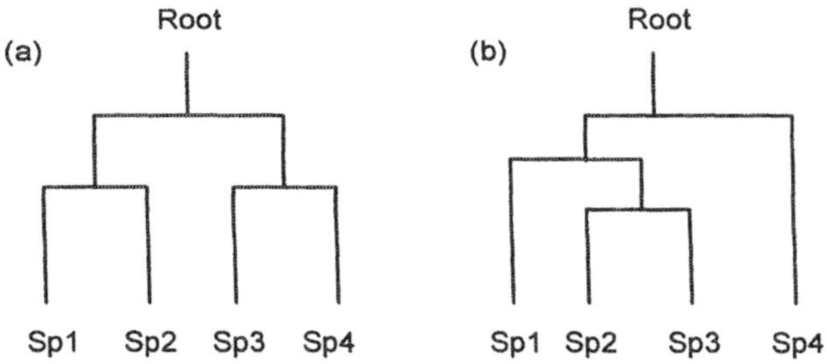

Fig. 7. Horizontal gene transfer: the two possible gene trees.

These mechanisms of horizontal gene transfer can introduce sequences of DNA that have little homology with the remaining DNA of the recipient cell. If the donor DNA and the recipient chromosome share some homologous sequences, the donor sequences can be stably incorporated into the recipient chromosome by homologous recombination. If the homologous sequences are located near sequences that are absent in the recipient, the recipient may acquire an insertion from another strain of unrelated bacteria; such insertions can be of any size.

3.2. Hybridization

Hybridization is another example of reticulate evolution. In Fig. 8, two lineages (Root-Species 2 and Root-Species 3) recombine to create a new species (Species 4). If the new species have the same number of chromosomes as the parent species, the process is called *diploid hybridization*. When it has the sum of the number of its parents' chromosomes, it is called *polyploid hybridization*. The three main mechanisms of hybridization are the following:

- *Autopolyploidization* is a speciation event involving the doubling of the chromosomes within a single species. It produces a bifurcating speciation event in a phylogenetic tree.
- *Allopolyploidization* is a type of hybridization between two species, when an offspring acquires the complete diploid chromosome complements of the two parents. In this case the parents do not need to have the same number of chromosomes. Allopolyploidization results in instantaneous speciation because any backcrossing to the diploid parents is likely to produce a sterile triploid offspring.
- *Diploid hybrid speciation* is a normal sexual event taking place between parents from different but related species. In nearly all cases, the two parents need to have the same number of chromosomes. In this case, successful backcrossing to the parents is possible, so the hybrids have to be isolated from the parents to become new species.

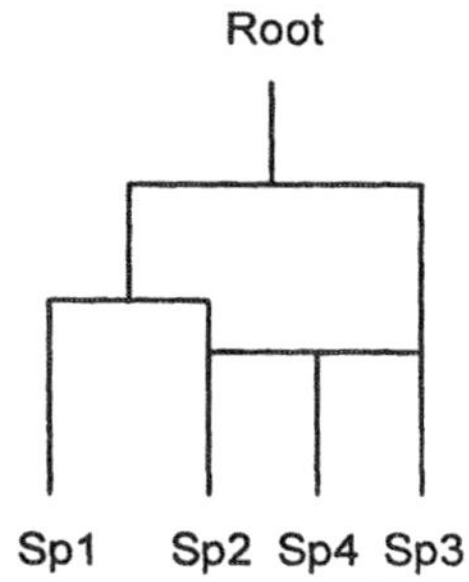

Fig. 8. Hybridization.

In sexually reproducing organisms, hybridization may lead to an entirely female hybrid population. It can sometimes reproduce either by parthenogenesis, or by gynogenesis, forming a new species consisting only of females. Gynogenesis, found among fish, amphibians and reptiles, is a mode of reproduction that allows a unisexual female hybrids population to reproduce, using the sperm from a related bisexual ancestor species to stimulate the development of the eggs (Dawley 1989).

Consider the problem of modeling reticulate evolution after diploid hybrid speciation. In normal diploid organisms, each chromosome consists of a pair of homologs. In the process of diploid hybridization, the hybrid inherits one of the two

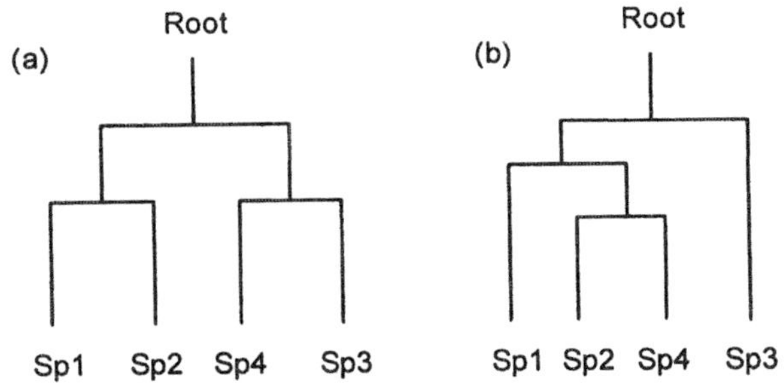

Fig. 9. Hybridization: two possible gene trees for the hybridization event shown in Fig. 8.

homologs for each chromosome from each of its two parents. Since the genes from both parents are contributed to the hybrid, the evolution of genes inherited from each parent can be represented on separate trees inside a network model. Classical phylogenetic analysis of the four species involved in a hybrid speciation event (Fig. 8) will produce either the tree in Fig. 9a or the one in Fig. 9b.

Hybridization is very common in plants, fish, amphibians and reptiles, and is virtually absent in other groups, particularly in birds, mammals, and most arthropods. The latter groups are only occasionally affected by hybrid speciation. They usually produce triploids which can only reproduce by asexual modes.

3.3. Homoplasy

Homoplasy is the development of organs or other bodily structures within different species, which resemble each other and have the same functions, but did not have a common ancestral origin. These organs arise via convergent evolution and are thus analogous, not homologous to each other. For example, the wings of insects, birds and bats, which are all used for flying, are homoplastic (meaning: similar in form and structure, but not in origin). As shown in Fig. 10, the wings of birds and bats are structurally different: the bird wing (a) is supported by digit number 2, the bat wing (b) by digits 2-5.

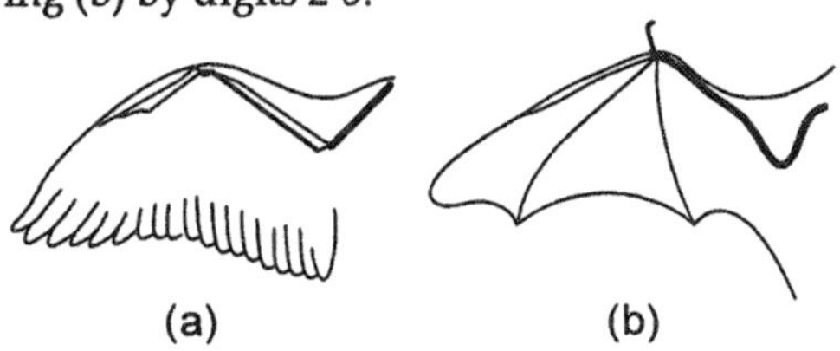

Fig. 10. The wings of birds and bats.

one another, the addition of reticulation branches to a tree produces a reticulogram (i.e. reticulated cladogram) which describes the data better than a tree would do.
Fig. 11, from Makarenkov and Legendre (2000). is an example of a reticulogram built for the primates data originally considered by Hayasaka et al. (1998). First, a distance matrix over 12 species of primates was computed on the base of protein-coding mRNA (898 bases). The phylogenetic tree was constructed from the distance matrix using the neighbor-joining method (Saitou and Nei 1987). The NJ tree is represented by solid lines in Fig. 11. Four groups of primates were found in the phylogeny. The reticulogram building algorithm (Makarenkov and Legendre 2000). added 5 reticulation branches (dashed lines) to the primate phylogeny. From the mathematical point of view, each reticulation branch improved the least-squares fit of the distance matrix, compared to the classical phylogenetic tree. From the biological point of view, the reticulation branches are long and they are formed between distant groups, so, they most likely represent homoplasy. For example, consider *Tarsius*: its position in the phylogeny of primates is uncertain (E. Douzery, personal communication). *Tarsius* is clustered with *Lemur catta* in the NJ

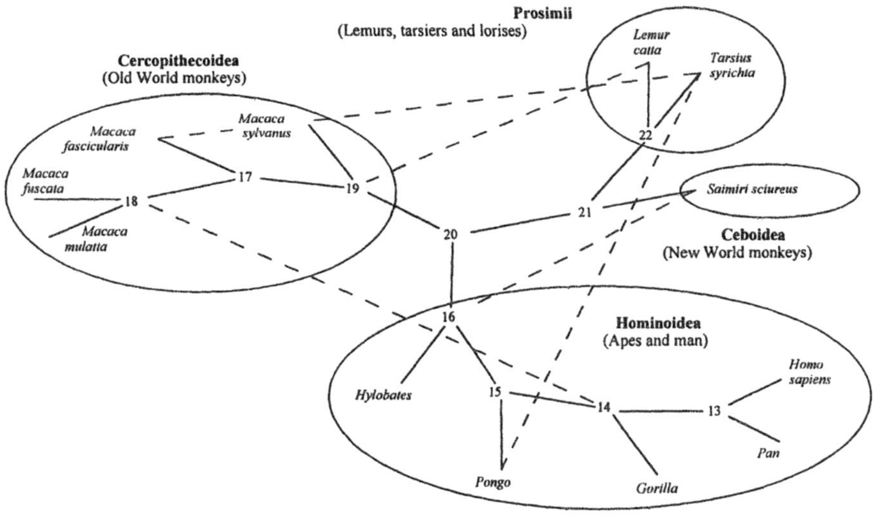

Fig. 11. Reticulogram representing homoplasy among primates (Makarenkov and Legendre[2] 2000, Fig. 2).

phylogenetic tree (solid lines), but it is also close to **Hominoidea** (reticulation branch between *Tarsius* and *Pongo*) and **Cercopithecoidea** (reticulation branch between *Tarsius* and *Macaca fascicularis*). Thus, modeling phylogenetic relationships among primates with reticulograms allowed the authors to depict alternative evolutionary features, homoplasy in this case, which cannot be represented by means of a classical tree model.

3.4. Genetic Recombination

Recombination refers to any process that gives rise to new combinations of genetic material, such as the reassortment of parental genes through crossing over during meiosis, which leads to the formation of gametes. Recombination creates reticulate evolution within lineages. Homologous chromosomes become paired during the prophase of meiosis, as shown diagrammatically in Fig. 12a. In crossing over, two homologous chromosomes swap a portion of their genetic material (Fig. 12b). After separation, each member of a pair of homologues contains parts of its partner's genetic material (Fig. 12c).

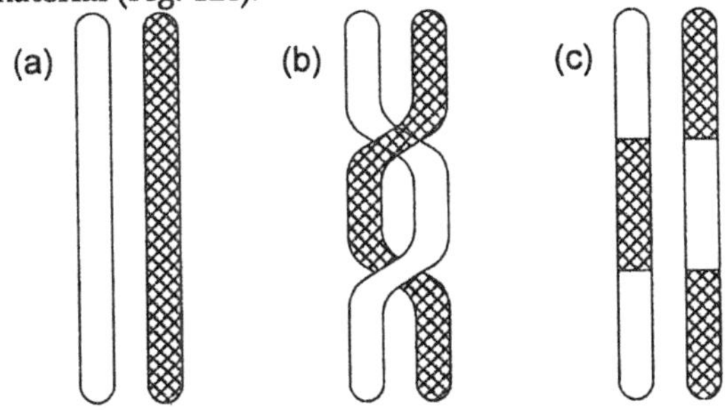

Fig. 12. Homologous chromosomes exchanging genetic material (their central portions) by crossing over.

The exchange of genetic material between homologous chromosomes, called *homologous genetic recombination* (also known as *general recombination* or *general homologous recombination*), may occur at any part of a chromosome. This event can take place in bacteriophage recombination, in recombination following bacterial conjugation, and during the formation of plasmid multimers. Site-specific recombination involves the exchange of genetic material at very specific sites only. Examples include the integration of a bacteriophage lambda into a host chromosome to form a prophage and the rearrangement of chromosomal DNA prior to expressing antibody genes.

Recombination has an important influence on genomes and on the genetic structure of populations. It affects biological evolution at many different levels and explains a considerable amount of genetic diversity in natural populations of sexually-reproducing species. In general, genes located in regions of the genome with low levels of recombination have low levels of polymorphism. Recombination reshuffles the existing variation and even creates new gene variants at the amino acid level. It shapes the genetic structure of natural populations (Anderson and Kohn 1998; Feil et al. 2001). and the action of natural selection (Marais et al. 2001).

Many applications in biology today are based on the estimation of phylogenetic trees. Since recombination leads to mosaic genes, where different regions may have different phylogenetic histories, it is important to take this process into account during the tree reconstruction. A number of statistical methods for the detection of recombination in DNA sequences are available. Their detailed description can be found in Posada and Crandall (2001a). who estimated the performance of 14 different algorithms dealing with recombination.

with low levels of recombination have low levels of polymorphism. Recombination reshuffles the existing variation and even creates new gene variants at the amino acid level. It shapes the genetic structure of natural populations (Anderson and Kohn 1998; Feil et al. 2001). and the action of natural selection (Marais et al. 2001).

Many applications in biology today are based on the estimation of phylogenetic trees. Since recombination leads to mosaic genes, where different regions may have different phylogenetic histories, it is important to take this process into account during the tree reconstruction. A number of statistical methods for the detection of recombination in DNA sequences are available. Their detailed description can be found in Posada and Crandall (2001a). who estimated the performance of 14 different algorithms dealing with recombination.

4. ALGORITHMS AND SOFTWARE FOR DETECTING RETICULATE EVOLUTION

In this section we discuss the algorithms and related software that have been created for the detection and visualization of patterns of reticulate evolution. The web page (http://evolution.genetics.washington.edu/phylip/software.html) supported by J. Felsenstein contains a comprehensive list of phylogeny reconstruction tools, which includes 251 software packages and 29 servers (available on January 12, 2006). In this paper we focus on the software that include algorithms for building and visualizing reticulate phylogenies. For a review of network-like structures used to detect reticulate evolution, readers can also consult the papers by Posada and Crandall (2001b). and Linder et al. (2003 and 2004). A special section dedicated to reticulate evolution and related problems has been published by the Journal of Classification (Legendre 2000a). with contributions from Sneath, Smouse, Lapointe, Rohlf, and Legendre.

Reticulate evolution has long been neglected in phylogenetic analyses. The first methods for studying the mechanisms of reticulate evolution started to appear in the mid-1970s (Sneath et al. 1975; Sonea and Panisset 1976). Several tentative methods have been proposed for the identification of reticulate evolution in nucleotide sequences. They include displays of compatibility (Sneath et al. 1975). tests for clustering (Stephens 1985). a randomization approach (Sawyer 1989). and an extension of the parsimony method of phylogenetic reconstruction that allows recombination (Hein 1993). Rieseberg and Morefield (1995). developed a computer program, RETICLAD, allowing one to identify hybrids based on the expectation that they would combine the characters of their parents. However, this program can only find reticulation events between terminal branches of a tree. Rieseberg and Ellstrand (1993). showed examples where the program appears to work well. The popular method of split decomposition enables the representation of data in the form of a splitsgraph revealing the conflicting signals contained in the data (Bandelt and Dress 1992a, 1992b). In a splitsgraph, a pair of nodes may be linked by a set of parallel edges depicting alternative evolutionary hypotheses. Hallet and Lagergren (2001). showed how lateral gene transfer events can be detected by evaluating topological differences between species and gene trees. Bryant and Moulton (2002, 2004). introduced a network-inferring method, NeighborNet, allowing the reconstruction of planar phylogenetic networks. Each of these methods has features that make them useful for the analysis of particular types of data, and they all have a role to play in detecting and describing reticulate evolution. Legendre and Makarenkov (2002). and Makarenkov and Legendre (2004). proposed to use reticulograms for detecting reticulation events in evolutionary data. They developed a distance-based method to infer reticulate phylogenies. That method uses the topology of a phylogenetic tree as a supporting structure for building a reticulogram. The other network-inferring techniques considered in the present paper are the following: HGT detection of Boc and Makarenkov (2003). and Makarenkov et al. (2004, 2006). Statistical parsimony (Templeton et al. 1992). Netting (Fitch 1997). Median networks (Bandelt et al. 1995 and 2000). Median-joining networks (Foulds et al. 1979; Bandelt et al. 1999). Molecular-variance parsimony (Excoffier and Smouse 1994). Pyramids (Diday and Bertrand 1986). and Weak hierarchies (Bandelt and Dress 1989).

4.1. Horizontal Gene Transfer Detection (Hallet and Lagergren)

Hallet and Lagergren (2001). and Addario-Berry et al. (2003). developed a model of horizontal gene transfer which compares the evolution of a set of gene trees to a species

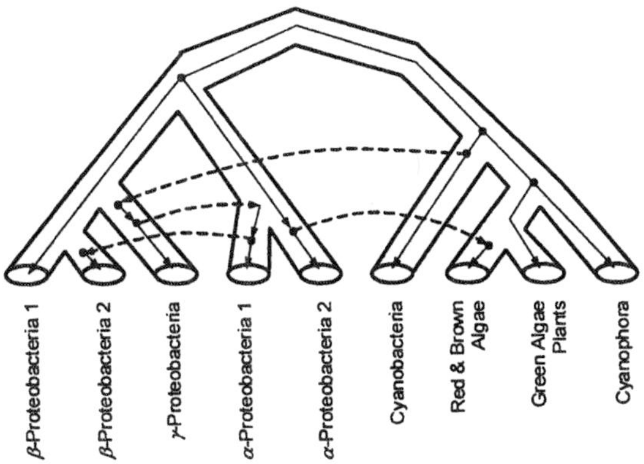

Fig. 13. Horizontal gene transfer scenario of the *rbcL* gene identified by Hallet and Lagergren (2001).

tree. The algorithm proceeds by mapping given gene trees into the species tree. A number of constraints are introduced in the model to make this mapping biologically meaningful. If a multiple copy of a gene appears in the species tree, the algorithm recognizes it as a possible lateral gene transfer. A scenario of lateral transfer of the *rbcL* gene is presented in Fig. 13 (example taken from Hallet and Lagergren 2001). This model also includes an activity parameter α that defines the number of genes allowed to be simultaneously active.

The algorithm is implemented in the Lateral Transfer software available at: http://cgm.cs.mcgill.ca/~laddar/lattrans/. This program also includes an option allowing one to seek scenarios under a combined lateral transfer/gene duplication model.

4.2. Horizontal Gene Transfer Detection (Boc and Makarenkov)

Two models for detection of horizontal gene transfer have been considered by Boc and Makarenkov (2003). Makarenkov et al. (2004, 2006). Both models use a distance approach and are based on the reconciliation of the topologies of the gene and species phylogenetic trees built for the same set of species.

The first model (Boc and Makarenkov 2003; Makarenkov et al. 2004). assumes partial gene transfer; it is based on the computation and optimization of the minimum path-length distances in a directed network (Fig. 14a). In this model, the phylogenetic tree is transformed into a connected and directed graph in which a pair of species can be linked by several paths. The second model (Makarenkov et al. 2006). assumes complete transfer: the species phylogenetic tree is gradually transformed into the gene phylogenetic tree by adding to it a horizontal gene transfer in each step. During this transformation, only a tree topology is taken into account and modified (Fig. 14b). Though the second model is less general, a fast and effective algorithm has been described to solve the problem. Moreover, two criteria, one metric and the other topological, can be combined in the optimization procedure (Makarenkov et al. 2006). Both models produce scenarios of horizontal transfers of the given gene. According to Makarenkov et al. (2006). the use of the topological

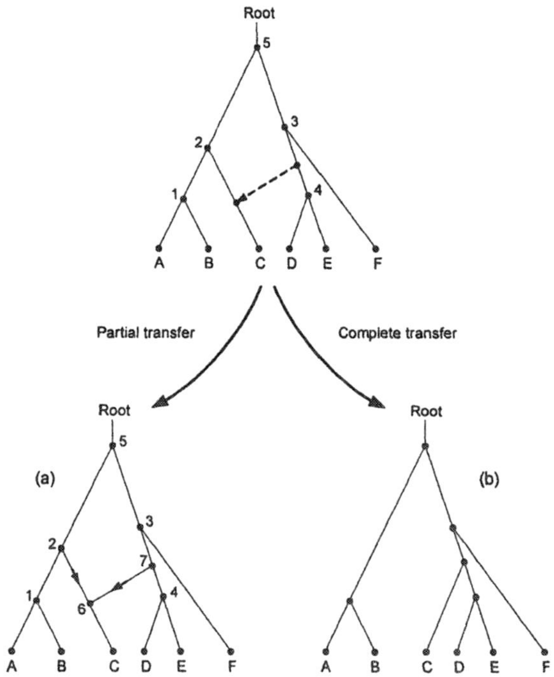

Fig. 14. Two evolutionary models, assuming that either a partial (a, model 1) or a complete (b, model 2) horizontal gene transfer has taken place. In the first case, only a part of the gene is transferred and the tree is transformed into a directed network, whereas in the second, the donor gene replaces the homologous gene of the host and the initial tree is transformed into a different phylogenetic tree.

criterion, which is the Robinson and Foulds (1981). topological distance, enables a better detection of gene transfers compared to the metric criterion (least-squares function); one of the considered examples concerned the well-known *rbcL* dataset from Delwiche and Palmer (1996). Among the recent developments in the field of HGT detection techniques, a validation procedure (bootstrapping) for gene transfer have been designed to measure the reliability of an individual transfer as well as that of a whole gene transfer scenario; see Makarenkov et al. (2006) for more detail. These methods were included in the T-REX package (Makarenkov 2001). which provides users with a friendly visualization support. T-REX is available at the following URL: http://www.trex.uqam.ca.

The main steps of the HGT detection algorithm (model 1) described in Boc and Makarenkov (2003). and Makarenkov et al. (2004). are the following. The algorithm first identifies the topological differences between the species and gene phylogenies. Then, it uses a least-squares optimization procedure to find where horizontal gene transfers between branches of the species tree may have taken place. A species phylogenetic tree T whose leaves are labeled according to the set of n taxa must have been constructed before starting the HGT detection algorithm. Tree T can be inferred from sequence or distance data using an appropriate tree fitting method. The tree should be explicitly rooted; the position of the root is important in this model. Likewise, a gene tree T_1 must have been inferred using a similar procedure; the leaves of T_1 are labeled according to the same set of n taxa labels as in the species tree T. Without loss of generality, the method assumes that T and T_1 are binary trees whose internal nodes are all of degree 3 and whose number of branches is $2n$-3.

If the topologies of T and T_1 are identical, the algorithm concludes that the evolution of the gene followed that of the species, and no horizontal gene transfers between branches of the species tree have taken place. However, if the two phylogenies are topologically different, it may be due to horizontal gene transfers. In this case, the gene tree T_1 can be mapped into the species tree T by fitting, by least squares, the branch lengths of T to the pairwise distances in T_1 [details on this least-squares fitting technique are available in Barthélemy and Guénoche (1991). and Makarenkov and Leclerc (1999)].

The goal of the next step is to determine the order of addition of HGT branches to the tree, considering all possible HGT connections between pairs of branches in T. There are $(2n-3)(2n-4)$ possibilities for the addition of the first HGT branch. This is the maximum number of different directed inter-branch connections in a binary phylogenetic tree with n leaves. The HGT connection providing the largest contribution to the decrease of the least-squares coefficient Q is the most probable case, in the least-squares sense, of horizontal gene transfer. That connection is added to the tree, transforming T into a network. After the first HGT branch has been added to T, all its branches, including the new HGT branch, are reassessed to fit optimally the inter-leaf distances from the gene tree T_1. Then, the best second, third, and so forth, HGT branches are added to T in the same way. Starting from the second HGT branch, addition of any new HGT connection takes into account all previously added HGTs. The algorithm stops when a predetermined number of HGT branches have been added to T. The phylogenetic network obtained in this way represents the best possible scenario, according to least squares, of horizontal transfer of the gene under study.

The following strategy was adopted to estimate the value of the least-squares coefficient Q for a given HGT branch (a,b). First, the algorithm lists all pairs of taxa such that the path between them can include the new HGT branch (a,b); this is controlled by a number of biological rules incorporated into the model. Second, the algorithm lists the pairs of taxa for which the minimum path-length distance may decrease after the addition of the branch (a,b). Third, the algorithm looks for the optimal value l of the length of branch (a,b), keeping fixed the lengths of all the other tree branches; see below. Fourth, all tree branch lengths are reassessed one at a time to improve the fit.

The set $A(a,b)$ of all pairs of taxa, such that the minimum path-length distances between them may change if the HGT branch (a,b) is added to the tree T (Fig. 15), is found as follows: $A(a,b)$ is the set of all pairs of taxa ij such that:

$$Min\{d(i,a) + d(j,b); d(j,a) + d(i,b)\} < d(i,j), \quad (1)$$

where $d(i,j)$ is the minimum path-length distance between taxa i and j in T; vertices a and b are located in the center of branches (x,y) and (z,w), respectively.

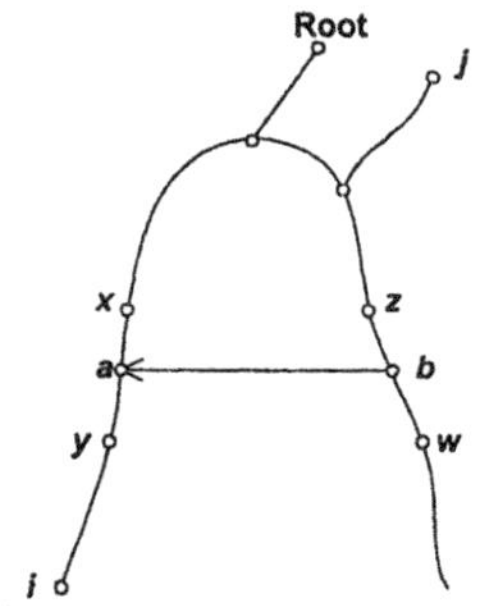

Fig. 15. The minimum path-length distance between taxa i and j can be affected by the addition of a new branch (a,b) representing the horizontal gene transfer between branches (z,w) and (x,y) in the species tree.

The following function is used:

$$dist(i,j) = d(i,j) - Min\{d(i,a) + d(j,b); d(j,a) + d(i,b)\}. \quad (2)$$

Thus, $A(a,b)$ is the set of all leaf pairs ij such that $dist(i,j) > 0$. The least-squares objective function to be minimized, with l used as an unknown variable, is formulated as follows:

$$Q(ab,l) = \sum_{dist(i,j)>l}(Min\{d(i,a)+d(j,b); d(j,a)+d(i,b)\}+l-\delta(i,j))^2 + \sum_{dist(i,j)\le l}(d(i,j)-\delta(i,j))^2, \quad (3)$$

where $\delta(i,j)$ is the minimum path-length distance between taxa i and j in the gene tree T_1. The function $Q(ab,l)$, measures the gain in fit when a new HGT branch (a,b) of length l is added to the species tree T. When the optimal value (i.e. the one that minimizes the function Q) of a new branch (a,b) is found, this computation is followed by an overall polishing procedure for all branch lengths in T. To reassess the length of any branch of T, one can use Equations 1, 2, and 3, assuming that the lengths of all the other branches are fixed. These computations are repeated for all pairs of branches in the species tree T. After all pairs of branches in T have been reassessed, only the HGT corresponding to the smallest value of Q is retained for addition to T. This algorithm requires $O(kn^4)$ operations to produce a HGT scenario with k HGT branches.

4.3. Reticulogram Reconstruction and the T-REX Package

In this subsection, we discuss the method for inferring connected and undirected reticulated networks (also called *reticulograms* or *reticulated networks*) from matrices of evolutionary distances between species. This method was used in several biological problems and turned up to be relevant for detecting hybrids, homoplasy and HGT, as well as biogeographic networks; see the papers by Makarenkov and Legendre (2000 and 2004). Legendre and Makarenkov (2002). and Makarenkov et al. (2004). The method is distance-based and works according to the following scheme: first, it infers a phylogenetic tree from a distance matrix using one of the existing tree fitting algorithms. Supplementary branches, called *reticulation branches*, are then added to the tree structure, one at a time, each one minimizing a least-squares or weighted least-squares loss function. The addition of reticulation branches stops when the minimum of a special goodness-of-fit function is reached. Four such functions have been proposed; each one takes into account the value of the least-squares criterion as well as the total number of branches of the reticulated network under construction. This algorithm requires $O(kn^4)$ time to add k reticulation branches to a phylogenetic tree with n leaves.

We will now describe the main features of this technique and show how it can be applied to study the evolution of a group of honeybees of the genus *Apis*. Let δ be a distance function used to estimate phylogenetic distances between the elements of the set X containing n taxa, and T a phylogenetic tree inferred from δ by means of an appropriate tree reconstruction method. Let d be an expression of the distances in T between the taxa of X (i.e. pairwise distances between the leaves of T). A reticulated network comprises more branches and thus uses more parameters than a phylogenetic tree. As in all statistical models, more parameters mean better fit, but fewer degrees of freedom and a loss of simplicity. A special cost criterion should be used to estimate how many reticulation branches have to be added to a network. The authors proposed four goodness-of-fit criteria to determine when to stop adding branches to a reticulogram (Makarenkov and Legendre 2004). When the exact number of reticulation branches is unknown, as it is often the case in evolutionary problems, one can stop the addition of new branches when the minimum of the selected criterion is reached.

The total number of nodes in a binary unrooted phylogenetic tree with n leaves is $2n-2$; this includes $n-2$ intermediate nodes and n terminal nodes (leaves, taxa). The maximum number of undirected branches one might place in a reticulated network inferred from a binary phylogenetic tree with n leaves is $(2n-2)(2n-3)/2$. Here we counted all possible connections between leaves, between nodes, and between leaves and nodes. However, any metric distance can be represented by a complete graph with $n(n-1)/2$ branches between the leaves. Thus, any of these two limits, $(2n-2)(2n-3)/2$ or $n(n-1)/2$, can be considered as the maximum possible number of branches in a reticulated network. If the latter limit is considered, the number of degrees of freedom of a reticulated network with N branches is $n(n-1)/2 - N$.

It would be reasonable to consider a penalty function opposing the loss in degrees of freedom to the gain in fit. The first proposed goodness-of-fit function is called Q_1:

$$Q_1 = \frac{\sqrt{\sum\limits_{i \in X} \sum\limits_{j \in X} (d(i,j) - \delta(i,j))^2}}{n(n-1)/2 - N} = \frac{\sqrt{Q}}{n(n-1)/2 - N}. \tag{4}$$

The numerator of this function is the square root of Q, which is the sum of squared differences between the values of the given distance δ and the corresponding reticulation estimates d. Interestingly, as was confirmed by a simulation study carried out by Legendre and Makarenkov (2002), function Q_1 usually has only one minimum over the interval $[2n-3, n(n-1)/2]$ of possible values of N. This minimum defines a stopping rule for addition of new branches to the reticulate phylogeny.

The least-squares function itself may be used as the numerator for a goodness-of-fit measure. Thus, one can consider a slightly different criterion, called Q_2, which usually adds more reticulation branches to the network than Q_1:

$$Q_2 = \frac{\sum\limits_{i \in X} \sum\limits_{j \in X} (d(i,j) - \delta(i,j))^2}{n(n-1)/2 - N} = \frac{Q}{n(n-1)/2 - N}. \tag{5}$$

One can also consider the Akaike Information Criterion (AIC) which is a useful and well-known statistic (Akaike 1987). A model with a minimum value of AIC may be chosen to be the best-fitting solution among several competing models. In our algorithm, the Akaike rule would select the model that minimizes the following quantity:

$$\text{AIC} = \frac{Q}{(2n-2)(2n-3)/2 - 2N}. \tag{6}$$

Another popular statistical estimator, the Minimum Description Length (MDL) criterion introduced by Rissanen (1978), can be also used as stopping rule for the reticulogram construction algorithm. The MDL criterion, which is closely related to the AIC statistics, is computed as follows:

$$\text{MDL} = \frac{Q}{(2n-2)(2n-3)/2 - N\log(N)}. \tag{7}$$

The reticulogram in Fig. 16 represents the evolutionary relationships within a group of honeybees. Makarenkov et al. (2004). applied the method for detection of reticulate evolution to the DNA sequence data of six species of honeybees (genus *Apis*). The DNA sequences (677 bases) considered in this work were taken from the SPLITSTREE package (Huson 1998). The bee phylogenetic tree was reconstructed by neighbor-joining (NJ; Fig. 16, full lines), and by maximum likelihood (ML which produced the same tree topology as NJ). The tree was validated by bootstrapping (Felsenstein, 1985) using 100 replicates for ML, and 1000 replicates for NJ. The phylogeny clearly separated two groups of bees, with the species *A. mellifera, A. dorsata,* and *A. cerana* forming the first group and species *A. andreniformis, A. florae,* and *A. koschevnikovi* the second group. The bootstrap support for the group separation branch was 88% for NJ and 89% for ML.

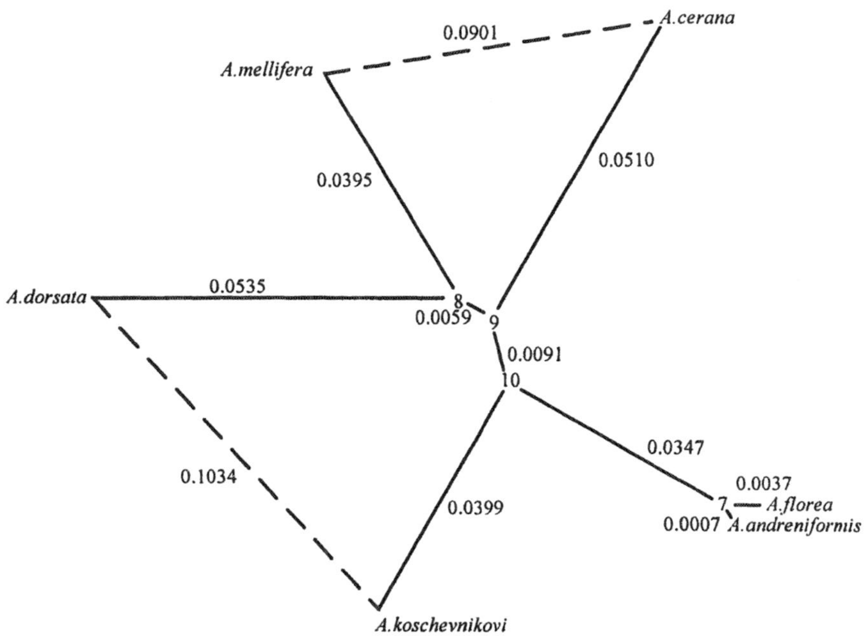

Fig. 16. Reticulogram representing the possible evolution of *Apis* honeybees.

The reticulogram construction algorithm was then applied to the phylogenetic tree provided by NJ. The goodness-of-fit function Q_2 was chosen as the stopping rule for addition of new branches. Two reticulation branches (dashed lines in Fig. 16) were added to the phylogenetic tree by the algorithm. The minimum of the goodness-of-fit function Q_2 was reached at the second step of the algorithm, decreasing the value of Q_2 from 0.000024 to 0.000020, whereas the value of the least-squares loss function Q dropped from 0.000143 to 0.000078. The decrease of Q after addition of only two reticulation branches was dramatic for these data. The gain in fit was 27.3% (Q = 0.000104) after the addition of the first branch, linking bees *A.*

mellifera and *A. cerana*, and the total gain was 45.5% (Q = 0.000078) after the addition of the second branch, linking species *A. dorsata* and A. *koschevnikovi*. These results indicate the relevance of the reticulogram model for the honeybee data, where reticulation branches bring to light conflicting features that are embedded in the phylogenetic tree. The poor bootstrap support (57% and 54% for NJ and ML, respectively) obtained for the branch linking nodes 8 and 9 of the tree is an indication of a close relationship between *A. mellifera* and *A. cerana*.

How should the reticulation branches be interpreted? The first reticulation branch linking *A. mellifera* and *A. cerana* is only about twice the length of the branches of the tree. It may be interpreted as a possible hybridization event involving the ancestors of the two species which occurred during the evolutionary process. This reticulation branch shows that the two species are genetically closer to each other than it is represented by the phylogenetic tree. Fig. 16 depicts what may have happened during evolution: a recent ancestor of *A. cerana* may have hybridized with one of the recent ancestors of *A. mellifera* to produce the modern *A. mellifera* bee. Or, conversely, a recent ancestor of *A. mellifera* may have hybridized with one of the recent ancestors of *A. cerana* to produce the modern *A. cerana* species. This hypothesis is in agreement with the belief, based on biological and behavioral data, that *A. mellifera* and *A. cerana* have shared a close common ancestor in relatively recent times (Milner 1996). The other reticulation branch, linking the species *A. dorsata* and *A. koschevnikovi*, also reveals that the relationship between these two species is closer than depicted by the phylogenetic tree.

The reticulogram reconstruction algorithm has been implemented in the T-REX (*tree and reticulogram reconstruction*) package (Makarenkov 2001) available for the Windows and Macintosh platforms and as a free web server. The program includes a number of popular algorithms for the reconstruction of phylogenetic trees and reticulograms from a distance matrix. Phylogenetic trees can also be inferred from data matrices containing missing values. T-REX provides a window with the tree or reticulogram fitting statistics and a window with the tree or reticulogram drawing. For tree reconstruction, the program includes six methods for fitting a tree metric (distance representable by a tree with non-negative branch lengths) to a distance matrix: the ADDTREE method of Sattath and Tversky (1977). the Neighbor-Joining (NJ) method of Saitou and Nei (1987). the BioNeighbor-Joining (BioNJ). method of Gascuel (1997a). the Unweighted Neighbor-Joining (UNJ) method of Gascuel (1997b). the Circular order reconstruction method of Makarenkov and Leclerc (1997). and Yushmanov (1984). and the Weighted least-squares method (MW) of Makarenkov and Leclerc (1999). Four fitting methods are offered for reconstruction of phylogenies from partial distance matrices (i.e. matrices containing missing values): the Triangle method of Guénoche and Leclerc (2001). the Ultrametric procedure for missing values estimation of De Soete (1984). and Landry and Lapointe (1997). the Additive procedure for missing values estimation of Landry and Lapointe (1997). and the Modified weighted least-squares method MW* of Makarenkov and Lapointe (2004). With the reticulogram inferring option, the program first computes a phylogenetic tree using one of the six available tree-fitting algorithms. Then, at each step of the reticulogram building procedure, a reticulation branch minimizing the least-squares or weighted least-squares loss function is added to the network. When the horizontal gene transfer option is selected, the program maps the gene tree into the species tree following the procedures by Boc and Makarenkov (2003). and Makarenkov et al. (2006).

4.4. Statistical Parsimony

The statistical parsimony method was developed by Templeton et al. (1992). It estimates the maximum number of differences among haplotypes which are caused by single substitution events. This estimation is complemented with a 95% statistical confidence. Multiple substitutions at a single site are neglected. The maximum number of differences is called the parsimony limit. The algorithm initially connects haplotypes differing by one change, then those differing by two, by three, and so on. The algorithm stops when either all the haplotypes are connected in a network or the parsimony connection limit is reached. Since the statistical parsimony method connects haplotypes with small differences, it shows the similarities rather than the dissimilarities between the haplotypes and provides an empirical assessment of deviations from parsimony. This method enables the identification of putative recombinants by looking at the spatial distribution, in the sequence, of the homoplasies defined by the network.

This method is implemented in the TCS Java computer program which estimates gene genealogies including multifurcations and/or reticulations. The corresponding software is described in the paper by Clement et al. (2000). It is available at the following web site: http://inbio.byu.edu/Faculty/kac/crandall_lab/tcs.htm. An example of the network generated by statistical parsimony for the *Apis* honeybees of Fig. 16 is shown in Fig. 17.

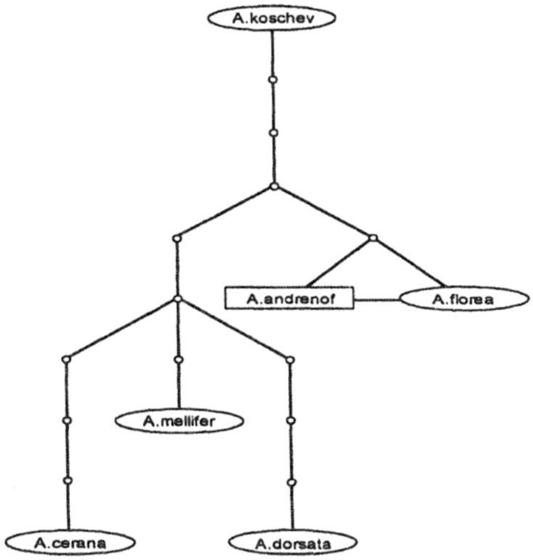

Fig. 17. Phylogenetic network for the *Apis* honeybees, generated by the TCS program.

4.5. Netting

This distance-based method (Fitch 1997). generates all the equally most parsimonious trees for a given data set and connects the leaves (sequences) into a single network. First, the algorithm connects the pair of sequences having the largest similarity. Then, it connects the joined sequences with the sequence having the largest similarity to them. This connection is made in such a way that the three pairwise differences are satisfied. Thus, the patristic distance between two sequences is necessarily equal to the number of differences. A new connection is added to the network if homoplasy is encountered. Gaps and invariant positions are not considered in the analysis. Since the method tends to satisfy all distances among haplotypes, the number of dimensions may be high and the representation of the network may become difficult.

4.6. Median Network

In the median-network method (Bandelt et al. 1995; Bandelt et al. 2000). sequences are first transformed into binary data, whereas constant sites are excluded from the analysis. Each split is encoded as a binary character taking values 0 and 1. Sites supporting the same split are clustered into one site which is then weighted by the number of clustered sites. Thus, this method represents haplotypes as binary vectors. Consensus or median vectors are estimated for each triplet of vectors until the median network is derived. With more than 30 haplotypes, the resulting median networks are very difficult to display due to the presence of high-dimensional hypercubes. Luckily, the size of a median network can be reduced using predictions from coalescence theory. All the most parsimonious trees are represented in a median network. Initially designed for the analysis of mtDNA data, median networks can be built for other kinds of data, as long as the data are binary or can be reduced to that form.

4.7. Molecular-variance Parsimony

The molecular-variance parsimony method developed by Excoffier and Smouse (1994). uses population statistics to select an optimal network. The algorithm generates a number of minimum-spanning trees which are translated into matrices of patristic distances among haplotypes. These matrices are used to compute some of relevant population statistics such as: squared patristic distances among haplotypes, geographic partitioning of populations, and functions of haplotype frequencies. The algorithm chooses the optimal trees by minimizing the molecular variance. This method makes explicit use of the sample haplotype frequencies and geographic subdivisions, and presents the solution in the form of a set of optimal networks.

Excoffier, Schneider, and Roessli have released the ARLEQUIN package, the program for carrying out the population genetics analysis. ARLEQUIN contains a number of useful methods including estimation of gene frequencies, testing of linkage disequilibrium, and analysis of diversity between populations. Another relevant feature of this program consists in its ability to compute a variety of evolutionary measures including the Jukes and Cantor (1969). Kimura 2-parameter (1981), and Tamura and Nei (1984). distances with or without correction for gamma-distributed rates of evolution. ARLEQUIN also computes minimum spanning tree networks. The executable for Windows, MacOS and Linux, Java source code, and a comprehensive documentation for this software are available at the following web site: http://acasun1.unige.ch/arlequin.

4.8. Median-joining Network

The median-joining network algorithm (Bandelt et al. 1999; Foulds et al. 1979). starts by combining the minimum-spanning trees within a single network. Using a parsimony criterion, the procedure adds to the network median vectors representing missing intermediates. Median-joining networks can be used to analyze large datasets and multistate characters. This technique is extremely fast and is able to process thousands of haplotypes in reasonable time. It can also be applied to amino acid sequences. However, the method cannot cope with recombinations, which restricts its application to the population level.

Röhl, Forster and Bandelt have written the NETWORK 4.1 program, the software for inferring median-joining networks from non-recombining DNA, STR, amino acid, and RFLP data. The networks can be constructed using either the reduced median network or the median-joining network method. Windows and DOS executables of the program are freely available at: http://www.fluxus-engineering.com/sharenet.htm. An example of the reduced median-joining network presented in Fig. 18 was calculated using NETWORK 4.1. This network was inferred for the dataset of *Apis* honeybees from Fig. 16.

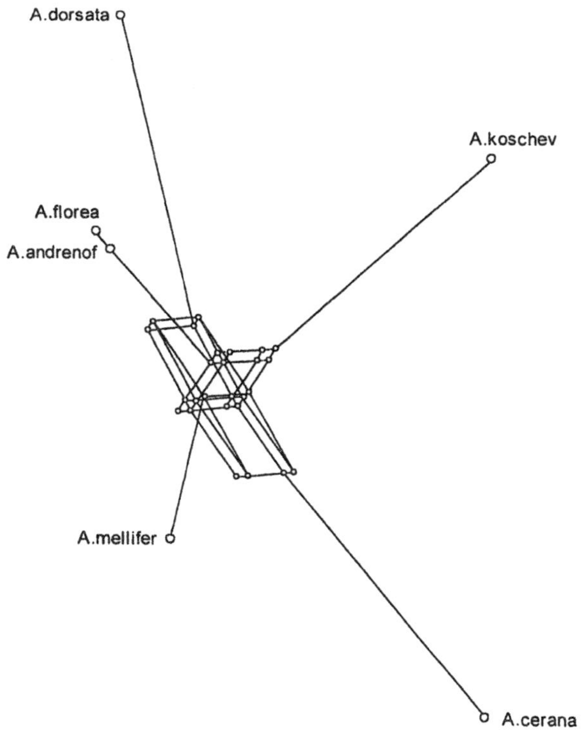

Fig. 18. Median-joining network for the *Apis* honeybees, generated by NETWORK 4.1.

4.9. Split Decomposition

Bandelt and Dress (1992a). designed the technique of split decomposition which transforms evolutionary distances into a sum of weakly compatible splits. There exist a number of algorithms for carrying out the split decomposition. The most popular is implemented in the SPLITSTREE program by Huson (1998).

We recall some basic definitions related to the split decomposition and splitsgraphs. Let X be a set of taxa. A split $S = \{B, B'\}$ is defined as a partition of X into two nonempty sets B and B' such that $B \cup B' = X$. For instance, any branch in a phylogenetic tree introduces a split consisting of all the taxa found on one side (set B) and on the other (set B') of this branch. A set S of splits is called *weakly compatible* if, for any three splits S_1, S_2, and S_3 from S and all $B_i \in S_i$ (i = 1, 2 and 3), at least one of the four intersections:

$$B_1 \cap B_2 \cap B_3,\ B_1 \cap B'_2 \cap B'_3,\ B'_1 \cap B_2 \cap B'_3, \text{ or } B'_1 \cap B'_2 \cap B_3$$

is empty (see Bandelt and Dress 1992a, b). A *splitsgraph* representing a weakly compatible split system S is a graph $G(S) = (V, E)$ whose vertices $v \in V$ are labeled by the set of taxa in X and whose edges (i.e. branches) $e \in E$ are straight line segments representing the splits in S (Fig. 19). In such a graph, each split $\{B, B'\}$ in S is depicted by a group of parallel branches of equal lengths, so that deleting all branches in such a group splits the graph into exactly two parts, one containing all vertices labeled by the taxa in B and the other containing all vertices labeled by the taxa in B'. This method requires an accurate estimation of pairwise distances. Any deviation from the optimal conditions leads to too many splits returned by the method.

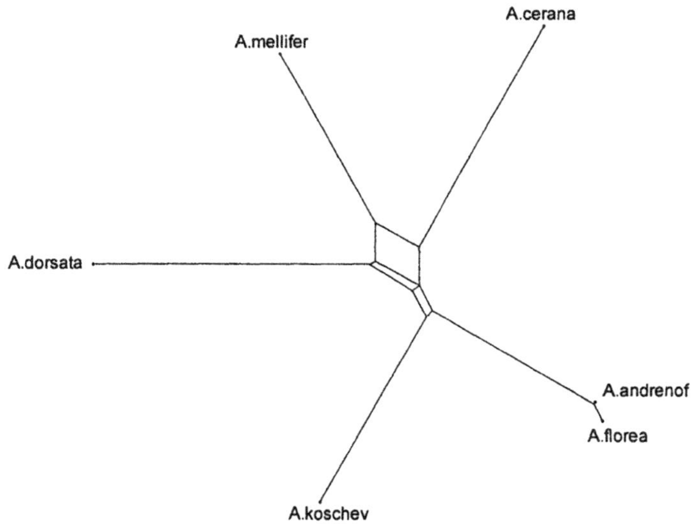

Fig. 19. SPLITSTREE network for the *Apis* honeybees.

The split decomposition method is fast, which means that a reasonable number of haplotypes can be analyzed. It can be applied to nucleotide or protein data. The program suuports the inclusion of models of the nucleotide substitution or amino acid replacement. The method is also suitable for bootstrap evaluation. Fig. 19 represents a splitsgraph built for the dataset of *Apis* honeybees using the LogDet (Steel 1994). evolutionary model selected to compute distances between species.

The SPLITSTREE package, which includes the split decomposition method, is available at: http://www-ab.informatik.uni-tuebingen.de/software/splits/welcome_en.html. The more recent SPLITSTREE 4.0 version includes also the NeighborNet method (Bryant and Moulton 2002, 2004). discussed in the next paragraph.

4.10. NeighborNet

NeighborNet (Bryant and Moulton 2002 and 2004). is a network construction and data representation method that combines the principles of the neighbor-joining and split decomposition techniques. Similarly to neighbor-joining, NeighborNet uses data agglomeration: taxa are combined into progressively larger and larger overlapping clusters.

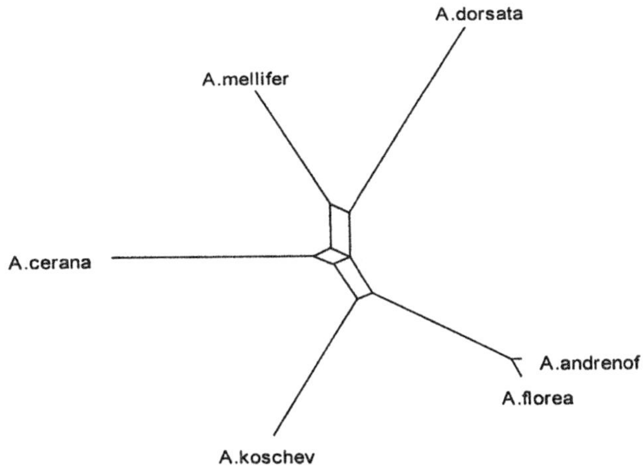

Fig. 20. NeighborNet network for the *Apis* honeybees, generated by SPLITSTREE 4.0.

This strategy has paid dividends in the tree building business with algorithms such as NJ (Saitou and Nei 1987). and BioNJ (Gascuel 1997a). The NeighborNet method can be used to represent multiple phylogenetic hypotheses simultaneously, or to detect complex evolutionary processes like recombination, lateral gene transfer or hybridization. NeighborNet tends to produce networks that are generally more resolved than those built by split decomposition. More precisely, NeighborNet generates a weighted circular split system rather than a hierarchy or a tree, which can subsequently be represented by a planar splitsgraph; for more detail see Bryant and Moulton (2002, 2004). In such graphs, bipartitions or splits of the taxa are represented by classes of parallel lines; conflicting signals or incompatibilities appear as boxes. The method runs in $O(n^3)$ time, for n species, and is well suited for the preliminary analysis of large phylogenetic data sets and for carrying out intensive validation techniques such as bootstrapping. A NeighborNet network for the *Apis* honeybee data is shown in Fig. 20. The LogDet (Steel 1994). evolutionary model was selected to compute distances between species. The NEIGHBORNET package, created by D. Bryant, implementing the method for the Linux and MacOS X platforms is available at the following website: http://www.mcb.mcgill.ca/~bryant/NeighborNet/. As mentioned in the previous paragraph, this method is also available in the SPLITSTREE 4.0 package.

4.11. Pyramids

The Pyramids method was introduced by Diday and Bertrand (1986). Its theoretical description can also be found in Diday (1984 and 1986). The pyramidal clustering model generalizes hierarchies by allowing non-disjoint classes at a given level instead of partitions. The classical hierarchical methods reconstruct a set of the non-overlapping, nested clusters. In contrast to them, pyramids represent a set of clusters that may overlap, with no need for them to be nested. Pyramids can be useful for depicting reticulation events among species. The method infers a pyramid by an agglomerative bottom-up algorithm. It is based on the computation of a Robinsonian dissimilarity matrix between species under study (set X). This means that X admits an ordering such that for any triplet (i, j, k) the dissimilarity value d_{ik} must be larger than or equal to the maximum of d_{ij} and d_{jk}.

The software, running on the Sun, Linux and Unix platforms, carrying out the Pyramids method, is available at the following website: http://195.221.65.10:1234/Pyramids. Fig. 21 shows a pyramid constructed for the *Apis* honeybee data. It was generated using the on-line software available at: http://bioweb.pasteur.fr/seqanal/interfaces/pyramids.html.

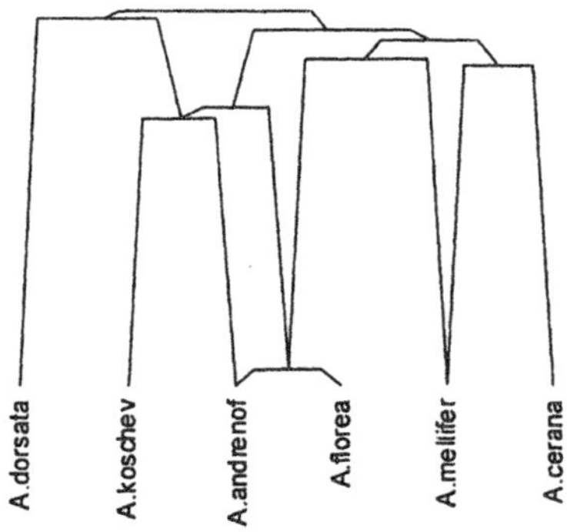

Fig. 21. Pyramid topology representing evolution of the *Apis* honeybees.

4.12. Weak Hierarchies

The method of Weak Hierarchies was introduced by Bandelt and Dress (1989). The method first uses the similarity matrix to infer a dendrogram (strong clusters), and then adds to it weak clusters representing supplementary inter-species relationships. Consequently, a weak hierarchy is an extension of dendrograms that includes both the weak and strong clusters. A subset C of the set X is regarded as a weak cluster if any two objects *a*, *b* in C are more similar to each other than any other object *x* from X-C is similar to either *a* or *b*.

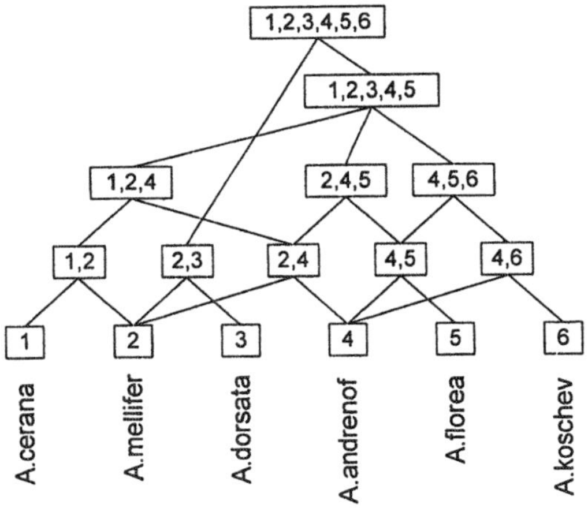

Fig. 22. Weak hierarchy representing the relationships among the *Apis* honeybees.

The mathematical definitions presented by Bandelt and Dress (1989). are as follows. Let *S* be a similarity function on a set *X* of objects. This function perfectly corresponds to a dendrogram if and only if it satisfies the ultrametric inequality (8):

$$S(a,b) \geq Min\{S(a,x), S(b,x)\}, \text{ for all } a, b, x \in X. \tag{8}$$

However, the ultrametric inequality is rarely satisfied for similarity measures encountered in reality. For an arbitrary similarity measure *S*, a subset C of the set *X* is called a strong cluster if it satisfies the inequality (9):

$$S(a,b) > Max\{S(a,x), S(b,x)\}, \text{ for all } a, b \in C \text{ and } x \in X\text{-}C. \tag{9}$$

If all objects in a subset C satisfy inequality (10), C is called a weak cluster:

$$(a,b) > Min\{S(a,x), S(b,x)\}, \text{for all } a, b \in C \text{ and } x \in X\text{-}C. \tag{10}$$

As pointed out by Bandelt and Dress (1989). potential applications of this method include fitting of dendrograms with few additional non-nested clusters and simultaneous representation of families of multiple dendrograms. Figure 21 shows a weak hierarchy for the *Apis* honeybee data also considered in the previous sections. Programs for computing weak hierarchies are available from either H-J. Bandelt (upon request) or V. Makarenkov (the C source code of the program is available at: http://www.info2.uqam.ca/~makarenv/software/Weak_Hierarchies.cpp).

5. CONCLUSION

Phylogenies can be estimated using distance-based, maximum parsimony, maximum likelihood, and Bayesian approaches. Methods and software for phylogenetic tree inferring have been developed since the seminal paper by Cavalli-Sforza and Edwards (1964). who described a tree reconstruction method for continuous characters. A standard format for representing phylogenies in computer-readable form, called the *Newick Standard*, was adopted by an informal committee convened during the Society for the Study of Evolution conference in Durham, New Hampshire, on June 26, 1986; see http://evolution.genetics.washington.edu/phylip/newicktree.html for more details. This format has enhanced the portability of results among computer packages and greatly facilitated the life and work of evolutionary biologists.

Patterns of reticulate evolution have been found in a variety of evolutionary contexts: lateral gene transfer, allopolyploidy, hybridization, as well as mechanisms operating at the micro-evolutionary level. These patterns can be modelled and analysed using methods of reticulate network reconstruction. Homoplasy can also be modelled using reticulate networks. Contrary to the tree inferring, the network building methods are still in their infancy. More refined methods need to be developed to address a variety of situations and research issues. Some of these issues have to be translated into mathematical and statistical form, requiring the help of mathematicians and statisticians. Development of new methods will involve collaboration between evolutionary biologists and computer scientists, as it has been the case for some of the presently available algorithms and models. The new and existing methods will have to be tested against carefully annotated benchmark data, representing different types of reticulate patterns, which should be made available to researchers in a remotely accessible repository. These methods should also be statistically validated and tested against simulated evolutionary data. The development of adequate simulation benchmarks should be discussed at length among evolutionary biologists. Software developers should also get together and develop a common format for the representation of reticulated networks, inspired by the *Newick* format mentioned in the previous paragraph. For the time being, many biologists conducting phylogenetic analysis still interpret their results in a conservative way, while the emerging field of reticulate evolution is trying to gain some level of confidence in the new methods.

REFERENCES

Addario-Berry L, Hallett M and Lagergren J (2003). Towards identifying lateral gene transfer events. Pac Symp Biocomput 8:279–290.

Anderson JB and Kohn LM (1998). Genotyping, gene genealogies and genomics bring fungal population genetics above ground. Trends Ecol Evol 13:444–449.

Atchley WR and Fitch WM (1991). Gene trees and the origins of inbred strains of mice. Science 254:554–558.

Atchley WR and Fitch WM (1993). Genetic affinities of inbred mouse strains of uncertain origin. Mol Biol Evol 10:1150–1169.

Atesson K (1999). The performance of Neighbor-Joining methods of phylogenetic reconstruction. Algorithmica 25: 251-278.
Aude JC, Diaz-Lazcoz Y, Codani JJ and Risler JL (1999). Application of the pyramidal clustering method to biological objects. Comput. Chem. 23:303-315.
Bandelt H-J and Dress AWM (1989). Weak hierarchies associated with similarity measures - an additive clustering technique. Bull Math Biol 51(1):133-166.
Bandelt H-J and Dress AWM (1992a). Split decomposition: a new and useful approach to phylogenetic analysis of distance data. Mol Phylogenet Evol 1:242–252.
Bandelt H-J and Dress AWM (1992b). A canonical decomposition theory for metrics on a finite set. Adv Math 92:47-65.
Bandelt H-J, Forster P, Sykes BC and Richards MB (1995). Mitochondrial portraits of human populations using median networks. Genetics 141:743–753.
Bandelt H-J, Forster P and Rohl A (1999). Median-joining networks for inferring intraspecific phylogenies. Mol Biol Evol 16:37-48.
Bandelt H-J, Macaulay V and Richards M (2000). Median networks:speedy construction and greedy reduction, one simulation, and two case studies from human mtDNA. Mol Phylogenet Evol 16:8–28.
Barthélemy J-P and Guénoche A (1991). Trees and Proximity Representations. Wiley, New York.
Baudry E, Solignac M, Garnery L, Gries M, Cornuet JM and Koeniger N (1998). Relatedness among honeybees *Apis mellifera* of a drone congregation. Proc R Soc Lond B 265:2009-2014.
Boc A and Makarenkov V (2003). New efficient algorithm for detection of horizontal gene transfer events. In: Algorithms in Bioinformatics, Springer, WABI 2003, pp 190-201.
Bryant D and Moulton V (2002). NeighborNet: an agglomerative method for the construction of planar phylogenetic networks. Algorithms in Bioinformatics: Second International Workshop, WABI 2002, Rome, Italy, September 17-21, pp 375 - 391.
Bryant D and Moulton V (2004). Neighbor-Net: an agglomerative method for the construction of phylogenetic networks. Mol Biol Evol 21:255-265.
Camin JH and Sokal RR (1965). A method for deducing branching sequences in phylogeny. Evolution 19: 311-326.
Cavalli-Sforza LL and Edwards AWF (1964). Analysis of human evolution. In: Genetics Today: Proc XI Int Congr Genet, pp 923-933.
Cheung B, Holmes RS, Easteal S and Beacham IR (1999). Evolution of class I alcohol dehydrogenase genes in catarrhine primates: gene conversion, substitution rates, and gene regulation. Mol Biol Evol 16:23-36.
Clement M, Posada D and Crandall KA (2000). TCS: a computer program to estimate gene genealogies. Mol Ecol 9:1657-1660.
Crandall KA (1995). Intraspecific phylogenetics: Support for dental transmission of human immunodeficiency virus. J Virol 69:2351-2356.
Dawley RM (1989). An introduction to unisexual vertebrates. In: RM Dawley and JP Bogart, eds. Evolution and Ecology of Unisexual Vertebrates. Albany, New York: New York State Museum, Bulletin 466, pp 1-18.
Delwiche CF and Palmer JD (1996). Rampant horizontal transfer and duplication of rubisco genes in Eubacteria and plastids. Mol Biol Evol 13:873-882.
De Soete G (1984). Additive-tree representations of incomplete dissimilarity data. Qual Quant 18:387-393.
Diday E (1984). Une representation des classes empiétantes : les pyramides. Research report INRIA 291.
Diday E (1986). Orders and overlapping clusters by pyramids. In: J.De Leeuw *et al.*, ed. Multidimensional Data Analysis Proc., DSWO Press, Leiden.
Diday E and Bertrand P (1984). An extension of hierarchical clustering: the pyramidal representation. In: ES Gelsema and LN Kanal eds., Pattern Recognition in Practice, Amsterdam, North-Holland, pp 411-424.
Doolittle WF (1999). Phylogenetic classification and the universal tree. Science 284:2124-2128.
Edwards AWF (1972). Likelihood. Oxford Univ. Press, Oxford, UK, pp 252.
Excoffier L and Smouse PE (1994). Using allele frequencies and geographic subdivision to reconstruct gene trees within a species:molecular variance parsimony. Genetics 136:343–359.

Farris JS (1970). Methods for computing Wagner trees. Syst Zool 19:83-92.

Farris JS (1977) Phylogenetic analysis under Dollo's Law. Syst Zool 26:77-88.

Feil EJ, Holmes EC, Bessen DE, Chan M-S, Day NPJ, Enright MC, Goldstein R, Hood DW, Kalia A, Moore CE, Zhou J and Spratt BG (2001). Recombination within natural populations of pathogenic bacteria: short-term empirical estimates and long-term phylogenetic consequences. Proc Natl Acad Sci USA 98:182-187.

Felsenstein J (1981). Evolutionary trees from DNA sequences: a maximum likelihood approach. J Mol Evol 17:368-376

Felsenstein J (1985). Confidence limits on phylogenies: an approach using the bootstrap. Evolution 39:783-791.

Felsenstein J (1997). An alternating least-squares approach to inferring phylogenies from pairwise distances. Syst Zool 46:101-111.

Felsenstein J (2003). Inferring Phylogenies. Sinauer Assoc pp 664.

Felsenstein J. (2004). PHYLIP (http://evolution.genetics.washington.edu/phylip.html - software download page and software manual) - PHYLogeny Inference Package.

Fitch WM (1971). Toward defining the course of evolution: Minimum change for a specific tree topology. Syst Zool 20:406-416.

Fitch WM (1997). Networks and viral evolution. J Mol Evol 44:65–75.

Fitch DHA, Mainone C, Goodman M and Sligh-Tom JL (1990). Molecular history of gene conversions in the primate fetal y-globin genes. J Biol Chem 265:781-793.

Foulds LR, Hendy MD and Penny D (1979). A graph theoretic approach to the development of minimal phylogenetic trees. J Mol Evol 13:127–149.

Gascuel O (1997a). BIONJ:an improved version of the NJ algorithm based on a simple model of sequence data. Mol Biol Evol 14:685-695.

Gascuel O (1997b). Concerning the NJ algorithm and its unweighted version, UNJ. In: B Mirkin, F R McMorris, F Roberts and A Rzhetsky, eds. Mathematical hierarchies and Biology. DIMACS Series in Discrete Mathematics and Theoretical Computer Science. Providence, RI: American Mathematical Society, pp 149–170.

Guénoche A and Leclerc B (2001). The triangles method to build X-trees from incomplete distance matrices. RAIRO Oper Res 35:283--300.

Guindon S and Gascuel O (2003). A simple, fast and accurate method to estimate large phylogenies by maximum-likelihood. Syst Biol 52:696-704.

Guttman DS and Dykhuizen DE (1994). Clonal divergence in *Escherichia coli* as a result of recombination, not mutation. Science 266:1380-1383.

Hallet MT and Lagergren J (2001). Efficient algorithms for lateral gene transfer problems. In: Proceedings of the 5th Ann Int Conf Compt Mol Biol (RECOMB 01), New York, ASM Press. pp 149-156.

Hatta M, Fukami H, Wang W, Omori M, Shimoike K, Hayashibara T, Ina Y and Sugiyama T (1999). Reproductive and genetic evidence for a reticulate evolutionary history of mass-spawning corals. Mol Biol Evol 16:1607-1613.

Hayasaka K, Gojobori T and Horai S (1998). Molecular phylogeny and evolution of primate mitochondrial DNA. Mol Biol Evol 5:626-644.

Hein J (1993). A heuristic method to reconstruct the history of sequences subject to recombination. J Mol Evol 36:396-405.

Hillis DM (1996). Inferring complex phylogenies. Nature 383:130–131.

Huelsenbeck JP, Ronquist F, Nielsen R and Bollback JP (2001). Bayesian inference of phylogeny and its impact on evolutionary biology. Science 294:2310–2314.

Hugall A, Stanton J and Moritz C (1999). Reticulate evolution and the origins of ribosomal internal transcribed spacer diversity in apomictic meloidogyne. Mol Biol Evol 16:157-164.

Huelsenbeck JP and Ronquist FR (2001). MRBAYES: Bayesian inference of phylogenetic trees. Bioinf 17:754–755.

Huson DH (1998). SplitsTree: a program for analyzing and visualizing evolutionary data. Bioinf 141:68-73.

Jukes TH and Cantor CR (1969). Evolution of protein molecules. In: H. N. Munro, eds. Mammalian Protein Metabolism, Academic Press, New York, pp 21-132.

Kim JH (1996). General inconsistency conditions for maximum parsimony: effects of branch lengths and increasing numbers of taxa. Syst Biol 45:363–374.
Kim J and Warnow T (1999). Tutorial on phylogenetic tree estimation. In: Proc. 7th Int'l Conf. on Intelligent Systems for Molecular Biology (ISMB99).
Kimura M (1981). Estimation of evolutionary distances between homologous nucleotide sequences. Proc Natl Acad Sci USA 78:454-458.
Koeniger G, Koeniger N, Mardan M and Wongsiri S (1993). Variance in weight of sexuals and workers within and between 4 Apis species (*A. florea, Apis dorsata, Apis cerana* and *Apis mellifera*). Asian Apicult 1:106-111.
Landry PA and Lapointe FJ (1997). Estimation of missing distances in path-length matrices: problems and solutions. In: B Mirkin, FR McMorris, F Roberts, A Rzhetsky eds., Mathematical hierarchies and Biology, DIMACS Series in Discrete Mathematics and Theoretical Computer Science, Amer Math Soc, Providence, RI, pp 209-224.
Lapointe F-J (2000). How to account for reticulation events in phylogenetic analysis: a comparison of distance-based methods. J Classif 17:175-184.
Larget B and Simon DL (1999). Markov chain Monte Carlo algorithms for the Bayesian analysis of phylogenetic trees. Mol Biol Evol 16:750–759.
Legendre P (Guest Editor) (2000a). Special section on reticulate evolution. J Classif 17:153–195.
Legendre P (2000b). Biological applications of reticulation analysis. J Classif 17:191–195.
Legendre P and Makarenkov V (2002). Reconstruction of biogeographic and evolutionary networks using reticulograms. Syst Biol 51:199–216.
Li W-H (1997). Molecular Evolution. Sunderland, Massachusetts: Sinauer Assoc, pp 487.
Li S, Pearl DK and Doss H (2000). Phylogenetic tree construction using Markov chain Monte Carlo. J Am Stat Assoc 95:493–508.
Linder CR, Moret BME, Nakhleh L and Warnow T (2003). Network (reticulate) evolution: biology, models, and algorithms. A tutorial presented at the Ninth Pacific Symposium on Biocomputing (PSB 2004).
Linder CR, Moret BME, Nakhleh L and Warnow T (2004). Reconstructing networks part II: computational aspects. A tutorial presented at the Ninth Pacific Symposium on Biocomputing (PSB 2004).
Makarenkov V and Leclerc B (1997). Tree metrics and their circular orders:some uses for the reconstruction and fitting of phylogenetic trees. In: B Mirkin, F R McMorris, F Roberts and A Rzhetsky, eds. Mathematical hierarchies and Biology. DIMACS Series in Discrete Mathematics and Theoretical Computer Science. Providence, RI: American Mathematical Society, pp 183–208.
Makarenkov V and Leclerc B (1999). An algorithm for the fitting of a tree metric according to a weighted least-squares criterion. J Classif 16:3–26.
Makarenkov V and Leclerc B (2000). Comparison of additive trees using circular orders. J Comput Biol 7:731–744.
Makarenkov V and Legendre P (2000). Improving the additive tree representation of a dissimilarity matrix using reticulations. In: HAL Kiers, J-P Rasson, PJF Groenen and M Schader, eds. Data Analysis Classification and Related Methods. Berlin: Springer, pp 35–40.
Makarenkov V (2001). T-Rex: reconstructing and visualizing phylogenetic trees and reticulation networks, Bioinf 17:664–668.
Makarenkov V and Legendre P (2004). From a phylogenetic tree to a reticulated network. J Comput Biol 11:195–212.
Makarenkov V, Legendre P and Desdevises Y (2004). Modeling phylogenetic relationships using reticulated networks. Zool Scrip 33:89–96.
Makarenkov V, Boc A and Diallo AB (2004). Representing lateral gene transfer in species classification. Unique scenario. In: Classification, Clustering, and Data Mining Applications, IFCS 2004, Chicago: Springer, pp 439-446.
Makarenkov V and Lapointe F-J (2004). A weighted least-squares approach for inferring phylogenies from incomplete distance matrices. Bioinformatics 20:2113-2121.
Makarenkov V, Boc A, Delwiche CF and Philippe H (2006). A new efficient method for detecting horizontal gene transfers: Modeling partial and complete gene transfer scenarios, submitted.
Marais G, Mouchiroud D and Duret L (2001). Does recombination improve selection on codon usage? Lessons from nematode and fly complete genomes. Proc Natl Acad Sci USA 98:5688–5692.

Mau B, Newton MA and Larget B (1997). Bayesian phylogenetic inference via Markov chain Monte Carlo methods. Mol Biol Evol 14:717-724.

McDade L (1995) Hybridization and phylogenetics. In PC Hoch and AG Stephenson, eds., Experimental and Molecular Approaches to Plant Biosystematics, Monographs in Systematic Botany from the Missouri Botanical Garden. pp 305–331.

Milner A (1996). An introduction to understanding honeybees, their origins, evolution and diversity. Available via Bibba electronic journal, URL:<http://www.bibba.com>.

Nei M and Kumar S (2000). Molecular Evolution and Phylogenetics. Oxford Univ. Press, New York, pp 333.

Nesbø CL, L'Haridon S, Stetter KO and Doolittle WF (2001). Phylogenetic analyses of two "archaeal" genes in *Thermotoga maritima* reveal multiple transfers between archaea and bacteria. Mol Biol Evol 18:362-375.

Odorico DM and Miller DJ (1997). Variation in the ribosomal internal transcribed spacers and 5.8s rDNA among five species of *Acropora* (*cnidaria; scleractinia*): Patterns of variation consistent with reticulate evolution. Mol Biol Evol, 14:465-473.

Posada D and Crandall KA (1998). Modeltest: testing the model of DNA substitution. Bioinf 14, 817-818.

Posada D and Crandall KA (2001a). Evaluation of methods for detecting recombination from DNA sequences: Computer simulations. Proc Natl Acad Sci USA 98(24):13757-13762.

Posada D and Crandall KA (2001b). Intraspecific gene genealogies: trees grafting into networks. Trends Ecol Evol 16 (1):37-45.

Rannala B and Yang Z (1996). Probability distribution of molecular evolutionary trees: a new method of phylogenetic inference. J Mol Evol 43:304–311.

Rieseberg LH and Ellstrand NC (1993). What can morphological and molecular markers tell us about plant hybridization? Crit Rev Plant Sci 12:213-241.

Rieseberg LH and Morefield JD (1995). Character expression, phylogenetic reconstruction, and the detection of reticulate evolution. In: PC Hoch and AG Stephenson, eds., Experimental and Molecular Approaches to Plant Biosystematics. Monographs in Systematic Botany from the Missouri Botanical Garden 53, pp 333-354.

Robertson DL, Hanh BH and Sharp PM (1995). Recombination in AIDS viruses. J Mol Evol 40:249-259.

Robinson DR and Foulds LR (1981). Comparison of phylogenetic trees. Math Biosci 53:131-147.

Rohlf FJ (1963). Classification of *Aedes* by numerical taxonomic methods (Diptera: Culicidae). Ann Entomol Soc Am 56:798–804.

Rohlf FJ (2000). Phylogenetic models and reticulations. J Classif 17(2):185-189.

Saitou N and Nei M (1987). The neighbor-joining method:a new method for reconstructing phylogenetic trees. Mol Biol Evol 4, 406–425.

Sattath S and Tversky A (1977). Phylogenetic similarity trees. Psychometrika 42:319-345.

Sawyer S (1989). Statistical tests for detecting gene conversion. Mol Biol Evol 6:526-536.

Schmidt HA and von Haeseler A (2003). Maximum-Likelihood Analysis Using TREE-PUZZLE. In A.D. Baxevanis, D.B. Davison, R.D.M. Page, G. Stormo, and L. Stein (eds.) Current Protocols in Bioinformatics, Unit 6.6, Wiley and Sons, New York.

Smouse PE (2000). Reticulation inside the species boundary. J Classif 17:165-173.

Sneath PHA, Sackin MJ and Ambler RP (1975). Detecting evolutionary incompatibilities from protein sequences. Syst Zool 24:311-332.

Sneath PHA (2000). Reticulate evolution in bacteria and other organisms: how can we study it? J Classif 17:159-163.

Sonea S and Mathieu LG (2000). Prokaryotology – A coherent view. Presses de l'Université de Montréal, Montréal.

Sonea S and Panisset M (1976). Pour une nouvelle bactériologie. Rev Can Biol 35:103-167.

Sonea S and Panisset M (1981). Introduction à la nouvelle bactériologie. Presses de l'Université de Montréal, Montréal and Masson, Paris, pp 127.

Stace CA (1984). Plant taxonomy and biosystematics. Edward Arnold, London, pp 272.

Steel MA (1994). Recovering a tree from the leaf colorations it generates under a Markov model. Appl Math Lett 72:19-24.

Stephens JC (1985). Statistical methods of DNA sequence analysis: detection of intragenic recombination or gene conversion. Mol Biol Evol 2:539-556.

Studier JA and Keppler KJ (1988). A note on the neighbor-joining algorithm of Saitou and Nei. Mol Biol Evol 5:729–731.

Swofford DL, Olsen GL, Waddell PJ and Hillis MD (1996). Phylogenetic Inference. In: D. M. Hill ed. Molecular Systematics. Sinauer, pp 407-514.

Swofford DL (2001). PAUP: Phylogenetic analysis using parsimony and other methods. Version 4.0d8. Champaign, Illinois: Illinois Natural History Survey.

Tajima F and Nei M (1984). Estimation of evolutionary distance between nucleotide sequences. Mol Biol Evol 1:269.

Templeton AR, Crandall KA and Sing CF (1992). A cladistic analysis of phenotypic associations with haplotypes inferred from restriction endonuclease mapping and DNA sequence data. III. Cladogram estimation. Genetics 132:619-633.

Walter SJ, Campbell CS, Kellogg EA and Stevens PF (1999). Plant systematics. A phylogenetic approach. Sinauer Associates. Inc. Sunderland, Massachusetts, USA, pp 576.

Whelan S Lio P and Goldman N (2001). Molecular phylogenetics:state-of-the-art methods for looking into the past. Trends Genet 17:262–272.

Xia X and Xie Z (2001). DAMBE: Data analysis in molecular biology and evolution. Journal of Heredity 92:371-373.

Yang ZH and Rannala B (1997). Bayesian phylogenetic inference using DNA sequences:a Markov chain Monte Carlo method. Mol Biol Evol 14:717–724.

Yushmanov SV (1984). Construction of a tree with p leaves from 2p-3 elements of its distance matrix (in Russian). Matematicheskie Zametki 35:877-887.

Applied Mycology
and Biotechnology
An International Series
Volume 6. Bioinformatics

4

Issues in Comparative Fungal Genomics

Tom Hsiang[1] and David L. Baillie[2]
[1] Department of Environmental Biology, University of Guelph, Guelph, Ontario, N1G 2W1, Canada (thsiang@uoguelph.ca); [2] Department of Molecular Biology and Biochemistry, Simon Fraser University, Burnaby, B.C., V5A 1S6, Canada (baillie@sfu.ca).

Biologists face an overwhelming richness of nucleotide and protein sequence data. By the middle of 2005, there were almost 300 complete genomes that were publicly accessible. Most of these were archeal or bacterial since prokaryotic genomes are much smaller than eukaryotic genomes. Among eukaryotes, fungi, particularly yeasts, have some of the smallest genome sizes and hence represent the highest number of complete or almost complete genomes sequenced. By mid-2005, there were over 43 fungal genomes that were completely or almost completely sequenced and publicly accessible. What are the relationships among fungi and between fungi and other organisms? What type of genes and pathways are required for pathogenicity and other fungal lifestyles? Researchers are addressing these types of questions with data from high-throughput genomic sequencing. This review examines some recent uses of fungal genomic data in comparative genome analyses. Comparative genomics can facilitate research into the following areas: evolution, phylogenetics, targeted drugs, gene discovery, and gene function. Each of these is discussed as well as the availability and ownership of the genomic data, and the concepts of homology (homologs, orthologs, paralogs) and similarity.

1. INTRODUCTION

By the middle of 2005, there were almost 300 complete genomes that were publicly accessible (http://www.genomesonline.org). Most of these (87%) were archeal or bacterial since prokaryotic genomes range in size from 1 to 5 Mb (Fraser et al. 2000), and are much smaller than eukaryotic genomes, which range in size from 10 Mb to over 3 Gb. Among eukaryotes, fungi, particularly yeasts, have some of the smallest genome sizes (10 to 50 Mb) and hence represent the highest number of complete or almost complete genomes sequenced. By mid-2005, there were over 43 fungal genomes that were completely or almost completely sequenced and publicly accessible (Table 1). Most of these were released since 2003 (84%), but many of them (56%) are considered "posted" but not "published" (Hyman 2001). In addition to

Corresponding author: T. Hsiang

publicly accessible genomes, there are privately-held complete or almost complete fungal genomic data, including *Cochliobolus heterostrophus* and *Gibberella fujikuroi* by Syngenta Biotechnology at the Research Triangle Park, NC (Turgeon et al. 2002), and *Aspergillus niger* sequenced by Gene Alliance (an alliance of five German Companies) for DSM Food Specialties (Heerlen, The Netherlands).

In 2000, the Fungal Genome Initiative (FGI) was formed to discuss and prioritize fungal genome sequencing. The FGI is a partnership between the fungal research community and the Broad Institute (which evolved from the Whitehead Institute/MIT Center for Genome Research in 2004). In February 2002, the FGI released the First White Paper on fungal species targeted for sequencing. Of the 15 fungi selected, the National Human Genome Research Institute in the U.S.A. agreed to fund the costs of sequencing seven, which have been completed or are almost completed. In June 2003, the FGI released the Second White Paper which contains a list of 44 fungal sequencing targets, with an emphasis on 10 major clusters of related species (*Penicillium, Aspergillus, Histoplasma, Coccidioides, Fusarium, Neurospora, Candida, Schizosaccharomyces, Cryptococcus,* and *Puccinia*). In July 2004, the FGI released the Third White paper which contains a list of four more target fungal species: *Schizosaccharomyces octosporus, Schizosaccharomyces japonicus, Trichophyton rubrum* and *Batrachochytrium dendrobatidis*. Copies of the White Papers, and more details on the status of these projects can be found at http://www.broad.mit.edu/annotation/fungi/fgi/history.html.

Other sequencing centers which have been responsible for release of fungal genomes include the U.S. Department of Energy Joint Genome Institute (http://www.jgi.doe.gov), The Wellcome Trust Sanger Institute (http://www.sanger.ac.uk), The Institute for Genomic Research (http://www.tigr.org), The Stanford Genome Technology Center (http://www-sequence.stanford.edu), The Génolevures Consortium (http://cbi.labri.fr/Genolevures), Genoscope (http://www.genoscope.cns.fr), The University of Paris (http://www.igmors.u-psud.fr), and Washington University (http://genome.wustl.edu). Funding for these projects has usually been obtained from government sources.

Recent reviews on fungal genomics have concentrated on food industry applications (Hofmann et al. 2003), pathogenicity (Yoder and Turgeon 2001; Lorenz 2002; Mitchell et al. 2003; Tunlid and Talbot 2002; Bos et al. 2003), antifungal drug discovery (Firon and d'Enfert 2002; Jiang et al. 2002; Parkinson 2002), uncovering human genes with fungal homologs (Zeng et al. 2001), yeast comparative genomics (Piskur and Langkjaer 2004, Liti and Louis 2005), and fungal genomics from an agricultural perspective (Yarden et al. 2003). Bennett and Arnold (2001) published an excellent broad overview of fungal genomics. There is also a recent review of fungal genomics targeted toward a general audience (Thacker 2003). The current review has evolved from a previous one (Hsiang and Baillie 2004), and the purpose is to provide an update on developments in comparative fungal genomics. Comparative genomics can facilitate research into phylogenetics, targeted drugs, gene discovery, and gene function. Each of these aspects is discussed in the following sections, beginning with

Table 1. Alphabetical listing of fungal genomes, showing year of first release, source, size, and current version. Information for this table was compiled from web searches, http://www.genomesonline.org (Bernal et al., 2001) and www-genome.wi.mit.edu/annotation/fungi/fgi/status.html.

First release	Species and strain	Genome source and publication[1]	Size[2]	File date and version[3]
2004	*Ashbya gossypii* ATCC 10895	Syngenta AG & Basel University (Dietrich et al. 2004) GenBank NC_005782 to 88	9 Mb	2004.3.4
2001	*Aspergillus fumigatus* AF293	TIGR (unpublished) GenBank NC_007194 to 201	29 Mb	2004.3.17
2003	*Aspergillus nidulans* FGSC-A4	Broad Institute (unpublished) GenBank AACD01000000	31 Mb	2003.6.20 Release 3
2005	*Botrytis cinerea* B05.10	Syngenta and Broad Institute (unpublished) http://www.broad.mit.edu/annotation/fgi/	30 Mb	2005.4.26
2002	*Candida albicans* SC5314	Stanford Genome Tech. Center (Tzung et al. 2001) http://www-sequence.stanford.edu/group/candida	16 Mb	2002.5.24 Assembly 19
2004	*Candida glabrata* CBS 138	Génolevures (Dujon et al. 2004). GenBank NC_005967 to NC_006036	12 Mb	2004.7.1
2004	*Candida guilliermondii* ATCC6260	Broad Institute (unpublished) GenBank AAFM01000000	12 Mb	2004.12.28 Assembly 1
2004	*Candida lusitaniae* ATCC 42720	Broad Institute (unpublished). GenBank AAFT01000000	16 Mb	2004.9.30 Assembly 1
2004	*Candida tropicalis* MYA-3404	Broad Institute (unpublished) GenBank AAFN01000000	30 Mb	2004.9.30 Assembly 1
2004	*Chaetomium globosum* CBS 148.51	Broad Institute (unpublished) GenBank: AAFU01000000	36 Mb	2004.12.10 Assembly 1
2004	*Coccidioides immitis* RS	Broad Institute (unpublished) GenBank AAEC01000000	29 Mb	2004.3.11 Assembly 1
2003	*Coprinus cinereus* Okayama 7	Broad Institute (unpublished) Gen Bank AACS01000000	38 Mb	2003.6.1 Assembly 1
2005	*Cryptococcus neoformans* JEC 21	TIGR (Loftus et al. 2005). GenBank NC_006670 to 94	21 Mb	2005-01-13
2003	*Cryptococcus neoformans* serotype A, strain H99	Broad Institute (unpublished) GenBank AACO01000000	20 Mb	2003.5.2 Assembly 1
2004	*Cryptococcus neoformans* Serotype B, strain R265	Broad Institute (unpublished) GenBank AAFP01000000	20 Mb	2004.8.18 Assembly 1
2003	*Cryptococcus neoformans* serotype D, strain B3501A	Stanford Genome Tech. Center (Loftus et al. 2005) www-sequence.stanford.edu/group/C.neoformans	18.5 Mb	2004.06.23 Assembly 040623
2004	*Debaryomyces hansenii* CBS 767	Génolevures (Dujon et al. 2004) GenBank NC_006043 to 49	12 Mb	2004.7.1
2001	*Encephalitozoon cuniculi* GB-M1	Genoscope (Katinka et al. 2001) GenBank NC_003229-42	3 Mb	2001.11.15
2003	*Fusarium graminearum* PH-1	Broad Institute (unpublished) GenBank AACM01000000	40 Mb	2003.10.03 Release 2
2003	*Fusarium verticillioides* 7600	Broad Institute (unpublished) http://www.broad.mit.edu/annotation/fgi/	36 Mb	2003.6.1 Assembly 2
2004	*Kluyveromyces lactis* NRRL Y-1140	Génolevures (Dujon et al. 2004). GenBank NC_006038 to 42	11 Mb	2004.7.1
2002	*Magnaporthe grisea* 70-15	Broad Institute (Dean et al. 2005). GenBank AACU01000000	40 Mb	2002.09.17 Release 2
2003	*Neurospora crassa* OR74A	Broad Institute (Galagan et al. 2003) GenBank AABX01000000	40 Mb	2005.2.17 release 7
2002	*Phanerochaete chrysosporium* RP-78	US DOE Joine Genome Inst. (Martinez et al. 2004) GenBank AADS00000000	36 Mb	2005.2.15 Release 2
2003	*Phytopthora infestans* T30-4	Broad Institute (unpublished). NCBI Trace Repository (http://www.ncbi.nlm.nih.gov/Traces)	237 Mb	2003.12.8
2003	*Phytophthora ramorum* UCD Pr4	US DOE Joint Genome Inst. (unpublished). http://genome.jgi-psf.org/ramorum1	65 Mb	2004.5.27 Release 1

2003	*Phytophthora sojae* P6497	US DOE Joint Genome Inst. (unpublished). http://genome.jgi-psf.org/sojae1	95 Mb	2004.05.27 Release 1
2004	*Podospora anserina* S Mat+	University of Paris (unpublished). http://podospora.igmors.u-psud.fr	34 Mb	2004.1.23 Assembly 1
2004	*Rhizopus oryzae* RA 99-880	Broad Institute (unpublished) GenBank AACW01000000	40 Mb	2004.12.28 Release 1
2003	*Saccharomyces bayanus* MCYC 623	Washington University (Cliften et al. 2003) http://genome.wustl.edu/	12 Mb	2003.03.28
2003	*Saccharomyces castellii* NRRL Y-12630	Washington University (Cliften et al. 2003) http://genome.wustl.edu/	12 Mb	2003.04.07
1997	*Saccharomyces cerevisiae* S288C	SGD, Stanford (Mewes et al. 1997a). GenBank NC_001133 to 48	12 Mb	2005.8.1 Version 5
2004	*Saccharomyces cerevisiae* RM11-1a	Broad Institute (unpublished) GenBank AAEG01000000	12 Mb	2004.9.10 Assembly 1
2003	*Saccharomyces kudriavzevii* IFO 1802	Washington University (Cliften et al. 2003) http://www.genetics.wustl.edu	12 Mb	2003.04.07
2003	*Saccharomyces kluyveri* NRRL Y-12651	Washington University (Cliften et al. 2003) http://genome.wustl.edu/	12 Mb	2003.04.07
2003	*Saccharomyces mikatae* IFO 1815	Broad Institute (Kellis et al. 2003). http://www.broad.mit.edu/annotation/fgi/	12 Mb	2003.03.28
2003	*Saccharomyces paradoxus* NRRL Y-17217	Broad Institute (Kellis et al. 2003). http://www.broad.mit.edu/annotation/fgi/	12 Mb	2003.03.28
2002	*Schizosaccharomyces pombe* 972h	Sanger Institute (Wood et al. 2002) GenBank NC_003421 to 24	14 Mb	2005.6.20 Version 2
2005	*Sclerotinia sclerotiorum* 1980	Broad Institute (unpublished) http://www.broad.mit.edu/annotation/fgi/	38 Mb	2005.4.13 Assembly 1
2005	*Stagonospora nodorum* SN15	Broad Institute (unpublished) GenBank AAGI00000000	37 Mb	2005.1.17 Release 1
2003	*Trichoderma reesei* QM9414	US DOE Joint Genome Inst. (unpublished) GenBank AAIL01000000	35 Mb	2003.7.18 Release 1
2003	*Ustilago maydis* 521	Broad Institute (unpublished) GenBank AACP01000000	20 Mb	2004.4.1 Release 2
2004	*Yarrowia lipolytica* CLIB99	Génolevures (Dujon et al. 2004) GenBank NC_006067 to 72	21 Mb	2004.7.1

Genome source: in addition to the GenBank accession numbers listed, sequence data from the Broad Institute can also be obtained directly from the FTP site (ftp://ftp.broad.mit.edu/pub/annotation/ fungi/). [2] Size: Estimated size of the genome provided by the source; if no estimate is given, then the data file size is listed. [3] File date and version: the date of the most recent release (year.month.day) is provided as well as the current version. In general, "Release" or "Version" refer to a version of the released sequence data, and "Assembly" refers to the process of joining sequence reads into continguous consensus sequences with the final goal of complete chromosomal sequences.

the availability and ownership of the genomic data, as well as the concepts of homology and similarity.

2. OWNERSHIP OF THE GENOMIC DATA

In 1991, the US National Human Genome Research Institute (NHGRI) and the US Department of Energy developed a data release policy whereby publicly funded sequencing projects should release their data within 6 months. In 1996, the International Human Genome Research Consortium adopted the "Bermuda Principles" with a policy of release of assembly data within 24 hr of generation. In early 2003, NHGRI issued a revision of release policies, reaffirming the 1996 Principles, as well as adding that sequence traces should be in a public trace archive within one week of production, and that whole genome assemblies should be deposited as soon as possible in public databases after the data has passed set quality evaluation criteria. In essence, the current policies state that publicly funded sequencing projects should release their data without restrictions, while sequence users should provide proper citation of the data source and keep in mind that the sequence generators would like to publish their own analyses of the sequence data (Dennis 2003). The full NHGRI report can be found at http://www.genome.gov/10506537.

Users of publicly available draft sequence data should consider that sequence generators require time from release of the first draft until the full sequence is sufficiently accurate for a full genome publication. For example, for the human genome, the first draft released in 2001 was considered 90% accurate while the completed version from 2003 was considered 99% accurate; however, this last 9% required as much time, effort and expense as the first 90% (International Human Genome Consortium 2004).

Situations have occurred where sequence generators felt that their prerogative to first publish using their data has been pre-empted by other researchers who have analyzed and published on the sequence data before full genome release in a peer-reviewed publication (Bell 2000, Hyman 2001, Marshall 2002). An Editorial in the journal Nature reaffirmed that journals will likely accept good research involving whole-genome analyses without restrictions on authorship, since that is in the best interests of science (Anon 2003). A response to the Editorial in Nature by several prominent bioinformatics researchers (Salzberg et al. 2003) asserts further that publicly funded genome sequence data should be available for use without restrictions.

3. HOMOLOGY

Comparative genomics involves comparisons of sequences to search for homologs. Homology is defined as similarity by descent. It is a qualitative measure rather than quantitative, since sequences are either homologous or not homologous (Doyle and Gaut 2000, Fitch 2000). In much of the molecular biology literature, homology is commonly used as a synonym for similarity, such as in a statement where two genes are said to be 75% homologous. It might be true that 75% of a gene shares common descent with another gene, while the remaining 25% does not, but this is usually not the intended meaning (Doyle and Gaut 2000). Instead of saying, "the two genes are 75% homologous", the statement should read, "the two genes are homologous with 75% similarity".

For quantitative assessments of relationships, the terms identity and similarity are often used, but the usage has been inconsistent. For nucleotides, both identity and similarity are used to refer to the occurrence of the same nucleotide at the same (homologous) position. For protein sequences, identity has the same usage as that for nucleotides, but similarity also includes matches with amino acids of similar triplet coding and similar chemical characteristics. For example, in the commonly used program, CLUSTALX (Jeanmougin et al. 1998), three characters are used in the multiple alignment to show conservation at each site: 'star' indicates positions which have a single, fully conserved residue; 'colon' indicates that one of the following strong groups is fully conserved (STA, NEQK, NHQK, NDEQ, QHRK, MILV, MILF, HY, and FYW); and 'period' indicates that one of the following weaker groups is fully conserved (CSA, ATV, SAG, STNK, STPA, SGND, SNDEQK, NDEQHK, NEQHRK, FVLIM, and HFY).

Various computer programs such as FASTA (Fast Alignment from Pearson 1990) or BLAST (Basic Local Alignment Search Tool from Altschul et al. 1990) have been used to assess the matches between a query sequence and a subject sequence. The output contains identity values for nucleotide or protein comparisons to indicate the percent matches between the query sequence and the matching database sequence. For protein searches, similarity values are also given in the output. For example, a BLASTP analysis (protein query vs. protein database) of the *S. cerevisiae* glucosidase protein YIL099W (549 amino acids) results in the following match with the *N. crassa* glucosidase protein NCU01517: Identities = 145/469 (30%), Positives = 224/469 (47%). This means that the 549 amino acid query sequence has a 469 amino acid portion which matched a sequence in the database, and in 469 amino acid portion, 145 positions were identical, and a further 79 (= 224 - 145) amino acids were similar. In this example, the 30% identical residues and an additional 17% similar residues resulted in 47% sequence similarity as indicated by 'Positives'.

Sequence similarity does not necessarily denote functional similarity. However, the more similar two sequences are, and by implication, the more recent the shared common ancestor, the more likely the retention of similar function (Webber and Ponting 2004). Structural similarity combined with sequence similarity increases the probability of homology (Webber and Ponting 2004) and of functional similarity.

What level of sequence identity or similarity is required to establish homology? For protein sequences, it is often said that 25% to 30% identity across a large segment is enough to call homologous. However, protein sequences may be homologous, yet not share statistically significant similarity (Pearson 1997), and conversely, protein sequences may share significant similarity in particular domains, yet not be truly homologous. A statistic often used as a criterion for homology is the expect value (e-value), which refers to: "the number of hits one can expect to see just by chance when searching a database of a particular size" (www.ncbi.nlm.nih.gov/BLAST/blast_FAQs.shtml). E-value accounts for both the percent similarity and the length over which the matching occurs, such that very high similarity over only a very short stretch of sequence does not result in a strong e-value. Just as with probability values, lower e-values indicate more significant matching than higher e-values.

In many studies, e-values of 10^{-20} or less have been considered a strong match, while e-values less than 10^{-5} have often been used as the criterion for homology (e.g. Keon et al. 2000; Kruger et al. 2002; Thomas et al. 2001; Thomas et al. 2002). Pearson (1998) states that an e-value of 0.02 could be used for inferring homology with only a 2% chance of a false positive. Some researchers consider e-values less than 10^{-1} to represent biological significance of the match, and have used the e-value as a measure of statistical significance (Pertsemlidis and Fondon 2002). By increasing the e-value in a BLAST analysis, the chances are increased of detecting evolutionarily distant homologs, and some strategies for homologous gene detection involve increasing e-values above 1. However, by increasing the e-value, the chances are also increased of finding false positives. Another consideration is that e-value is directly proportional to the size of the database, such that a match against a local database, which is probably much smaller than the full GenBank database (www.ncbi.nlm.nih.gov/Genbank), will necessarily give a much higher e-value than for the exact same match as found in the GenBank database.

A further complication is that there are several distinct types of homologs: orthologs, paralogs and xenologs (Fitch 2000). Orthology is the relationship between homologous genes found in different organisms where the single ancestral gene was present in the most recent ancestor of the different organisms. Paralogy is the relationship between homologous genes which arose by gene duplication, such as members of a gene family found within the same organism. Xenology describes the relationship between two homologous genes found in different organisms where one gene was derived by lateral gene transfer into another organism. In phylogenetic analyses, if paralogs or xenologs are used in the place of orthologs, a phylogeny could result that is correct for the genes, but not for the organisms (Fitch 2000). The difficulty is that it is sometimes not possible to distinguish between these different types of homologs with the data available (Blattner et al. 1997).

4. COMPARATIVE GENOMICS

Insights into biology and evolution have been gained from studies of comparative genomics (Koonin et al. 2000, Hardison 2003) among bacteria (Fraser et al. 2000; Alekshun 2001; Fraser et al. 2002; Mira et al. 2002; Parkhill et al. 2003; Thomson et al. 2003) or eukaryotes (Rubin et al. 2000, Philip et al. 2005, Philippe et al. 2005) such as phytoplankton (Fuhrman 2003), higher plants (Bennetzen 2002; Hall et al. 2002; Schmidt 2002; Shimamoto and Kyozuka 2002; Pertea and Salzberg 2002; Resier et al. 2002; Kirst et al. 2003; Yu et al. 2005), protozoa (El-Sayed et al. 2005) or animals (Ureta-Vidal et al. 2003; Bofelli et al. 2004; Enard and Paabo 2004; Ptak et al. 2005). Through such comparisons, many secrets of a genome be revealed. For example, the tiger pufferfish (*Fugu rubripes*) was the second vertebrate genome sequenced after humans (Aparicio et al. 2002), and researchers were able to calculate the number of predicted genes conserved in both species or unique to either vertebrate. Genes conserved in these two divergent species after over 400 million years of evolution may have important functions. Although only one-ninth of the size of the human genome, the pufferfish genome has the same number of predicted genes, but with less repetitive DNA and shorter introns (Hedges and Kumar 2002).

The mouse genome was released shortly after that, and while slightly smaller that the human genome, 99% of human genes were found to have a homolog in mouse (Mouse Genome Sequencing Consortium 2002). During comparison of the two genomes, more predicted human genes were uncovered (Mouse Genome Sequencing Consortium 2002), and among the genes exclusive to mouse, many are involved in the sense of smell. Interestingly, among 33 pseudogenes uncovered in the completed sequence of the human genome, 10 may have been involved in olfactory reception (International Human Genome Consortium 2004). These pseudogenes are thought to have recently acquired one or more mutations that caused them to be nonfunctional, and among the 33, five were found to be still functional in chimpanzees (International Human Genome Consortium 2004). Increasing the number of genomes compared also increases the likelihood of detecting conserved sequences which are functional (Bofelli et al. 2004).

Chimpanzees (*Pan troglodytes*) are the closest relative to humans having diverged 5 to 7 million years ago, and the comparative genome analysis was released in late 2005 (Chimpanzee Sequencing and Analysis Consortium 2005). The sequences that can be directly compared between the two genomes are almost 99% identical, but when insertions and deletions are also considered, the similarity is closer to 96%. Compared to other mammals, certain classes of genes were found to be evolving more quickly in humans and chimpanzees including ones related to sound perception, nerve signal transmission, sperm production, and ion transport. More than 50 genes found in the human genome were not found in the chimpanzee genome.

5. PHYLOGENETICS

Complete-genome comparative analyses may also provide more definitive answers on phylogenetic assignments of organisms. Wolf et al. (2001) used different methods of tree construction based on complete genome data from diverse taxa of bacteria, and concluded that there were two primary prokaryotic domains. Datasets from the genomes of seven *Saccharomyces* species consisting of a few or a small number of genes often gave rise to conflicting topologies, whereas combined analysis of 8 or more genes yielded a tree with moderate bootstrap support (all branches over 70%), and a combined analysis of 20 or more genes yielded a single fully resolved tree with over 95% bootstrap support at all branches (Rokas et al. 2003). The implication of this research is that a larger number of genes is required in phylogenetic analyses to give more resolution.

Although full genome comparisons should seem to be able to settle questions in systematics, there are several issues that need consideration and further investigation. Soltis et al. (2004) demonstrated that even when whole genomes are used, if the number of taxa used is low, incorrect phylogenetic reconstructions can be obtained. A major controversy in metazoan systematics is the relationship among vertebrates, arthropods and nematodes. The Coelomata hypothesis argues that arthropods and vertebrates are more closely related because they have a true body cavity, while the Ecdysozoa hypothesis places arthropods as a sister group to nematodes. Using available genomic data, two research groups came to different conclusions, with Philip et al. (2005) supporting the Coelomata hypothesis while Philippe et al. (2005) supported the Ecdysozoa hypothesis. This demonstrates that even when starting with the same or similarly large sets of sequence data, different conclusions can be obtained depending on the analyses.

For species where multiple genomes have been sequenced or studied, researchers have found significant intraspecific variability (Bergthorsson and Ochman 1995). For bacterial species, these differences can as large as 11% for *Salmonella enterica* (McClelland et al. 2001) and 10% for *Pseudomonas aeruginosa* (Spencer et al. 2003). For *P. aeruginosa*, Spencer et al. (2003) concluded that loss, gain or rearrangements of large blocks of DNA were responsible for the significant intraspecific variability. The normal nucleotide substitution rate of 0.5% leads to some divergence between genomes (Spencer et al. 2003), and between any two humans, there is an average of 0.1% difference (Maher 2003). However, humans are different from most other species in having such a narrow genetic range, approaching that of asexually-reproducing species such as *Mycobacterium tuberculosis*, where variation is expected to be low (Kato-Maeda et al. 2001). For fungi, there may also be variable chromosome numbers (Covert 1998) and chromosome lengths (Plummer et al. 1993; Zolan 1995; Plummer et al. 1995; Dewar et al. 1997), in addition to variations in gene sequences between genomes of the same species. These factors could give rise to tremendous differences in genomic sequences, and the use of a particular genome in a phylogenetic assay could lead to biased results if the genome were not representative of the species.

A further consideration is that although genomes are said to be completely sequenced, they still contains gaps and usually exclude multiple copies of ribosomal genes and highly repetitive sequences. For example, the completed version of the human genome still contains 341 gaps that require new technology to complete (International Human Genome Consortium 2004). For the fungal genomes presented in Table 1, the statistics range from 90% to 100% complete. If gene absence or presence is used as an indicator of evolutionary relatedness (Huson and Steel 2004), then the occurrence of gaps and missing information in genomes could have a large effect on the results.

6. UNIQUE TARGET SITES IN PESTS

One of the major purported uses of microbial comparative genomics has been the discovery of antimicrobial target sites. By comparing the genomes of the host and of the pathogen, or of the pathogen and a species similar to the pathogen but non-pathogenic, insights can be gained into target sites for antimicrobial activity including novel fungicide target sites. Hsiang and Baillie (2005) found 17 uniquely fungal genes in their analyses of 14 fungal genomes compared with 2 genomes each of plants, animals and bacteria. They pointed out that seven of these 14 genes were already listed in U.S. patents dealing with antifungal drug discovery. Kessler et al (2002) compared 3000 cDNA sequences from *A. fumigatus* against genomes of three yeasts: *Saccharomyces cerevisiae*, *Schizosaccharomyces pombe*, and *Candida albicans*. They found that 49% of the clones did not have a match at e-value $\leq 10^{-5}$, and concluded that these could be *A. fumigatus*-specific genes that could be used as potential candidates for novel antifungal targets specific to this fungus. Caution must be taken with this approach to antimicrobial research, since many agricultural pesticides which turned out to have strong non-target effects often affected sites in the host or other non-target organisms which were not homologous to the target site in the pest. For example, the insecticide DDT which affects the nervous system in insects turned out to also cause egg-shell thinning in birds, but the mechanism of action is not the same (Mellanby 1992). Similarly, many human therapeutic drugs turned out to have side-effects which are not related to their target sites. Despite these limitations, a major direction in the use of microbial sequences is to identify specific targets for inhibitor-based drug design (Wu et al. 2003). By searching for gene families that may be important in parasitic or pathogenic activities, and by comparing the presence of these genes in other organisms, specific targets for chemical inhibition may be identified. Many researchers have mentioned this issue as a strength of comparative genomics, and claim that it may be able to pinpoint novel target sites in pathogens which are absent in the host (e.g. Kessler et al. 2002). A more comprehensive method of characterizing pharmacological targets may involve phylogenomics, where the evolutionary analyses of potential target sites are also considered (Searls 2003).

7. GENE PREDICTION AND GENE FUNCTION

While gene sequences are likely to be very accurate, with the level of error estimatable based on the sequencing procedure used, annotation involves interpretation of the sequence and is often subject to error (Parkhill 2002), particularly if the annotation is automated (Nierman et al. 2005). Gene prediction algorithms are based first on finding open reading frames (ORF) larger than a given size (usually 100 aa), which have a start and stop codon in the same reading frame, and then determining whether the coding sequence has properties such as G+C content similar to known coding sequences in that organism (Parkhill 2002). In addition to similarity searches to assign function, there are non-similarity methods such as physically proximity and frequent co-occurrence (Parkhill 2002). Cliften et al. (2001) used comparative sequence analysis to identify conserved functional elements in several *Saccharomyces* genomes to predict genes. Kellis et al. (2003) compared the genomes of four *Saccharomyces* species (*S. cerevisiae, S. paradoxus, S. mikatae* and *S. bayanus*), and found a high degree of synteny across the genomes. By examining regulatory motifs and analyzing conservation of predicted gene sequences, they concluded that the proteome of *S. cerevisiae* could be reduced by approximately 500 predicted genes.

Once gene sequences are identified, how is function determined? Lockhart and Winzeler (2000) claim that "guilt by association" can allow for many groups of sequences to be simultaneously classified, since strong correlations between expression profiles may indicate similar functional assignments. Uetz et al. (2000) applied this concept in their two-hybrid analysis of protein interactions in yeast, and were able to identify interactions between proteins of known and unknown function, and shed light both on the existence of the interactions and on the possible roles of the proteins with undescribed function. Date and Marcotte (2003) extended this by using phylogenetic profiles to analyze pairwise coinheritance of genes within genomes to predict thousands of functional linkages and identify large-scale cellular systems. Nardone et al. (2004) describe how the use of conserved non-coding regulatory regions in cross-species comparisons can give insights into homologous transcriptional regulation.

The annotation of gene functions is a major bottleneck in genomics (Pallen 2002), and is one reason for the delay between genome release and publication (Table 1). Most genes have not yet been characterized. For example, although ~4000 of ~6000 predicted genes in yeast have been annotated (Cherry et al. 1998), it is not known how many of these annotations are accurate. In 2005, 8 years after this first eukaryotic genome was released in 1997, still only 66% of the 6591 open reading frames in *S. cerevisiae* were considered verified and characterized (http://www.yeastgenome.org/cache/genomeSnapshot.html), while 22% were uncharacterized and 12% were considered dubious.

When analysis of the first draft of the human genome was published (International Human Genome Consortium 2001), they estimated 30,000 to 40,000 protein-coding genes. Before the draft was released, estimates ranged up to 120,000

(Liang et al. 2000), and other estimates based on the draft gave 65,000 to 75,000 transcriptional units (Wright et al. 2001). In 2004, the International Human Genome Consortium published on the completed human genome, and revised the estimate down to 22,287 gene loci with a total of 34,214 transcripts. Imanishi et al. (2004) investigated the function of 19,574 protein-coding human genes that were derived from experimental evidence, and were able to assign 50.1% of them to a functional group.

Predicted genes are often given a functional annotation that is derived from the BLAST hit with the lowest e-value, but this assignment of function makes the assumption that sequence similarity is equivalent to functional similarity, and, as discussed above, this is not always the case. Once an erroneous annotation is provided, it may become propagated throughout different databases and the original evidence may become difficult to track down (Pallen 2002). For example, Bridge et al. (2003) examined over 200 fungal ribosomal RNA sequences from publicly available databases, and concluded that 20% appeared to be misidentified, dubious or chimeric with 38% not linked to traceable material.

Comparative genomics provides a major route for the study of functional genomics. We may discover what is occurring in one organism because the same thing happens in another organism. Since model organisms such as *Saccharomyces cerevisiae* for fungi, *Arabidopsis thaliana* for plants, and *Caenorhabditis elegans* for nematodes, are among the best studied organisms in their respective taxa and have been completely sequenced, determination of gene function in one of these more easily manipulated organisms often gives insight into homologous functions in higher or larger organisms. Rehm (2001) discusses some methods involved in sequence analyses including functional assignment of genes. There are attempts to classify genes from a variety of organisms into functional classes such as GO (Gene Ontology)(Gene Ontology Consortium 2000), COG (Cluster of Orthologous Genes) (Rashidi and Buehler 2000; Tatusov et al. 2000), MIPS (Martinsreid Institute for Protein Sequences) (Mewes et al. 1997b), and InterPro (InterPro Consortium 2001).

For genes without known function, one method to determine function is by gene knockout (Capecchi 1989). Prior to this breakthrough technique, researchers had already developed gene transfer technology in mice in the early 1980's, but they could neither control nor predict where the transgene would be inserted into the genome of the target organism (Pray 2002). Using homologous recombination, Cappechi (1989) demonstrated that the transgene could be precisely aimed at a target site in the genome and the replacement of a specific gene with an inactive or mutated allele would knock out the function of this gene (Pray 2002). Other more recent methods for assessing gene function include RNA interference (RNAi) (Fire et al. 1998) and Targeted Induced Local Lesions in Genomes (TILLING) (Till et al. 2003).

Gene expression technologies are developing rapidly, and RNA detection includes standard procedures such as northern blots, RT-PCR (reverse transcription of RNA followed by PCR), cDNA sequencing, differential display, and more recently derived procedures such as microarray analyses (Lockhart and Winzeler 2000), serial analysis of gene expression (SAGE, Velculescu et al. 1995) and analyses of expressed sequence tags (ESTs) (Soanes et al. 2002). ESTs are the fastest growing segment in

GenBank, and Jongeneel (2001) presents a good overview of searching for genes in EST databases. These technologies for establishing gene function and expression are still developing, but the technologies for genomic sequencing have advanced at a far greater rate, and unexplored or lightly explored sequence data are accumulating exponentially.

8. COMPARATIVE GENOMICS BETWEEN FUNGI AND OTHER ORGANISMS

A genome represents the complete set of genes of an organism. This set includes all the instructions for maintenance, defense, growth and reproduction of the organism, and while a smaller genome is less expensive to maintain, it lacks the genetic flexibility of larger genomes (Fuhrman 2003). With greater complexity and larger genome sizes, the proportion of genes in a genome which can be found in other genomes in publicly available databases decreases. For prokaryotes, ~70% of the genes in any genome may be identified in other organisms, perhaps also reflecting the greater number of prokaryotic genomes available (Braun et al. 2000). For *S. cerevisiae*, which has one of the smallest eukaryotic genomes, more than 60% of the genes have a match in at least one other organism (Braun et al. 2000). However, for more complex eukaryotes such as *Caenorhabditis elegans* or *Arabidopsis thaliana*, the proportion of genes that have a match in other organisms is much smaller (Braun et al. 2000). Zeng et al. (2001) found almost 1000 human proteins with higher similarity to homologs in fungal genomes than in other animals, such as *C. elegans* or *Drosophila melanogaster*, and concluded that functional genomics with human genes should involve yeasts and higher fungi.

A massive comparative study of the genomes of *D. melanogaster, C. elegans,* and *S. cerevisiae* was conducted by over 50 researchers (Rubin et al. 2000) representing a wide array of agencies. They found that the two animal genomes had nonredundant protein sets which were similar in size and twice that of yeast, and that the multidomain proteins and signaling pathways in the animals were more complex than those of yeast. Another massive comparative genomics study (Thomas et al. 2003) compared a large genomic region in 13 vertebrate species including human, other primates, cat, dog, cow, pig, chicken, rodents, and fishes. Their analysis supported the closer phylogenetic relationship of primates to rodents than to the other mammals listed. They identified DNA segments that were conserved across a wide range of species but apparently not coding for any proteins. Non-coding DNA can represent a large part of the genome of an organism, such as 98% of the DNA in *Homo sapiens*, but some of this non-coding DNA actually contains hidden genes that work through RNA (Gibbs 2003). Roy and Gilbert (2005) examined the pattern of intron conservation in eukaryotes using seven fully sequenced genomes. They found that modern introns generally are very old and that 40% of the introns found in animals, plants and fungi date to their common ancestor.

There are also attempts using comparative genomics to distinguish between genes of the pathogen and that of the most in mixed libraries. Hsiang and Goodwin (2003) used the complete genomes of a plant and a fungal pathogen to assess the origin of ESTs from fungal-infected plant tissues. In trials with pure fungal or pure plant sequences, they showed that their method was better able to place the taxonomic origin of the sequences than a comparison with the GenBank NR database, and explained that since so many more plant genes have been investigated than fungal genes, a best match to a plant sequence from GenBank did not necessary ensure that the query sequence was of plant origin. Xu et al. (2003) used a similar method involving computational subtraction with human genome sequences to remove the human component from a cDNA library of virus-infected human tissue (27,840 sequences). They then designed primers for the remaining 32 non-matching sequences, and attempted to amplify these sequences from infected and non-infected tissues. Twenty-two were found to amplify from uninfected tissues, leaving 10 sequences, and all 10 of these sequences were found to match viral sequences (Xu et al. 2003). A major advantage of studying a human disease is that complete genomic data may be available for both the host and the pathogen, while for plant diseases, it is rare to have complete genomic sequences for both the host and pathogen. Furthermore, for fungal plant diseases, both the host and pathogen are eukaryotes and hence their sequences may be more difficult to distinguish, unlike human diseases where the important pathogens are mostly bacterial or viral.

9. FUNGAL COMPARATIVE GENOMICS

Fungal comparative genomics can be used to address many very fundamental questions in biology and evolution. As noted by Goswami and Kistler (2004), comparative genomics can give insights into evolution of gene clusters (Ward et al. 2002) and gene family expansions and extinctions (Kroken et al. 2003), and gene prediction using a reading frame conservation test (Kellis et al. 2003). Comparative analyses will also provide information on gene dispersion and loss, genome rearrangements, the acquisition of species-specific genes, and other mechanisms which should be applicable to eukaryotes in general (Goffeau 2004). Because of the greater number of fungal genomes currently available and soon to become available, comparative genomics with fungi should continue to be at the leading edge of the field of eukaryotic comparative genomics.

The complete genome sequences of particular fungal species also allows a full inventory of genes that might be related to sexual reproduction, particularly in species that are considered to be asexual. For example, the presence of certain types of mating genes in the genomic sequence of *Aspergillus fumigatus* suggested that it is able to mate and undergo meiosis (Paoletti et al. 2005). Similarly, Wong et al. (2003) found through datamining, the presence of genes involved mating and meiosis in the presumed asexual yeast, *Candida glabrata*. Although mating type genes can be found by using degenerate primers (e.g. Hsiang et al. 2003), such attempts have not always proven successful. The availability of complete genomic sequences provides the opportunity to datamine genomes for the presence of genes that might be involved in reproduction.

Yeast comparative genomics continues to be a highly active area of research (Grunfelder and Winzeler 2002, Dujon et al. 2004, Kellis et al. 2004, Piskur and Langkjaer 2004, Rokas and Carroll 2005, Fabre et al. 2005). Among the 43 genomes listed in Table 1, 18 are species of yeasts. Yeasts generally have smaller genome sizes than filamentous fungi, and were among the earliest genomes sequenced. For genera such as *Candida* and *Saccharomyces,* multiple species have been sequenced which allows for evolutionary comparisons within genera and between these two genera which diverged over 100 million years ago (Berbee and Taylor 2001, Heckman et al. 2001).

Other recent studies in fungal comparative genomics include a survey of *Aspergillus* species (Archer and Dyer 2004), since the genomes of four *Aspergillus* species have been sequenced (but not all are publicly available). Tekaia and Latge (2005) compared *A. fumigatus* to other fungal genomes and concluded that based on the presence of certain types of genes and enzymatic machinery, that *A. fumigatus* is a saprophyte and opportunistic invader of humans. Nierman et al. (2005) reviewed the progress on comparative genomics among *Aspergillus* species, and stated that the species are distantly related (compared to congeneric taxa among plants or animals), and that only 50% of each genome can be aligned with the corresponding region of the other genomes.

10. FUNGAL COMPARATIVE GENOMICS - EVOLUTIONARY BIOLOGY

Cliften et al. (2003) compared the genomes of six *Saccharomyces* species to find functional non-protein-coding sequences, such as gene regulatory elements. These are generally difficult to recognize because they are often short, degenerate and can be distant from the genes they control. By finding these "phylogenetic footprints", the authors were able to revise the catalog of yeast predicted genes, and to identify motifs that may be targets of transcriptional regulatory proteins. Schoch et al. (2003) inventoried the kinesin gene families in three filamentous fungi, *Botryotinia fuckeliana, Cochliobolus heterostrophus,* and *Gibberella moniliformis,* and compared these to two yeasts, *Saccharomyces cerevisiae* and *Schizosaccharomyces pombe*. They found that the filamentous species contained a constant set of 10 kinesins in nine subfamilies while the yeasts had much fewer kinesins. Kellis et al. (2004) compared the genomes of *S. cerevisiae* and *Kluyveromyces waltii* and concluded that *S. cerevisiae* arose from an ancient whole-genome duplication.

Zelter et al. (2004) looked for homologs of yeast calcium signalling machinery in *Neurospora crassa* and *Magnaporthe grisea* in a comparative genomics study. They found a greater number of homologs for various calcium signalling genes in the filamentous fungi than in yeast, and speculated that there was greater complexity in the filamentous forms because of their more complex cellular organization and possibly greater range of external signals in their natural habitats.

Dietrich et al. (2004) compared *S. cerevisiae* to the genome of *Ashbya gossypii*, a filamentous, ascomycetous, plant pathogen with a very small genome size (9.2 Mb). They found, using BLAST and FASTA, that 95% of the *A. gossypii* genes showed homology with *S. cerevisiae* genes, with percent identity values from 19% to 100%. Among *A. gossypii* genes, 90% showed homology and synteny with *S. cerevisiae* genes, 5% showed homology but not synteny, and 5% did not show homology, but were considered to be real genes because of the presence of homologues in other species. Through these comparisons, they found evidence that *S cerevisiae* resulted from a whole genome duplication or fusion of two related species.

Nielsen et al. (2004) examined intron loss and gain in four ascomycete species (*Magnaporthe grisea, Neurospora crassa, Fusarium graminearum,* and *Aspergillus nidulans*). Since the time of their divergence from the most recent comment ancestor over 300 million years ago, there have been up to 250 intron gains and 350 intron losses in each lineage, and the authors suggest that intron gain has been a major driving force in the evolution of fungi.

Fungi are good model organisms for the study of evolutionary biology using comparative genomics. First, the number of fungal genomes that have been sequenced is greater than that for other major eukaryotic taxa. Second, the relatively small and compact fungal genomes facilitate computational analyses. Third, ascomycetous yeast species alone cover the evolutionary range comparable to the entire phylum of chordates (Hedges and Kumar 2003).

11. FUNGAL COMPARATIVE GENOMICS - FUNGAL BIOLOGY

Papp et al. (2003) used genomic sequences of *S. cerevisiae* to search for paralogs (e-value $\leq 10^{-2}$) to identify gene family size. Then they compiled a list of interacting protein pairs which did not belong to the same gene family, and found that out of almost 7000 pairs, over 4300 had the two members with the same-sized gene families. They also found that members of large gene families were rarely involved in complexes, and supported the assertion that dominance is a by-product of physiology and metabolism rather than the result of selection to mask the effects of deleterious mutations (Papp et al. 2003).

Tzung et al. (2001) compared *C. albicans* with *S. cerevisiae* to assess whether genes important for sexual reproduction and meiosis might be present in *C. albicans*. The complete repertoire of genes related to sexual reproduction was not found, leading to the suggestion that *C. albicans* has alternative mechanisms of genetic exchange. Fungi are known to undergo asexual recombination under the parasexual cycle (Pontecorvo 1956), and the presence of homologs to genes involved in vegetative incompatibility suggests that this may be a method by which *C. albicans* generates genetic variation (Tzung et al. 2001).

Wagner (2000) examined the ability of *S. cerevisiae* to compensate for mutations and concluded that interactions among unrelated genes are the major cause of robustness against mutations. Gu et al. (2003) continued this line of research by studying a near complete set of single-gene-deletion mutants of *S. cerevisiae* with functional annotations. They found that for genes with paralogs, there was a greater probability of functional compensation than for singleton genes (Gu et al. 2003). They estimated for *S. cerevisiae*, that of the gene deletions which resulted in no phenotypic change, 25% were because of compensation by duplicate genes, and at least some of the remaining were because of alternative pathways.

Yoder and Turgeon (2001) compared the occurrence of selected protein families in genomes of selected pathogenic and saprophytic fungi, and concluded that the plant pathogens *Cochliobolus sativus, Fusarium graminearum,* and *Botrytis cinerea* have more genes dedicated to secondary metabolism than do saprophytes such as *Neurospora crassa, Ashbya gossypii,* and *S. cerevisiae*. They found that the three plant pathogenic fungi were rich in peptide synthetases and polyketide synthases, some of which are known to be virulence factors (Kroken et al. 2003) , whereas the saprophytes encoded few or none of these proteins. Yarden et al. (2003) contend that searches for differences between plant pathogenic fungi and nonpathogenic ones can be confounded when orthologous genes are present in both types of organisms, but the orthologous pathways may not be; hence, direct comparisons of presence or absence may be an oversimplification.

Gardiner and Howlett (2005) used previously characterized genes involved in sirodesmin biosynthesis in *Leptosphaeria maculans* to uncover a cluster of 12 genes putatively involved in gliotoxin production in *Aspergillus fumigatus*. The gliotoxin-related genes were identified by comparative genomics, since both gliotoxin and sirodesmin are epipolythiodioxopiperazine toxins. Further experimental work quantified gene expression using quantitative RT-PCR, and identified genes that were co-regulated and showed expression of timing correlated with gliotoxin production as measured by HPLC.

12. FUNGAL COMPARATIVE GENOMICS - ESSENTIAL FUNGAL GENES

Braun et al. (2000) conducted a whole genome comparison between *Saccharomyces cerevisiae* and *Neurospora crassa*. They found that *N. crassa*, with its larger genome, has more unique genes than *S. cerevisiae* by making comparisons with the GenBank protein database. The presence of a gene in *N. crassa* that could also be found in other organisms but not in *S. cerevisiae* was interpreted as gene loss from *S. cerevisiae*. They were also able to find genes in *N. crassa* that were not found in any non-fungal species in GenBank, and postulated that these were fungal-specific proteins (Braun et al. 2000).

Firon and d'Enfert (2002) reviewed some of the methods for identifying essential genes in fungal pathogens of humans, including transposon mutagenesis and post-transcriptional gene silencing. They contend that the characterization of genes essential for growth in fungal pathogens is an important step in development of novel antifungal drugs, as well as providing insights into biological diversity of fungi.

Decottignies et al. (2003) used a PCR-based gene deletion procedure on 100 genes of *S. pombe* and found that 17.5% of these deletions were of essential genes. They then compared 450 proteins from two yeasts (*S. cerevisiae* and *S. pombe*) with those of Metazoa, plants and prokaryotes in the GenBank nonredundant protein database, and estimated that 80% of the essential genes of *S. pombe* were shared with other eukaryotes, with half of these genes also found in prokaryotes, while only 10% of essential genes were fungal specific. Similar numbers were found for *S. cerevisiae*, with the criterion for homology at e-value $\leq 10^{-5}$. With a greater number and taxonomic range of fungal genomes being sequenced every year, our ability to uncover genes which are conserved across many fungal taxa will be enhanced. We may then be able to determine which genes are exclusively fungal that help make fungi distinctive from other organisms.

Strobel and Arnold (2004) compared cDNAs from the AIDs-related fungal pathogen *Pneumocystis carinii* to the saprophytes *Schizosaccharomyces pombe* and *Saccharomyces cerevisiae*. They identified 200 sequences shared with these other fungi and considered these to be essential genes. Because the cDNA library was thought to include half of all *P. carnii* genes, they then estimated the essential eukaryotic core to be approximately 400 genes.

Hsiang and Baillie (2005) searched for homologs of *Saccharmoyces cerevisiae* genes among 13 other fungal species. They found that out of the 6355 putative

Saccharomyces cerevisiae genes, 3340 were present in at least 12 other fungal genomes (at e-value $\leq 10^{-5}$). Of these 3340 genes, 938 had homologs in plants, animals and bacteria, while 17 were found to lack homologs in non-fungal species. These 17 core fungal genes did not seem to share peculiarities in GC content, codon usage patterns, or putative functional characteristics, and only one of these was considered to be essential from gene deletion studies.

13. FUNGAL COMPARATIVE GENOMICS - SMALL SCALE STUDIES

Bioinformatic tools are necessary to process the enormous amounts of genomic data that are generated. These tools include gene-matching algorithms, such as BLAST, and processing of output from such programs with computer scripts specifically written for these activities in languages, such as PERL (practical extraction and report language) (Tisdall 2003). As biologists, our goal in genomic studies is to enhance our understanding of the biology of the organisms and generally not just to catalogue the component parts (Lockhart and Winzeler 2000). Analytical tools are available to handle the masses of genetic data to generate results, but making biological interpretations from the results is a daunting task (Lockhart and Winzeler 2000). Most biologists do not consider themselves to be bioinformatics-enabled, but new computer programs should reduce the complexity of bioinformatic tools (Buckingham 2003). These tools are being directed toward the exponentially increasing amounts of genetic data, as well as toward categorizing the ever growing number of publications related to analysis and interpretation of such data (Buckingham 2003). These tools are generally freely available and can be downloaded from many websites on the Internet.

Many articles on comparative genomics studies have been written with a multitude of authors, arising from labs that may have both high-powered molecular biology and computational tools; however there is still a role for smaller research labs in comparative genomics. The fact that the massive computing power available to a super-computing center may be able to process all the data and make the sequence comparisons in one day, a task which may take several months for a smaller research program to conduct, doesn't outweigh the fact that the smaller research programs may come up with important novel ideas for an analysis which haven't been considered by the larger research programs. Although the learning curve can be quite steep for biologists, comparative genomic analyses can be conducted on common desktop computers using Windows, Mac, or Linux operating systems, and the results of these types of analysis can be very rewarding. Furthermore, genomes databases have been set up which allow users to search for homologs, and find current information on the annotation and physical location of loci in particular genomes. The January 2005 supplemental issue (Volume 33 Database Issue) of Nucleic Acids Research was devoted to descriptions of available genomic database resources.

14. CONCLUSION

This article has discussed just a few of the discoveries that are possible using comparative genomics, and certainly many more are possible. We encourage mycologists and plant pathologists to explore the use of the new tools of bioinformatics. After all, biologists do not usually hand over their data to statisticians for analysis and interpretation, but undertake the data analysis with the help of statisticians, since extensive training in biology is required to make many of the important biological interpretations from the results of statistical analyses of biological data. Similarly, with the ever-burgeoning amounts of sequence data, there is plenty for researchers to analyze to bring forth important discoveries of biological significance.

Acknowledgements: We gratefully acknowledge the research support provided by the Natural Sciences and Engineering Research Council of Canada.

REFERENCES

Alekshun MN (2001). Beyond comparison - antibiotics from genome data? Nature Biotech 19:1124-1125.

Altschul SF, Gish W, Miller W, Myers EW and Lipman DJ (1990) Basic local alignment search tool. J Mol Biol 215:403-10.

Anon (2003). Sacrifice for the greater good. Nature 421:875.

Aparicio S, Chapman J, Stupka E, Putnam N, Chia JM, Dehal P, Christoffels A, Rash S, Hoon S, Smit A etc. (2002). Whole-genome shotgun assembly and analysis of the genome of *Fugu rubripes*. Science 297:1301-1310.

Archer DB and Dyer PS (2004). From genomics to post-genomics in Aspergillus. Curr Op Microbiol 7: 499-504.

Bell E (2000) Publication rights for sequence data producers. Science 290:1696-1698.

Bennett JW and Arnold J (2001) Genomics for fungi. In: RJ Howard and NAR Gow, ed.The Mycota VIII: Biology of the fungal cell. Berlin: Springer-Verlag GmbH & Co, pp. 267-297.

Bennetzen J (2002) Opening the door to comparative plant biology. Science 296:60-63.

Berbee ML and JW Taylor (2001) Fungal molecular evolution: gene trees and geologic time. In: DJ McLaughlin, EG McLaughlin and PA Lemke, ed. The Mycota VII: Systematics and Evolution. Berlin: Springer-Verglab GmbH & Co, pp. 229-245.

Bergthorsson U and Ochman H (1995) Heterogeneity of genome sizes among natural isolates of *Escherichia colia*. J Bacteriol 10:5784-5789.

Bernal A, Ear U, Kyrpides N (2001) Genomes OnLine Database (GOLD): a monitor of genome projects world-wide. Nucl Acid Res 29:126-127.

Blattner FR, Plunkett G, Bloch CA, Perna NT, Burland V, Riley M, Collado-Vides J, Glasner JD, Rode CK, Mayhew GF, Gregor J, Davis NW, Kirkpatrick HA, Goeden MA, Rose DJ, Mau B and Shao Y (1997) The complete genome sequence of *Escherichia coli* K-12. Science 277:1453-74.

Bofelli D, Nobrega MA and Rubin EM (2004) Comparative genomics ate the vertebrate extremes. Nat Rev Genet 5:456-465.

Bos JIB, Armstrong M, Whisson SC, Torto TA, Ochwo M, Birch PRJ, and Kamoun S (2003) Intraspecific comparative genomics to identify avirulence genes from *Phytophthora*. New Phytol 159:63-72.

Braun EL, Halpern AL, Nelson MA and Natvig DO (2000) Large-scale comparison of fungal sequence information: mechanisms of innovation in *Neurospora crassa* and gene loss in *Saccharomyces cerevisiae*. Genome Res. 10:416-430.

Bridge PD, Roberts PJ, Spooner BM and Panchal G (2003) on the reliability of published DNA sequences. New Phytol 160:43-48.

Buckingham S (2003) Programmed for success. Nature 425:209-215.

Capecchi MR (1989) Altering the genome by homologous recombination. Science 244:1288-1292.

Cherry M, Adler C, Ball C, Chervitz SA, Dwight SS, Hester ET, Jia Y, Juvik G, Roe T, Schroeder M, Weng S and Botstein D (1998) SGD: *Saccharomyces* Genome Database. Nucl Acid Res. 26:73-80.

Chimpanzee Sequencing and Analysis Consortium (2005) Initial sequence of the chimpanzee genome and comparison with the human genome. Nature 437:69–87.

Cliften PF, Hillier LW, Fulton L, Graves T, Miner T, Gish WR, Waterston RH Johnston M (2001) Surveying *Saccharomyces* genomes to identify functional elements by comparative DNA sequence analysis. Genome Res 11:1175-1186.

Cliften PF, Sudarsanam P, Desikan A, Fulton L, Fulton B, Majors J, Waterston R, Cohen BA Johnston M (2003) Finding functional features in *Saccharomyces* genomes by phylogenetic footprinting. Science 301:71-76.

Covert SF (1998) Supernumerary chromosomes in filamentous fungi. Curr Genet 33:311-319.

Date SV and Marcotte EM (2003) Discovery of uncharacterized cellular systems by genome-wide analyses of functional linkages. Nat Biotech 21:1055-1062.

Dean RA, Talbot NJ, Ebbole DJ, Farman ML, Mitchell TK, Orbach MJ, Thon M, Kulkarni R, Xu JR, Pan H, Read ND, Lee YH, Carbone I, Brown D, Oh YY, Donofrio N, Jeong JS, Soanes DM, Djonovic S, Kolomiets E, Rehmeyer C, Li W, Harding M, Kim S, Lebrun MH, Bohnert H, Coughlan S, Butler J, Calvo S, Ma LJ, Nicol R, Purcell S, Nusbaum C, Galagan JE and Birren BW. 2005. The genome sequence of the rice blast fungus *Magnaporthe grisea*. Nature 21:980-986.

Decottignies A, Sanchez-Perez I and Nurse P (2003) *Schizosaccharomyces pombe* essential genes: a pilot study. Genome Res 13:399-406.

Dennis C (2003) Draft guidelines ease restrictions on use of genome sequence data. Nature 421:877-878.

Dewar K, Bousquet J, Dufour J and Bernier L 1997. A meiotically reproducible chromosome length polymorphism in the ascomycete fungus *Ophiostoma ulmi* (sensu lato). Mol Gen Genet 255:38-44.

Dietrich FS, Voegeli S, Brachat S, Lerch A, Gates K, Steiner S, Mohr C, Pohlmann R, Luedi P, Choi S, Wing RA, Flavier A, Gaffney TD and Philippsen P (2004) The *Ashbya gossypii* genome as a tool for mapping the ancient *Saccharomyces cerevisiae* genome. Science 304:304-307.

Doyle JJ and Gaut BS (2000) Evolution of genes and taxa: a primer. Plant Mol Biol 42:1-23.

Dujon B, Sherman D, Fischer G, Durrens P, Casaregola S, Lafontaine I, De Montigny J, Marck C, Neuveglise C, Talla E, etc. (2004) Genome evolution in yeasts. Nature 430:35-44.

El-Sayed NM, Myler PJ, Blandin G, Berriman M, Crabtree J, Aggarwal G, Caler E, Renauld H, Worthey EA, Hertz-Fowler C, etc. (2005) Comparative genomics of trypanosomatid parasitic protozoa. Science 309:404-9

Enard W and Pääbo S (2004) Comparative primate genomics. Ann Rev Genomics Hum Genet 5:351-78.

Fabre E, Muller H, Therizols P, Lafontaine I, Dujon B and Fairhead C (2005) Comparative genomics in hemiascomycete yeasts: evolution of sex, silencing and subtelomeres. Mol Biol Evol 22:856-73.

Fire A, Xu S, Montgomery MK, Kostas SA, Driver SE and Mello CC (1998) Potent and specific genetic interference by double-stranded RNA in *Caenorhabditis elegans*. Nature 391:806-811.

Firon A and d'Enfert C (2002) Identifying essential genes in fungal pathogens of humans. Trends Microbiol 10:456-462.

Fitch WM (2000) Homology, a personal view on some of the problems. Trends Genet 16:227-231.

Fraser CM, Eisen JA and Salzberg SL (2000) Microbial genome sequencing. Nature 406:799-803.

Fraser CM, Eisen JA, Nelson KE, Paulsen IT and Salzberg SL (2002) The value of complete microbial genome sequencing (You get what you pay for). J Bacteriol 184:6403-6405.

Fuhrman J (2003) Genome sequences from the sea. Nature 424:1001-1002.

Galagan JE, Calvo SE, Borkovich KA, Selker EU, Read ND, Jaffe D, FitzHugh W, Ma LJ, Smirnov S, Purcell S, etc. (2003) The genome sequence of the filamentous fungus *Neurospora crassa*. Nature 422:859-868.

Gardiner DM and Howlett BJ (2005) Bioinformatic and expression analysis of the putative gliotoxin biosynthetic gene cluster of As pergillus fumigatus. FEMS Microbiology Letters 248:241-248.

Gene Ontology Consortium (2000) Gene Ontology: tool for the unification of biology. Nature Genet 25: 25-29.

Gibbs WW (2003) The unseen genome: gems among the junk. Sci Amer 289(5):46-53.

Goffeau A (2004) Evolutionary genomics: seeing double. Nature 430:25-26.

Goswami RS and Kistler C (2004) Heading for disaster: *Fusarium graminearum* on cereal crops. Molecular Plant Pathology 5:515-525.

Grunenfelder B and Winzeler EA. (2002) Treasures and traps in genome-wide data sets: case examples from yeast. Nat Rev Genet 3:653-661.

Gu Z, Steinmetz L.M, Gu X, Scharfe C, Davis RW and Li WH (2003) Role of duplicate genes in genetic robustness against null mutations. Nature 421:63-66.
Hall AE, Fiebig A, and Preuss D (2002) Beyond Arabidopsis genome: opportunities for comparative genomics. Plant Physiol 129:1439-1447.
Hardison RC (2003) Comparative Genomics. PLoS Biol 1(2):e58
Heckman DS, Geiser DM, Eidell BR, Stauffer RL, Kardos NL and Hedges SB (2001) Molecular evidence for the early colonization of land by fungi and plants. Science 293:1129-1133.
Hedges SB and Kumar S (2002) Vertebrate genomes compared. Science 297:1283-1285.
Hedges SB and Kumar S (2003) Genomic clocks and evolutionary timescales. Trends Genet 19:200-206.
Hofman G, McIntyre M and Nielsen J (2003) Fungal genomics beyond *Saccharomyces cerevisiae*. Curr Opin Biotech 14:226-231.
Hsiang T and Baillie DL (2004) Recent progress, developments and issues in comparative fungal genomics. Can J Plant Pathol 26:19-30.
Hsiang T and Baillie DL (2005) Comparison of the yeast proteome to other fungal genomes to find core fungal genes. J Mol Evol 60:475-483.
Hsiang T and Goodwin PH (2003) Distinguishing plant and fungal sequences in ESTs from infected plant tissues. J Microbiol Meth 54:339-351
Hsiang T, Chen F and Goodwin PH (2003) Detection and phylogenetic analysis of mating type genes of *Ophiosphaerella korrae*. Can J Bot 81:307-315.
Huson DH and Steel M (2004) Phylogenetic trees based on gene content. Bioinformatics 20:2044-2049.
Hyman RW (2001) Sequence data: posted vs. published. Science 291:827.
Imanishi T, Itoh T, Suzuki Y, O'Donovan C, Fukuchi S, Koyanagi KO, Barrero RA, Tamura T, Yamaguchi-Kabata Y, Tanino M, etc. (2004) Integrative annotation of 21,037 human genes validated by full-length cDNA clones. PLoS Biology 2: 856-875.
International Human Genome Consortium (2001) Initial sequencing and analysis of the human genome. Nature 409:860-921.
International Human Genome Consortium (2004) Finishing the euchromatic sequence of the human genome. Nature 431:931-945.
InterPro Consortium (2001) The InterPro database, an integrated documentation resource for protein families, domains and functional sites. Nucl Acids Res 29:37-40.
Jeannmougin F, Thompson JD, Gouy M, Higgins DG and Gibson TJ. (1998) Multiple sequence alignment with Clustal X. Trends Biochem Sci 23:403-5.
Jiang B, Bussey H and Roemer T (2002) Novel strategies in antifungal lead discovery. Curr Opin Microbiol 5:466-471.
Jongeneel V (2001) Searching the expressed sequence tag (EST) databases: panning for genes. Brief Bioinform 1:76-92.
Katinka MD, Duprat S, Cornillot E, Metenier G, Thomarat F, Prensier G, Barbe V, Peyretaillade E, Brottier P, Wincker P, Delbac F, El Alaoui H, Peyret P, Saurin W, Gouy M, Weissenbach J, Vivares CP (2001) Genome sequence and gene compaction of the eukaryote parasite *Encephalitozoon cuniculi*. Nature 414:401-402.
Kato-Maeda M, Rhee JT, Gingeras TR, Salamon H, Drenkow J, Smittipat N and Small PM (2001) Comparing genomes within the species *Mycobacterium tuberculosis*. Genome Res 11:547-554.
Kellis M, Birren BW, Lander ES (2004) Proof and evolutionary analysis of ancient genome duplication in the yeast *Saccharomyces cerevisiae*. Nature 428:617-624.
Kellis M, Patterson N, Endrizzi M, Birren B, and Lander E.S (2003) Sequencing and comparison of yeast species to identify genes and regulatory elements. Nature 423:241-254.
Keon J, Bailey A and Hargreaves J (2000) A group of expressed cDNA sequences from the wheat fungal leaf blotch pathogen, *Mycosphaerella graminicola* (*Septoria tritici*). Fung Genet Biol 29:118-133.
Kessler MM, Willins DA, Zeng Q, Del Mastro RG, Cook R, Doucette-Stamm L, Lee H, Caron A, McClanahan TK, Wang L, Greene J, Hare RS, Cottarel G and Shimer GH (2002) The use of direct cDNA selection to rapidly and effectively identify genes in the fungus *Aspergillus fumigatus*. Fung Genet Biol 36:59-70.
Kirst M, Johnson AF, Baucom C, Ulrich E, Hubbard K, Staggs R, Paule C, Retzel E, Whetten R and Sederoff R (2003) Apparent homology of expressed genes from wood-forming tissues of loblolly pine (*Pinus taeda* L.) with *Arabidopsis thaliana*. PNAS USA 100:7383-7388.

Koonin EV, Aravind L and Kondrashov AS (2000) The impact of comparative genomics on our understanding of evolution. Cell 101:573-576.

Kroken S, Glass NL, Taylor JW, Yoder OC, Turgeon BG (2003) Phylogenomic analysis of type I polyketide synthase genes in pathogenic and saprobic ascomycetes. PNAS USA 100:15670-15675

Kruger WM, Pritsch C, Chao S and Muehlbauer GJ (2002) Functional and comparative bioinformatic analysis of expressed genes from wheat spikes infected with *Fusarium graminearum*. Mol Plant-Microbe Interact 15: 445-455.

Liang F, Holt I, Pertea G, Karamycheva S, Salzberg SL, Quackenbush J (2000) Gene Index analysis of the human genome estimates approximately 120,000 genes. Nat Genet 25:239-240.

Liti G and Louis EJ (2005) Yeast genome evolution and comparative genomics. Ann Rev Microbiol 59:135-153.

Lockhart DJ and Winzeler EA (2000) Genomics, gene expression and DNA arrays. Nature 405:827-836.

Loftus BJ, Fung E, Roncaglia P, Rowley D, Amedeo P, Bruno D, Vamathevan J, Miranda M, Anderson IJ, Fraser JA, etc. (2005) The genome and transcriptome of *Cryptococcus neoformans*, a basidiomycete fungal pathogen of humans. Science 307:1321-1324.

Lorenz MC (2002) Genomic approaches to fungal pathogenicity. Curr Opin Microbiol 5:372-378.

Maher BA (2003) The 0.1% portrait of human history. The Scientist, June 30, 2003.

Marshall, E (2002) DNA sequencer protests being scooped with his own data. Nature, 295:1206-1207.

McClelland M, Sanderson KE, Spieth J, Clifton SW, Latreille P, Courtney L, Porwollik S, Ali J, Dante M, Du F, etc. (2001) Complete genome sequence of *Salmonella enterica* serovar Typhimurium LT2. Nature 413:852-846.

Mellanby K (1992) The DDT Story. British Crop Protection Council, Farnham, Surrey, UK.

Mewes HW, Albertmann K, Bahr M, Frishman D, Gkeissner A, Hani J, Heumann K, Kleine K, Maierl A, Oliver SG, Pfeiffer F and Zollner A. (1997a) Overview of the yeast genome. Nature 387(suppl):7-65.

Mewes HW, Albermann K, Heumann K, Liebl S and Pfeiffer F (1997b) MIPS: a database for protein sequences, homology data and yeast genome information. Nucl Acids Res 25:28-30.

Mira A, Klasson L and Andersson SGE (2002) Microbial genome evolution: sources of variability. Curr Opin Microbiol 5:506-512.

Mitchell TK, Thon MR, Jeong JS, Brown D, Deng J and Dean RA (2003) The rice blast pathosystem as a case study for the development of new tools and raw materials for genome analysis of fungal plant pathogens. New Phytol 159:53-61.

Mouse Genome Sequencing Consortium (2002) Initial sequencing and comparative analysis of the mouse genome. Nature 420:520-562.

Nardone J, Lee DU, Ansel KM. and Rao A (2004) Bioinformatics for the 'bench biologist': how to find regulatory regions in genomic DNA. Nat Immunol 5:768-774.

Nielsen CB, Friedman B, Birren B, Burge CB and Galagan JE (2004) Patterns of intron gain and loss in fungi. PLoS Biol 2:2234-2242.

Nierman WC, May G, Kim HS, Anderson MJ, Chen D and Denning DW (2005) What the *Aspergillus* genomes have told us. Medical Mycol 43:S3 - S5.

Pallen M (2002) From sequence to consequence: in silico hypothesis generation and testing. Meth Microbiol 33:27-48.

Paoletti M, Rydholm C, Schwier EU, Anderson MJ, Szakacs G, Lutzoni F, Debeaupuis JP, Latge JP, Denning DW and Dyer PS (2005) Evidence for sexuality in the opportunistic fungal pathogen *Aspergillus fumigatus*. Curr Biol 15:1242-1248.

Papp B, Pal C and Hurst LD (2003) Dosage sensitivity and the evolution of gene families in yeast. Nature 424:194-197.

Parkhill J (2002) Annotation of microbial genomes. Meth Microbiol 33:3-26.

Parkhill J, Sebaihia M, Preston A, Murphy LD, Thomson N, Harris DE, Holden MT, Churcher CM, Bentley SD, Mungall KL, etc. (200) Comparative analysis of the genome sequences of *Bordatella pertussis, Bordatella parapetussis*, and *Bordatella bronchiseptica*. Nat Genet 45:32-40.

Parkinson T (2002) The impact of genomics on anti-infectives drug discovery and development. Trends Microbiol 10:S22-S26.

Pearson WR (1990) Rapid and sensitive sequence comparison with FASTP and FASTA. Meth Enzymol 183:63-98.

Pearson WR (1997) Identifying distantly related protein sequences. CABIOS 13:324-332.

Pearson WR (1998) Empirical statistical estimates for sequence similarity searches. J Mol Evol 276:71-84.
Pertea M and Salzberg SL (2002) Computational gene finding in plants. Plant Mol Biol 48:39-48.
Pertsemlidis A and Fondon JW (2002) Having a BLAST with bioinformatics (and avoiding BLASTphemy). Genome Biol 2(10):1-10.
Philip GK, Creevey CJ and McInerney JO (2005) The Opisthokonta and the Ecdysozoa may not be clades: stronger support for the grouping of plant and animal than for animal and fungi and stronger support for the Coelomata than Ecdysozoa. Mol Biol Evol 22:1175-1184.
Philippe H, Lartillot N and Brinkmann H (2005) Multigene analyses of bilaterian animals corroborate the monophyly of Ecdysozoa, Lophotrochozoa and Protostomia. Mol Biol Evol 22:1246-1253.
Piskur J and Langkjær RB (2004) Yeast genome sequencing: the power of comparative genomics. Mol Microbiol 53:381-389.
Plummer KM and Howlett BJ (1993) Major chomosomal length polymorphisms are evident after meiosis in the phytopathogenic fungus *Leptosphaeria maculans*. Curr Genet 24:107-113.
Plummer KM and Howlett BJ (1995) Inheritance of chromosomal length polymorphisms in the ascomycete *Leptosphaeria maculans*. Mol Gen Genet 247:416-22.
Pontecorvo G (1956) The parasexual cycle in fungi. Ann Rev Microbiol 10:393-400.
Pray L (2002) Refining transgenic mice. The Scientist 16(13):34.
Ptak SE, Hinds DA, Koehler K, Nickel B, Patil N, Ballinger DG, Przeworski M, Frazer KA and Pääbo S (2005) Fine-scale recombination patterns differ between chimpanzees and humans. Nat Genet 37:429-434
Rashidi HH, and Buehler LK (2000) Bioinformatics Basics. Boca Raton: CRC Press.
Rehm BHA (2001) Bioinformatic tools for DNA/protein sequence analysis, functional assignment of genes and protein classification. Appl Microbiol Biotechnol 57:579-592.
Reiser L, Mueller LA and Rhee SY (2002) Surviving in a sea of data: a survey of plant genome data resources and issues in building data management systems. Plant Mol Biol 48:59-74.
Rokas A and Carroll SB (2005) More genes or more taxa? The relative contribution of gene number and taxon number to phylogenetic accuracy. Mol Biol Evol 22:1337-1344.
Rokas A, Willaims BL, King N and Carroll SB (2003) Genome-scale approaches to resolving incongruence in molecular phylogenies. Nature 425:798-804.
Roy SW and Gilbert W (2005) Complex early genes. PNAS USA 102:1086-1991.
Rubin GM, Yandell MD, Wortman JR, Gabor Miklos GL, Nelson CR, Hariharan IK, Fortini ME, Li PW, Apweiler R, etc. (2000) Comparative genomics of Eukaryotes. Science 287:2204-2215.
Salzberg S, Birney E, Eddy S and White O (2003) Unrestricted free access works and must continue. Nature 422:801.
Schmidt R (2002) Plant genome evolution: lessons from comparative genomics at the DNA level. Plant Mol Biol 48:21-37.
Schoch CL, Aist JR, Yoder OC and Turgeon BG (2003) A complete inventory of fungal kinesins in representative filamentous ascomycetes. Fungal Genet Biol 39:1-15.
Searls DB (2003) Pharmacophylogenomics: genes, evolution and drug targets. Nature Rev 2:613-623.
Shimamoto K and Kyozuka J (2002) Rice as a model for comparative genomics of plants. Ann Rev Plant Biol 53:399-419.
Soanes DM, Skinner W, Keon J, Hargreaves J and Talbot NJ (2002) Genomes of phytopathogenic fungi and the development of bioinformatic resources. Mol. Plant-Microbe Interact 15:421-427.
Soltis DE, Albert VA, Savolainen V, Hilu H, Qiu YL, Chase MW, Farris JS, Stefanovic S, Rice DW, Palmer JD and Soltis PS (2004) Genome-scale data, angiosperm relationships, and 'ending incongruence': a cautionary tale in phylogenetics. Trends Plant Sci 19:477-483
Spencer DH, Kas A, Smith EE, Raymond CK, Sims EH, Hastings M, Burns JL, Kaul R and Olson MV (2003) Whole-genome sequence variation among multiple isolates of *Pseudomonas aeruginosa*. J Bacteriol 185:1316-1325.
Strobel G and Arnold J (2004) Essential eukaryotic core. Evolution 58:441-446.
Tatusov RL, Galperin MY, Natale DA and Koonin EV (2000) The COG data-base: a tool for genome-scale analysis of protein functions and evolution. Nucl Acids Res 28:33-36.
Tekaia F, Latge JP (2005) *Aspergillus fumigatus*: saprophyte or pathogen? Curr Opin Microbiol 8:385-92.
Thacker PD (2003) Understanding fungi through their genomes. BioScience 53:10-15.

The *C. elegans* Sequencing Consortium (1998) Genome sequence of the nematode *C. elegans*: a platform for investigating biology. Science 282:2012-2018.
Thomas JW, Touchman JW, Blakesley RW, Bouffard GG, Beckstrom-Sternberg SM, Margulies EH, Blanchette M, Siepel AC, Thomas PJ, McDowell JC, Maskeri B, Hansen NF, Schwartz MS, Weber RJ, etc. (2003) Comparative analyses of multi-species sequences from targeted genomic regions. Nature 424:788-793.
Thomas SW, Glaring MA, Rasmussen SW, Kinane JT and Oliver RP (2002) Transcript profiling in the barley mildew pathogen *Blumeria graminis* by serial analysis of gene expression (SAGE). Mol Plant-Microbe Interact 15:847-856.
Thomas SW, Rasmussen SW, Glaring MA, Rouster JA, Christiansen SK and Oliver RP (2001) Gene identification in the obligate fungal pathogen *Blumeria graminis* by expressed sequence tag analysis. Fung Genet Biol 33:195-211.
Thomson N, Sebaihia M, Cerdeno-Tarraga A, Bentley S, Crossman L and Parkhill J (2003) The value of comparison. Nature Rev Microbiol 1:11-12.
Till BJ, Reynolds SH, Greene EA, Codomo CA, Enns LC, Johnson JE, Burtner C, Odden AR, Young K, Taylor NE, Henikoff JG, Comai L and Henikoff S (2003) Large-scale discovery of induced point mutations with high-throughput TILLING. Genome Res 13:524-530.
Tisdall J (2003) Mastering PERL for Bioinformatics. Cambridge, Massachusetts: O'Reilly & Associates.
Tunlid A, and Talbot NJ (2002) Genomics of parasitic and symbiotic fungi. Curr Opin Microbiol 5:513-519.
Turgeon BG, Kroken S, Lee BN, Bsaker SE, Amedeo P, Catlett N, Gunawardena U, Wagner E, Robbertse B, Wu J, Yoder OC, Glass NL and Taylor JW (2002) Comparative genomic analysis of fungal plant pathogens: secondary metabolites and mechanisms of pathogenesis. APS Symposium on Functional Genomics of Plant Pathogen Interactions, Milwaukee, Wisconsin, July 27-31, 2002.
Tzung KW, Williams RM, Scherer S, Federspiel N, Jones T, Hansen N, Bivolarevic V, Huizar L, Komp C, Surzycki R, Tamse R, Davis RW and Agabian N (2001) Genomic evidence for a complete sexual cycle in *Candida albicans*. PNAS 98:3249-3253.
Uetz P, Giot L, Cagney G, Mansfield TA, Judson RS, Knight JR, Lockshon D, Narayan V, Srinivasan M, Pochart P, Qureshi-Emili A, Li Y, Godwin B, Conover D, Kalbfleisch T, Vijayadamodar G, Yang M, Johnston M, Fields S and Rothberg JM (2000) A comprehensive analysis of protein-protein interactions in *Saccharomyces cerevisiae*. Nature 403:601-603.
Ureta-Vidal A, Ettwiller L and Birney E (2003) Comparative genomics: genome-wide analysis in metazoan eukaryotes. Nat Rev Genet 4:251-262
Wagner A (2000) Robustness against mutations in genetic networks of yeast. Nat Genet 24:355-361.
Ward TJ, Bielawski JP, Kistler HC, Sullivan E and O'Donnell K (2002) Ancestral polymorphism and adaptive evolution in the trichothecene mycotoxin gene cluster of phytopathogenic Fusarium. PNAS USA 99:9278–9283.
Webber C and Ponting CP (2004) Genes and homology. Curr Biol 14:R332-R333.
Wolf YI, Rogozin IB, Grishin NV, Tatusov RL and Koonin EV (2001) Genome trees constructed using five different approaches suggest new major bacterial clades. BMC Evol Biol 1:8.
Wong S, Fares MA, Zimmermann W, Butler G and Wolfe KH (2003) Evidence from comparative genomics for a complete sexual cycle in the 'asexual' pathogenic yeast *Candida glabrata*. Genome Biol 4:R10.
Wood V, Gwilliam R, Rajandream MA, Lyne M, Lyne R, Stewart A, Sgouros J, Peat N, Hayles J, Baker S, Basham D, Bowman S, Brooks K, Brown D, Brown S, Chillingworth T, Churcher C, etc. (2002) The genome sequence of *Schizosaccharomyces pombe*. Nature 415: 871-880.
Wright FA, Lemon WJ, Zhao WD, Sears R, Zhuo D, Wang J-P, Yang HY, Baer T, Stredney D, Spitzner J, Stutz A, Krahe R and Yuan B (2001) A draft annotation and overview of the human genome. Genome Biol 2(2): research0025.1-0025.18.
Wu Y, Wang X, Liu X and Wang Y (2003) Data-mining approaches reveal hidden families of proteases in the genome of malaria parasite. Genome Res 13:601-616.
Xu Y, Stange-Thomann N, Weber G, Bo R, Dodge S, David RG, Foley K, Beheshti J, Harris NL, Birren B, Lander E and Meyerson M (2003) Pathogen discovery from human tissue by sequence-based computational subtraction. Genomics 81:329-335.
Yarden O, Ebbole DJ, Freeman S, Rodriquez RJ and Dickman MB (2003) Fungal biology and agriculture: revisiting the field. Molec Plant-Microbe Interact 16:859-866.

Yoder OC and Turgeon BG (2001) Fungal genomics and pathogenicity. Curr Opin Pl Biol 4:315-321.
Yu J, Wang J, Lin W, Li S, Li H, Zhou J, Ni P, Dong W, Hu S, Zeng C, etc. (2005) The genomes of *Oryza sativa*: A history of duplications. PLoS Biol 3(2):38.
Zelter A, Bencina M, Bowman BW, Yarden O and Read ND (2004) A comparative genomic analysis of the calcium signaling machinery in *Neurospora crassa, Magnaporthe grisea, Aspergillus fumigatus* and *Saccharomyces cerevisiae*. Fung Genet Biol 41:827-841.
Zeng Q, Morales AJ and Cottarel G (2001) Fungi and humans: closer than you think. Trends Genet 17:682-684.
Zolan ME (1995) Chromosome-length polymorphism in fungi. Microbiol Rev 59:686-698.

ELSEVIER

Applied Mycology and Biotechnology
An International Series
Volume 6. Bioinformatics
2006 Published by Elsevier B.V.

Fungal Genomic Annotation

Igor V. Grigoriev[1], Diego A. Martinez[2] and Asaf A. Salamov[1]
[1]US Department of Energy Joint Genome Institute, Walnut Creek, CA 94598 (ivgrigoriev@lbl.gov, AASalamov@lbl.gov); [2]Los Alamos National Laboratory Joint Genome Institute, P.O. Box 1663 Los Alamos, NM 87545 (admar@lanl.gov).

Sequencing technology in the last decade has advanced at an incredible pace. Currently there are hundreds of microbial genomes available with more still to come. Automated genome annotation aims to analyze this amount of sequence data in a high-throughput fashion and help researches to understand the biology of these organisms. Manual curation of automatically annotated genomes validates the predictions and set up 'gold' standards for improving the methodologies used. Here we review the methods and tools used for annotation of fungal genomes in different genome sequencing centers.

1. INTRODUCTION

In recent years the power of DNA sequencing has dramatically increased, with dedicated centers running 24 hours a day 7 days a week able to produce as much as 2 gigabases of raw sequence or more a month. The researchers who work on a variety of fungi are fortunate, as most fungal genomes are under 50 megabases and produce high-quality draft assembly almost as easily as bacteria. This feature of fungal genomes is a key reason that the first sequenced eukaryotic genome was of the ascomycete *Saccharomyces cerevisiae* (Goffeau et al. 1996). As of the submission of this chapter, one can obtain draft sequences of more than 100 fungal genomes (Table 1) and the list is growing. While some are species of the same genus (e.g., Aspergillus has three members and more coming), there still remains a height of data that could confuse and bury a researcher for many years.

Large-scale fungal genome annotation and analysis started after the sequencing of the yeast *S. cerevisiae* was completed (Goffeau et al. 1996), followed by another yeast *Schizosaccharomyces pombe* (Wood et al. 2002). This period also saw the first filamentous fungi *Neurospora crassa* (Galagan et al. 2003), the first basidiomycete genome of *Phanerochaete chrysosporium* (Martinez et al. 2004) and, through the Phytophthora Genome Initiative (Waugh et al. 2000), the first oomycetes, *Phytophthora sojae* and *Phytophthora ramorum*, were sequenced (genome.jgi-psf.org/sojae1 and genome.jgi-psf.org/ramorum1). Large genome sequencing

Corresponding author: Diego A. Martinez

centers have begun to focus some of their sequencing capacity on the fungal kingdom. One such center, the Joint Genome Institute (JGI) (www.jgi.doe.gov), started the sequencing and annotation of fungi with the whiterot genome (*P. chrysosporium*) over two years ago and now has approximately 20 genomes in various stages of the sequencing and annotation pipeline. The JGI has also hosted three fungal annotation *jamborees* (see Section5.0) for *P. chrysosporium*, *Trichoderma reesei*, and the two *Phytophthora* genomes. Both the Broad Institute and the JGI are set to sequence members of the zygomycetes and the chytridiomycetes. In the 1990s there was a call for many other fungal genomes to be sequenced, and to heed this call, the Fungal Genome Initiative (FGI) (www.broad.mit.edu/annotation/fungi/fgi/) started a coordinated effort on targeted sequencing fungal genomes in a kingdom-wide manner; that is, by selecting a set of fungi that maximizes the overall value through a comparative approach. Currently, from the list of about 40 genomes, 20 were sequenced at the Broad Institute and gene models are available for 7 of those genomes. Unlike the Broad Institute's FGI, the JGI is sequencing individual fungi proposed by researchers world-wide and selected through the Community Sequencing Program (www.jgi.doe.gov/CSP/index.html) on the basis of the organism's scientific and economic importance and through the Department of Energy's microbial genomics program (microbialgenome.org).

The Génolevures Consortium is another large initiative on fungal genomics, focused on large-scale comparative genomics between *S. cerevisiae* and 14 other yeast species representative of the various branches of the Hemiascomycetous class. The consortium sequenced and manually curated the complete genome sequences of four yeast species: *Debaryomyces hansenii*, *Kluyveromyces lactis*, *Candida glabrata*, and *Yarrowia lipolytic*, as well as a number of random genomic libraries (Table 1) (Dujon et al. 2004, Sherman et al. 2004).

To combat the initial problem of making sense of the incredible amount of data, many sequencing centers offer resources to make genomic information more accessible and assist in stimulating research. Collectively these resources are termed annotation. In the field of genomics, the term annotation refers to two types of annotation. The first type, which is performed after assembly, is to locate genes and describe gene structure. This is often termed *structural annotation* or *gene modeling*. In bacteria, this process is relatively straightforward as prokaryotes utilize almost all of their DNA for coding. Of the prokaryotic genomes listed at NCBI (www.ncbi.nlm.nih.gov), the average percentage of coding DNA is 85.5% (C Stubben, personal communication, 2005). For eukaryotes of even small to medium genome sizes, this task can be quite challenging because of the complexity of eukaryotic gene structure and the amount of noncoding DNA. For comparison, the percentage of coding DNA in the whiterot Basidiomycete *P. chrysosporium* is approximately 45%. The second type of annotation is called *functional annotation*. Once the genes have been identified, an attempt is made to identify what the gene does for the cell in a biochemical, structural, signaling, etc. context. This discovery method relies largely on an analysis of the resulting protein.

Table 1. Non-exhaustive list of genomes and respective sequencing centers important to agriculture biotechnology. Also shown are current status and availability of information. For a complete list of genomes, please visit the GOLD database (www.genomesonline.org). †Also available in the MIPS Pedant genome database. *Indicates more than one strain from this species has been sequenced. This includes strains sequenced at the same institution.

Sequencing Center and Genome	Sequenced	Annotated	Published	References
European Consortium				
Saccharomyces cerevisiae†	+	+	+	Goffeau et. Al. 1996
Sanger Center				
Schizosaccharomyces pombe†	+	+	+	Wood et. al.
Broad Institute/ German Consortium				
Neurospora crassa†	+	+	+	Galagan et al. 2003
US DOE Joint Genome Institute				
Phanerochaete chrysosporium†	+	+	+	Martinez et. al. 2004
Phytophthora sojae	+	+		
Phytophthora ramorum†	+	+		
Trichoderma reesei	+	+		
Pichia stipitis	+			
Laccaria bicolor	+			
Nectria haematococca	+			
Glomus intraradices	+			
Postia placenta	+			
Aspergillus niger	+			
Mycosphaerella graminicola	+			
Sporobolomyces roseus				
Mycosphaerella fijiensis				
Piromyces sp.				
Melampsora larici-populina				
Batrachochytrium dendrobatidis				
Phycomyces blakesleenus				
Xanthoria parietina				
Trichoderma virens				
Phytophthora capsici				
Broad Institute				
Aspergillus nidulans†	+	+	+	Galagan et. al. 2005
Chaetomium globosum	+	+		
Fusarium graminearum†	+	+		
Magnaporthe grisea†	+	+	+	Dean et. al . 2005
Stagonospora nodorum	+	+		
Ustilago maydis†	+	+		
Phytophthora infestans				
Botrytis cinerea	+			

Candida guilliermondii	+			
Candida lusitaniae	+			
Candida tropicalis	+			
Coccidioides immitis†	+	+		
Coprinus cinereus†	+	+		
Cryptococcus neoformans (A)†*	+	+		
*Cryptococcus neoformans (B)**	+			
Fusarium verticillioides	+			
Rhizopus oryzae	+			
Saccharomyces cerevisiae RM11-1a	+			
Saccharomyces paradoxus†		+		
Sclerotinia sclerotiorum	+			
Uncinocarpus reesii	+			
Washington University, St Louis				
Alternaria brassicicola				
Saccharomyces kudriavzevii†	+	+		
Saccharomyces mikatae†	+	+		
Saccharomyces castellii†	+	+		
TIGR				
Aspergillus clavatus	+	+		
Aspergillus flavus	+			
Coccidioides posadasii	+	+		
Neosartorya fischeri	+	+		
Penicillium chrysogenum	+			
TIGR/ Univ. of Manchester / Sanger Centre / Institut Pasteur / Univ. of Salamanca / Nagasaki Univ				
Aspergillus fumigatus†	+	+	+	Nierman et al. 2005
TIGR / Stanford Genome Sequencing Center				
Cryptococcus neoformans(D)	+	+	+	Loftus et al. 2005
Stanford Genome Sequencing Center				
Candida albicans	+	+	+	Braun et al. 2005, Jones et al. 2004
Japanese Consortium				
Aspergillus oryzae	+	+	+	Machida et al. 2005
Genoscope				
Kluyveromyces thermotolerans†	+	+		
Kluyveromyces marxianus var marxianus†	+	+		
Kluyveromyces lactis†*	+	+		
Saccharomyces exiguus†	+	+		
Saccharomyces exiguus†	+	+		
Saccharomyces servazzii†	+	+		
Zygosaccharomyces rouxii†	+	+		

Debaryomyces hansenii var hansenii†*	+	+		
Yarrowia lipolytica†*	+	+		
Pichia angusta†	+	+		
Pichia sorbitophila†	+	+		
Candida glabrata†	+	+		
Candida tropicalis†	+	+		
Broad Institute/Genoscope				
Saccharomyces bayanus†*	+	+		
Washington University, St Louis/Genoscope				
Saccharomyces kluyveri†*	+	+		
Syngenta Biotechnology Inc				
Ashbya gossypii†	+	+	+	Dietrich et al. 2004

2. Gene Discovery in the Fungi

With more genomes, computational methods for genome annotation have evolved and different research groups and centers have developed various gene prediction methods and tools. Nevertheless, it appears that there are no completely automated methods to predict gene models in eukaryotes. Most of the eukaryotic gene predictors have been developed for the human genome or other higher eukaryotes and cannot be used for the annotation of a "random" genome without carefully tuning the parameters for gene prediction. Furthermore, gene modeling algorithms made for complex vertebrate genomes show a marked decrease in accuracy even when applied to other vertebrate genomes (Burset and Guigo 1996) and therefore will likely perform poorly on fungal genomes. Guigo et al. have also shown that gene prediction accuracy drops significantly for draft sequences (Burset and Guigo 1996). The methods that rely on open reading frame ORF compatibility across exons (e.g., Fgenesh (Salamov and Solovyev 2000)) suffer most. Others, such as Grail (Xu et al. 1997) and GeneWise (Birney et al. 2004), allow frameshifts, but then produce a mixture of real genes damaged by sequencing errors and potential pseudogene candidates. This is, however, a useful feature for finding pseudogenes (see Section3.3).

2.1 Gene Modelers

Genes in eukaryotic genomes can be predicted using a variety of different approaches, including *ab initio*, homology-based, EST-based, and synteny-based methods, the first two of which are the most used approaches, especially in the absence of ESTs or sequences of other closely related genomes. Overall, performance of *ab-initio* gene finding algorithms greatly depends on which species gene structures were used in the generation of modeling parameters. In general, the predicted models will be highly inaccurate if the genome that the gene finding algorithm is applied to is different in gene structure than the genome that the algorithm was trained on (Korf 2004, Salamov 2005). Therefore, one seeks to train a modeling algorithm on as much data from the genome that it is going to be run on.

Gene-specific parameters are generally subdivided into *content-based* and *signal-based*. Content-based parameters describe oligonucleotide compositions of coding, intronic and intergenic sequences, and also such characteristics as distributions of exon and intron lengths specific to a given genome, average number of exons per gene, etc. Many programs, such as GeneMark (Lukashin and Borodovsky 1998), Genscan (Burge and Karlin 1997), and Fgenesh (Salamov and Solovyev 2000),use 5th order Markov chain probabilities for describing oligonucleotide preferences of genomic sequences. Signal-based parameters describe the specific patterns of splice sites, branch points, polypyrimidine tracts and other functional signals that are important for mechanisms of splicing and transcription. They can be modeled by position weight matrices, weight array matrices (generalized multipositional weight matrices) or by some combined features of sequences, implemented for example through neural nets, discriminant functions and other techniques (Solovyev 2002).

Gene modeling parameters are tuned based on a collection of known gene structures for annotated genome. For genomic information, there should be at least several pieces of relatively large (> 50kb) genomic contig sequences, and this is usually available from early stages of genomic sequencing. All known genes from GenBank, full length cDNA, and EST data are then mapped to the genomic sequences, providing coding, intronic and information about splice sites. Exploratory data analysis is then performed, for example removing redundancy in sequences, removing some questionable EST mappings and estimating if enough data is available to make reliable parameter values. A subset of the above information is usually set aside to form a test set from known genes, where prediction accuracies with various methods and parameters can be obtained. From the above it is obvious that the quality of the parameters greatly depends on the number of available known gene structures for a given genome. For example, if the number of known genes is too small for the reliable estimation of the oligonucliotide composition parameters, it is better to use the parameters from other related species from which they were calculated, or at least from organisms with comparable GC content. For some functional signals, such as the TATA box, signal peptides, polyA signals and transcription start sites (TSS), often little species-specific information is known, and thus it is difficult to train them for specific genomes and only general available data may be used. The investigation of these elements is usually left to the end-user.

If a given genome has a sufficient number of known genes or full-length cDNAs, then all these parameters can be efficiently computed and implemented through existing gene-finding algorithms. This presents a problem for many newly sequenced genomes, including new fungal genomes, where there is a scarcity of high-quality information about gene structures. In such a situation, some glimpses about particular gene structures prevalent in a given genome can be inferred from EST data. EST collections are a significant source of data for annotation (Loftus 2003). They can be either mapped directly, or used in EST-based gene predictors like GrailEXP (Xu et al. 1997), Exonerate by ENSEMBL (Slater and Birney 2005) and EST_MAP (softberry.com). Another source of known genes comes from homology-based gene modeling programs such as GeneWise (Birney et al. 2004) or Fgenesh+

(softberry.com). Homology-based programs rely on close protein homologs, which retain similar exon-intron structures.

In recent years, there has been a trend to sequence and annotate genomes of closely related organisms, some even in the same genus. This rapid increase in the number of complete genomes of closely related organisms allows us to effectively use synteny-based gene prediction methods that predict genes in one genome on the basis of comparison with gene models in another. In the last few years a number of such methods have been developed ((Kellis et al. 2003), in yeast). Although in general they provide a reasonable quality of predicting exons, large-scale genome prediction suffers from chimerism, i.e., linking neighbor models into one long model. Therefore, application of these methods is often limited to correction of gene models. For example, in the annotation of *P. sojae* and *P. ramorum* genomes, Fgenesh2 (softberry.com) was used to correct orthologous gene models predicted by other methods if coverage of the alignment between the orthologs was higher in one protein than in another (Tyler et al., in preparation). Other examples of successful use of these methods include the annotation of two *Aspergillus* genomes by TIGR using TWAIN (Majoros et al. 2005) in combination with TigrScan (Majoros et al. 2004) and annotation of different serotypes of *Cryptococcus neomorphans* genomes using TwinScan (Flicek et al. 2003, Korf et al. 2001) followed by RT-PCR validation (Tenney et al. 2004).

Each gene prediction method has its own advantages and disadvantages. A number of benchmarks of different gene prediction methods on different sets of data have been published. Combining different methods can improve overall quality of gene models. Methods to select entire gene models (e.g., Bayesian framework (Pavlovic et al. 2002)) or assemble model fragments into *de novo* models (e.g., Combiner (Allen et al. 2004)) have been proposed. Annotation pipelines at JGI and the Broad Institute employ the first approach to combine several gene predictors, each of which by itself already maximizes use of available evidence.

2.2 Fungal Gene Structure

The G+C content of genomes is a feature of genomic organization that affects codon usage and other oligonucleotide preferences. Most gene modelers predict more accurately in low GC regions because they strongly rely on hexamer frequencies to discriminate between coding and noncoding regions (Burset and Guigo 1996). In fungal genomes the G+C content varies greatly from 33% for *Candida albicans* to 57% in *P. chrysosporium*. The number of exons per gene also varies greatly among diverse fungi, from the largely single-exon gene structure of *S.cerevisae* to the high proportion of multi-exon genes in *C. neoformans*. However, in comparison with metazoan genes, fungal genes have relatively short introns. For example, in *C. neoformans*, preliminary analysis has shown that introns have a very tight distribution around 68bp and therefore, when annotating this genome, authors explicitly coded this 'spiked' intron length distribution in the TWINSCAN program instead of the default geometric distribution used in the original program (Tenney et al. 2004). Kupfer et al. (Kupfer et al. 2004) provided the first comprehensive analysis of introns and splicing sites in five diverse fungi, which included the yeasts *S. cerevisae* and *S. pombe*; two well-studied Ascomycetes, *A. nidulans* and *N crassa*; and one Basidmycete, *C. neoformans*. Based on EST data they found that for all studied fungi more than 98% of all splice sites have the canonical 5'GT ... AG3' donor-acceptor pairs in agreement with vertebrate splice sites. On the other hand, they found that polypyrimidine tracts between the intron 3' end and the branch point are absent in a large fraction (31%-72%) of introns across all studied genomes. Their results also suggest that for some short introns, absent polypyrimidine tracts may be compensated by poly(T) tracts upstream of the branch point.

2.3 Validation of Gene Predictions

Validation of predicted gene models is an important part of automated annotation. It is not sufficient to determine an average accuracy of gene predictors on the test set of genes. Divergence of fungal genomes makes it impossible to use the same parameters for different genomes and therefore accuracy also varies from genome to genome.

Predicted gene models can be normally validated through either their expression or conservation. Evidence of gene expression can be collected from ESTs/cDNAs overlapping with a gene model, oligonucleotide probes placed on microarrays, or peptides from mass-spectrometry experiments aligned against genomic sequence. Conservation can be inferred from homology of a predicted protein and proteins from other organisms in either hand curated datasets like SwissProt (Boeckmann et al. 2003) or all the proteins in Genbank (www.ncbi.nlm.nih.gov). In addition, the percentage coverage of the alignment of the predicted protein and its best homolog serves as a measure of completeness of the predicted gene model especially in alignments between the orthologs. Independent of gene prediction, the alignment between genomic sequences of two or more closely related organisms can reveal islands of DNA conservation and suggest or confirm location of exons and nonconserved functionally important regions. For this reason the VISTA genome analysis tool (Mayor et al. 2000) became a standard feature of JGI genome annotation.

While the number of gene models supported by either of the aforementioned types of evidence describes overall quality of gene models, knowing the quality of every individual gene model is important for many biologists. Based on the same lines of evidence all genes are divided into more or less reliable predictions using gene-naming conventions. While the naming conventions vary from place to place, all genes can be divided into three major categories by their functional assignment: (1) higher confidence assignment based on strong homology to protein from GenBank or SwissProt (e.g., TIGR: "known/putative", Broad Institute: "known/conserved hypothetical/hypothetical, similar to"), (2) lower confidence assignment supported by ESTs (e.g., TIGR: "expressed") or weak homology (Broad Institute: hypothetical), and (3) *ab initio* gene predictions without homology or EST support (e.g., TIGR: "hypothetical," Broad Institute: "predicted").

Analysis of the aforementioned lines of evidence may help to elucidate an overpredicted portion of a gene set, i.e. *ab inito* gene models, without any additional support. On the other hand, a conservative approach to genome annotation can cause gene underprediction, which can be assessed given a "core" reference set of genes/functions. However, this is a challenging task. First, generation of such a set requires analysis of large collection of diverse genomes. Second, a lack of a "core" gene in a genome does not necessarily mean underprediction because of (1) the draft nature of genome sequence and a good chance of finding the gene in gaps or unassembled DNA reads, or (2) nonhomologous gene substitution, i.e., recruitment of a different protein to perform the same or similar function. Both of these tasks for the moment can be only addressed by a human curator.

3. FUNCTIONAL ANNOTATION

The promise of genomics to biology is not only to find genes but also to describe the function of each resulting protein. While this set of goals was originally that of the fields of genetics and cell and molecular biology, in the genomic era it takes on a new scope. Of the genomes from Table 1, 40 have been through the gene-modeling process, and several have at least preliminary functional annotations. While many biologists feel that manual annotation is best, and will volunteer to examine the staggering numbers of gene models that are predicted for their organism of interest, (e.g., the manual annotation of *C. albicans*) (Braun et al. 2005) and the continued annotation by the Munich Information Center for Protein Sequences (MIPS) (Mewes et al. 2004), there appears to be a need for a reliable automated functional annotation. The *N. crassa* genome alone contains 4,140 (40%) completely unknown genes. Automated annotation, however, has its problems. In Koonin and Galperin (Koonin and Galperin 2002) there are several humorous examples of automated annotation, of which we should be aware. Finally, we must also ask the question, "can we assign protein function by computational methods?"

3.1 Automated Methods

Most sequencing centers have turned to some form of first pass automated annotation to deal with the numbers of genomes that are being sequenced. This data is usually used by the community to attempt to find a function. We present here various approaches to discover gene function that are used in whole genome projects.

3.1.1 Homologous relationships and gene identity

The attempt to transfer gene function from a known protein to an unknown protein can be a difficult task, as evolution can change the context of what a gene does depending on the environment (Francino 2005) that the organism has been in since the time of speciation.

The general approach is to tease out evolutionary relationships by discovering orthologous and paralogous relationships between protein sequences in whole genomes. Orthologs are genes originating from a single ancestral gene in the last common ancestor of the compared genomes (Fitch 1970, Koonin 2005). Paralogs are genes within the same genome that arose from duplications. While conserved function of the proteins is not a part of the definition of orthology, it would reason that the amino acid conservation is due to functional conservation (Koonin 2005, Storm and Sonnhammer 2002). Such an approach is useful because it is less likely that paralogous genes that have fixed in the population have retained the same function and may have been recruited (Lynch and Conery 2000), thus making their function ambiguous.

The most widely accepted method for inferring orthology is through the analysis of phylogenetic trees. Many robust phylogenetic methods exist for recovering the orthologous relationships between genes from different organisms. These are especially useful for understanding more complicated relationships among groups of related genes, such as paralogs, which may appear as many one-to-one orthologs depending on the time of speciation since duplication. This is, however, usually a manually if not computationally intensive method for understanding related genes. Automation is thus required to efficiently process the quantity of sequences found in whole genomes. There has been some headway in automating phylogenetic analyses (Storm and Sonnhammer 2002, Zmasek and Eddy 2002), but they are still limited because of the complexities involved in building phylogenetic trees.

Because of the complexities and manual analysis involved in phylogenetics, most people use a method that relies on a sequence similarity method often called "mutual best hits" or "bidirectional best hits" to identify putative orthologs. This relationship is calculated with all the proteins in the genome. The logic in performing this is as follows: in two genomes A and B containing genes Xa and Xb, respectively, Xa and Xb are potential orthologs if there is no better alignment to Xa from genome B than Xb, and there is no better alignment to Xb from genome A than Xa (Lee et al. 2002, Overbeek et al. 1999). COGs (Tatusov et al. 2001) extends this approach by requiring that orthologs be from three genomes ("triangles" of proteins termed BeTs) to be considered orthogous, thus ensuring that the gene has persisted through time.

There is an unfortunate caveat to the usefulness of such techniques. In all genomes, there is a large fraction of genes whose function is unknown, for example, in the well-studied filamentous fungi there are 4,140 (41%) genes with no similarity to any protein in GenBank (Galagan et al. 2003). It is immediately apparent that there is a need to develop techniques to identify the function of many thousands of genes in a high-throughput manner.

3.2.1 Annotation in fungi with experimental data

With a dramatic increase in the number of unknown and hypothetical genes being produced from whole genome projects, there is a need to integrate the data from high-throughput experiments into the annotation process. The database for this organism is in the Saccharomyces Genome Database (Balakrishnan et al. 2005). One can access the data from a variety of microarray information for many of the approximately 6,000 genes predicted in this yeast. An approach of integrating data in the fashion of SGD will drive fungal research and assist in the search for the function of all the genes in a genome.

With transcriptomics and proteomics we are able to understand under what conditions and times mRNAs accumulate in the cell. The types of studies that appear in the literature for fungi are particularly useful for annotation, as they are often under conditions that are unique to the organism, and likely will give clues to many of the species or fungal-specific genes that are common in databases (Lorenz 2002, Rementeria et al. 2005). It is also possible to create a probe for every exon in the genome, so that the predicted structure of a gene can be verified with useful suggestions on how to correct some gene models (Sims et al. 2004).

Because most functioning genes create proteins it is also possible to describe them with proteomics. In fungi this is often identifying what proteins are secreted, as fungi are important degraders of biomass (Medina et al. 2005, Medina et al. 2004, Vanden Wymelenberg et al. 2005) have symbiotic relationships with roots of agriculturally important plants (Bestel-Corre et al. 2004) and protect plants from other soil-borne microbes (Grinyer et al. 2005, Grinyer et al. 2004). The majority of these studies are again targeting biological niches that are dominated by fungi, and are expected to involve fungal-specific genes.

3.3 Pseudogene Annotation

In all studied genomes, eukaryotic and prokaryotic, there are remnants of genes that are no longer transcriptionally active. These inactivated genes are called *Pseudogenes,* often preceded with the greek letter *psi.* There are two types of pseudogenes that are named for how they arise: *processed* and *nonprocessed.* Processed pseudogenes occur when a normal gene is transcribed, introns removed, and a DNA copy is made from the gene by the reverse-transcriptase enzyme of a retrotransposon. Processed pseudogenes usually do not appear to have introns or regulatory elements and can often have poly-A tails. In addition, this type of pseudogene usually contains disablements over the length, such as frameshifts and stop codons in the coding frame. The second type, nonprocessed pseudogenes, were once genes or were duplications of genes. Like processed pseudogenes they contain disablements; however, nonprocessed pseudogenes often have features that make them appear to be genes. This makes nonprocessed pseudogenes more difficult to identify and they can be listed erroneously as a transcribed gene.

In fungi there are previously described pseudogenes (Borsuk et al. 1988, Fink 1987, Gniadkowski et al. 1991, Metzenberg et al. 1985) which were discovered before the genomic era. The determination of pseudogenization was done by manual analysis. In the postgenomic era however, few researchers have the luxury to analyze the average 10,000 or so genes that may contain the hallmarks of pseudogenes. To keep up with the barrage of genomic data in fungi, it will be necessary to apply automated analyses in discovering pseudogenes. Such techniques have already been developed for humans (Zhang et al. 2003). In the yeast genomes, *S. cerevisiae* and *Schizosaccharomyces pombe*, there are 221 for the former (Harrison et al. 2002) and 33 (Wood et al. 2002) for the latter. For the larger filamentous fungus, *P. chrysosporium* (Martinez et al. 2004) no analysis of pseudogenes has been provided because of ambiguity in their discovery. This is also the case for *N. crassa, Magnaporthe grisea (Dean et al. 2005),*and *C. albicans* (Braun et al. 2005, Jones et al. 2004) likely because of the ambiguity of stop codons in draft genomes.

One of the key features of pseudogenes is the appearance of stop codons and frameshifts in the coding region. This is usually found by using GeneWise (Birney et al. 2004) which performs a sensitive alignment to a known gene in order to create a gene model, placing an "X" in the predicted amino acid sequence where a frame shift is likely to have occured, thus allowing the extension of the gene model beyond what could be a sequencing error. There are other criteria (Zhang and Gerstein 2004); however, the stop appears to be the strongest signal. This is the primary difficulty in finding pseudogenes for many genome projects. The data in whole genome shotgun is of the highest quality of sequencing; the error rate is usually 1 in 10,000 (Martinez et al. 2004) for draft genomes. This means that several hundred genes in each genome could contain frame shifts caused by sequencing error alone, Recently however, Torrents et al. (Torrents et al. 2003) has devised a novel technique in verifying pseudogenes that does not rely on the presence of stops. This method applies the Ka/Ks ratio test (rate of nonsynonymous vs. synonymous substitutions) to decide whether a gene is really a pseudogene. In a recent technique comparison from Zhang and Gerstein (Zhang and Gerstein 2004), with some alteration of parameters, the Torrents technique is able to predict the approximately 14,000 pseudogenes in the human genome that other methods were able to find. With the application of this technique, it now may be possible to identify pseudogenes in draft genomes.

4. ANNOTATION PIPELINES

The centers involved in fungal annotation use a system of steps in order to produce a final set of gene models and annotation, collectively called a *pipeline*. With this broad variety of methods and tools available for gene prediction it is interesting to understand the practical solutions that have been developed by these centers (Table 1). The overall workflow is similar between the different pipelines and includes a few major steps common to all. These common steps are (1) repeat masking, (2) mapping ESTs/known genes, (3) homologs, (4) gene modeling using different methods sequentially or in parallel and then combining them (see Figure 1), and (5) annotating produced sets of gene models using various domain prediction and homology searches.

The JGI and the Broad Institute both use a similar basic set of gene predictors (Fgenesh (Salamov and Solovyev 2000), Fgenesh+ (softberry.com), and GeneWise (Birney et al. 2004)), but in order to produce a nonredundant set of genes they combine them in a slightly different way. Broad Institute uses a prioritization system weighting various gene predictors on the amount and quality of information that exists and the performance of each algorithm. This system gives first priority to GeneWise models with >90% amino acid identity to the translated genome, the second to Fgenesh+ models with identity between 80% and 90%, and then selects the one with the best homology among Fgenesh, Fgenesh+ and GeneWise predictions. This is a sequential gene prediction procedure. JGI predicts all models independently, utilizing ESTs to correct and expand predicted gene models and add UTR regions, and fixes incomplete models by analysis of local genomic regions. The JGI treats all models equally (except known genes that have a higher weight). The JGI selection procedure analyzes each cluster or locus of overlapping models. The final gene model is chosen according to a hierarchy of criteria: (1) homology to other proteins, (2) EST support, and (3) length and completeness.

After gene models are predicted, each of them is translated and the predicted proteins are functionally analyzed in terms of functional domains and homologs. Functions are automatically assigned on basis of the best homology hit. Comparison with the specialized databases (e.g,, KEGG (Kanehisa et al. 2004)) and functional classification allows one to map the predicted proteins onto metabolic pathways, Gene Ontology and KOG (Tatusov et al. 2003) categories provide the user with multiple entry points into the annotation data. Although implementation of these steps varies, most of the pipeline utilizes Blast or Smith-Waterman searches to find all potential homologs, InterProScan (Mulder et al. 2005) or various domain-search

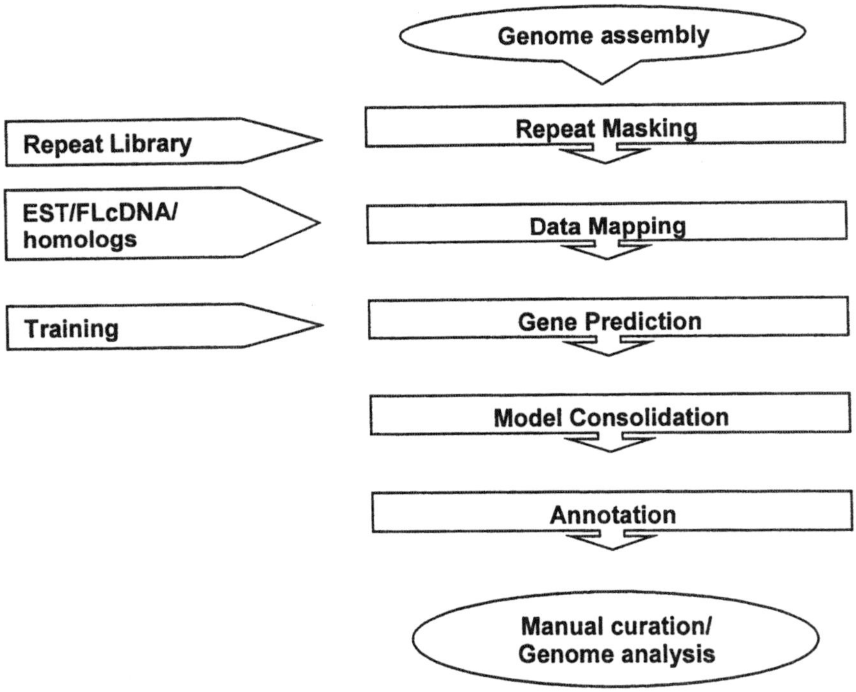

Fig. 1. Annotation pipeline workflow diagram

methods to predict domains, and public software (e.g., TMHMM (www.cbs.dtu.dk/services/TMHMM/), SignalP (Bendtsen et al. 2004), and TargetP (Emanuelsson et al. 2000)) for more specialized analysis.

In the CAAT-box package (Frangeul et al. 2004) used for annotation of yeast genomes (Dujon et al. 2004, Sherman et al. 2004), gene prediction and functional annotation are integrated with assembly process. However, genes in CAAT-box are identified simply as ORFs (similar to bacterial gene prediction) and while is acceptable for yeasts with low number of exons (a similar approach was taken for *S.cerevisiae* (Goffeau et al. 1996)) it cannot be used broadly for all yeast genomes or especially fungi in general. Even for yeasts the package was used as a first-pass tool combined with the use of GeneMark in the intragenic regions. A similar combination of tools was used in the annotation of *C. albicans* (Braun et al. 2005)

MIPS (Mewes et al. 2004) provides both structural and functional annotation for many of the genomes listed in Table 1. For all genomes housed at MIPS the automated functional annotation system Pedant (Frishman et al. 2003) is used. The Pedant system performs Blast against known proteins from GenBank and the Funcat database (Ruepp et al. 2004), as well as predicting domains using Interpro (Mulder et al. 2005) and other domain-specific databases. For the genomes *S. cerevisiae, N. crassa, Ustilago myadis* and *Magnaporthe grisea,* MIPS performs in-depth manual curation and verification of both gene structure (provided by the sequencing centers) and gene function.

5. MANUAL CURATION: IT TAKES A VILLAGE

Automated annotation and functional genomics methods have reduced the amount of work needed to turn the data in whole genome projects into useful information. There is however still some amount of error in the results in both automated functional and structural annotation (Bork 2000). To verify the calls made by automatic methods and to add the value of personal knowledge to the information presented, volunteers will manually curate the data. Such a resource currently exists or is under development for all known fungal genomes.

Community annotation usually begins with a conference, often termed "Jamboree," so named for the original *Drosophila melanogaster* genome annotation conference (Pennisi 2000). The jamboree serves several purposes. The volunteers that will be manually curating the information are trained how to use the specialized tools. Groups of genes are assigned to individuals, and they will then become the curator of that family of genes or pathways. The group of curators will then proceed to manually verify both automated gene calls as well as automated functional data using custom interfaces that connect to a relational database, usually via the web through a web browser.

Several of the fungal genomes listed in Table 1 are currently being curated or have been curated in this manner. The JGI uses custom software for functional annotation and the Apollo editor (Lewis et al. 2002) for updating gene structure features. The results can be viewed on the web, and include the genomes of the basidiomycete *P. chrysosporium,* the oomycete *Phytophthora* species *sojae* and *ramorum,* and the ascomycete *T. reesei.* The genome of *S cerevisiae* has one of the oldest databases available on line, the Saccharomyces Genome Database (www.yeastgenome.org). The Broad Institute, an important center for fungal genomes, is in the process of creating an interface for community annotation; however, their automated annotations are available (www.broad.mit.edu/annotation/). Other fungal genomes have employed the community annotation model, such as the *Aspergillus* (www.cadre.man.ac.uk) and *C. albicans* (Braun et al. 2005) communities.. The *Aspergillus* site uses the Ensembl (Hubbard et al. 2005) system, while the *C. albicans* annotation project used the Artemis system(Berriman and Rutherford 2003).

6. CONCLUSION

The genome of the yeast S. cerevisiae was completed and published nearly a decade ago. Further improvements in sequencing technology will provide a rapid explosion in the number of fungal genomes, which will result in a critical mass of data for fungal genomes and is essential for changing annotation strategy, as more genome sequences will provide a better understanding of the individual genomes. It is quite possible that someday soon acquiring the genome of the organism you wish to study will be another tool in the biology lab, akin to a centrifuge. Creating resources and perfecting methods to make sequence information accessible is key to making it useful. There exists a need, however, to be able to compare multiple fungal genomes at one time. Despite a number of rich information resources for individual species there is not a unified fungal genomics resource that allows one to quickly compare a newly sequenced genome against others and get an understanding of commonalities and specifics on all levels from individual genes to families and pathways to whole genome organization. On this front collaboration from all centers and researchers involved need to address the need to create a common interface and work together to produce the best available fungal genomic resource possible.

Acknowledgements: This work was performed under the auspices of the US Department of Energy's Office of Science, Biological and Environmental Research Program, and Lawrence Livermore National Laboratory under Contract No. W-7405-Eng-48, Lawrence Berkeley National Laboratory under contract No. DE-AC02-05CH11231, and Los Alamos National Laboratory under contract No. W-7405-ENG-36. We would like to thank our colleagues Gary Xie, Jean Challacombe, and Monica Misra for their critical review of this work.

REFERENCES

Allen JE, Pertea M and Salzberg SL (2004). Computational gene prediction using multiple sources of evidence. Genome Research 14 (1):142-148.

Balakrishnan R, Christie KR, Costanzo MC, Dolinski K, Dwight SS, Engel SR, Fisk DG, Hirschman JE, Hong EL, Nash R, Oughtred R, Skrzypek M, Theesfeld CL, Binkley G, Lane C, Schroeder M, Sethuraman A, Dong S, Weng S, Miyasato S, Andrada R, Botstein D and Cherry JM (2005). Saccharomyces Genome Database. http://www.yeastgenome.org/

Bendtsen JD, Nielsen H, von Heijne G and Brunak S (2004). Improved prediction of signal peptides: SignalP 3.0. Journal of Molecular Biology 340 (4):783-795.

Berriman M and Rutherford K (2003). Viewing and annotating sequence data with Artemis. Briefings in Bioinformatics 4 (2):124-132.

Bestel-Corre G, Dumas-Gaudot E and Gianinazzi S (2004). Proteomics as a tool to monitor plant-microbe endosymbioses in the rhizosphere. Mycorrhiza 14 (1):1-10.

Birney E, Clamp M and Durbin R (2004). GeneWise and genomewise. Genome Research 14 (5):988-995.

Boeckmann B, Bairoch A, Apweiler R, Blatter MC, Estreicher A, Gasteiger E, Martin MJ, Michoud K, O'Donovan C, Phan I, Pilbout S and Schneider M (2003). The SWISS-PROT protein knowledgebase and its supplement TrEMBL in 2003. Nucleic Acids Research 31 (1):365-370.

Bork P (2000). Powers and pitfalls in sequence analysis: The 70% hurdle. Genome Research 10 (4):398-400.

Borsuk P, Gniadkowski M, Bartnik E And Stepien PP (1988). Unusual Evolutionary Conservation Of 5s Ribosomal Rna Pseudogenes In Aspergillus-Nidulans Similarity Of The Dna Sequence Associated With The Pseudogenes With The Mouse Immunoglobulin Switch Region. Journal Of Molecular Evolution 28 (1-2):125-130.

Braun BR, van het Hoog M, Enfert C, Martchenko M, Dungan J, Kuo A, Inglis DO, Uhl MA, Hogues H, Berriman M, Lorenz M, Levitin A, Oberholzer U, Bachewich C, Harcus D, Marcil A, Dignard D, Iouk T, Zito R, Frangeul L, Tekaia F, Rutherford K, Wang E, Munro CA, Bates S, Gow NA, Hoyer LL, hler G, Morschh, user J, Newport G, Znaidi S, Raymond M, Turcotte B, Sherlock G, Costanzo M, Ihmels J, Berman J, Sanglard D, Agabian N, Mitchell AP, Johnson AD, Whiteway M

and Nantel A (2005). A Human-Curated Annotation of the Candida albicans Genome. PLoS Genetics 1 (1):e1.
Burge C and Karlin S (1997). Prediction of complete gene structures in human genomic DNA. Journal of Molecular Biology 268 (1):78-94.
Burset M and Guigo R (1996). Evaluation of gene structure prediction programs. Genomics 34 (3):353-367.
Dean RA, Talbot NJ, Ebbole DJ, Farman ML, Mitchell TK, Orbach MJ, Thon M, Kulkarni R, Xu J-R, Pan H, Read ND, Lee Y-H, Carbone I, Brown D, Oh YY, Donofrio N, Jeong JS, Soanes DM, Djonovic S, Kolomiets E, Rehmeyer C, Li W, Harding M, Kim S, Lebrun M-H, Bohnert H, Coughlan S, Butler J, Calvo S, Ma L-J, Nicol R, Purcell S, Nusbaum C, Galagan JE and Birren BW (2005). The genome sequence of the rice blast fungus Magnaporthe grisea. Nature 434 (7036):980-986.
Dujon B, Sherman D, Fischer G, Durrens P, Casaregola S, Lafontaine I, de Montigny J, Marck C, Neuveglise C, Talla E, Goffard N, Frangeul L, Aigle M, Anthouard V, Babour A, Barbe V, Barnay S, Blanchin S, Beckerich JM, Beyne E, Bleykasten C, Boisrame A, Boyer J, Cattolico L, Confanioleri F, de Daruvar A, Despons L, Fabre E, Fairhead C, Ferry-Dumazet H, Groppi A, Hantraye F, Hennequin C, Jauniaux N, Joyet P, Kachouri R, Kerrest A, Koszul R, Lemaire M, Lesur I, Ma L, Muller H, Nicaud JM, Nikolski M, Oztas S, Ozier-Kalogeropoulos O, Pellenz S, Potier S, Richard GF, Straub ML, Suleau A, Swennen D, Tekaia F, Wesolowski-Louvel M, Westhof E, Wirth B, Zeniou-Meyer M, Zivanovic I, Bolotin-Fukuhara M, Thierry A, Bouchier C, Caudron B, Scarpelli C, Gaillardin C, Weissenbach J, Wincker P and Souciet JL (2004). Genome evolution in yeasts. Nature 430 (6995):35-44.
Emanuelsson O, Nielsen H, Brunak S and von Heijne G (2000). Predicting subcellular localization of proteins based on their N-terminal amino acid sequence. Journal of Molecular Biology 300 (4):1005-1016.
Fink GR (1987). Pseudogenes In Yeast? Cell 49 (1):5-6.
Fitch Wm (1970). Distinguishing Homologous From Analogous Proteins. Systematic Zoology 19 (2):99-113.
Flicek P, Keibler E, Hu P, Korf I and Brent MR (2003). Leveraging the mouse genome for gene prediction in human: From whole-genome shotgun reads to a global synteny map. Genome Research 13 (1):46-54.
Francino MP (2005). An adaptive radiation model for the origin of new gene functions. Nature Genetics 37 (6):573-577.
Frangeul L, Glaser P, Rusniok C, Buchrieser C, Duchaud E, Dehoux P and Kunst F (2004). CAAT-Box, contigs-Assembly and Annotation Tool-Box for genome sequencing projects. Bioinformatics (Oxford) 20 (5):790-NIL_0758.
Frishman D, Mokrejs M, Kosykh D, Kastenmuller G, Kolesov G, Zubrzycki I, Gruber C, Geier B, Kaps A, Albermann K, Volz A, Wagner C, Fellenberg M, Heumann K and Mewes HW (2003). The PEDANT genome database. Nucleic Acids Research 31 (1):207-211.
Galagan JE, Calvo SE, Borkovich KA, Selker EU, Read ND, Jaffe D, FitzHugh W, Ma LJ, Smirnov S, Purcell S, Rehman B, Elkins T, Engels R, Wang SG, Nielsen CB, Butler J, Endrizzi M, Qui DY, Ianakiev P, Pedersen DB, Nelson MA, Werner-Washburne M, Selitrennikoff CP, Kinsey JA, Braun EL, Zelter A, Schulte U, Kothe GO, Jedd G, Mewes W, Staben C, Marcotte E, Greenberg D, Roy A, Foley K, Naylor J, Stabge-Thomann N, Barrett R, Gnerre S, Kamal M, Kamvysselis M, Mauceli E, Bielke C, Rudd S, Frishman D, Krystofova S, Rasmussen C, Metzenberg RL, Perkins DD, Kroken S, Cogoni C, Macino G, Catcheside D, Li WX, Pratt RJ, Osmani SA, DeSouza CPC, Glass L, Orbach MJ, Berglund JA, Voelker R, Yarden O, Plamann M, Seller S, Dunlap J, Radford A, Aramayo R, Natvig DO, Alex LA, Mannhaupt G, Ebbole DJ, Freitag M, Paulsen I, Sachs MS, Lander ES, Nusbaum C and Birren B (2003). The genome sequence of the filamentous fungus Neurospora crassa. Nature 422 (6934):859-868.
Galagan JE, Calvo SE, Cuomo C, Ma L-J, Wortman JR, Batzoglou S, Lee S-I, Basturkmen M, Spevak CC, Clutterbuck J, Kapitonov V, Jurka J, Scazzocchio C, Farman M, Butler J, Purcell S, Harris S, Braus GH, Draht O, Busch S, D'Enfert C, Bouchier C, Goldman GH, Bell-Pedersen D, Griffiths-Jones S, Doonan JH, Yu J, Vienken K, Pain A, Freitag M, Selker EU, Archer DB, Penalva MA, Oakley BR, Momany M, Tanaka T, Kumagai T, Asai K, Machida M, Nierman WC, Denning DW, Caddick M, Hynes M, Paoletti M, Fischer R, Miller B, Dyer P, Sachs MS, Osmani SA and Birren

BW (2005). Sequencing of Aspergillus nidulans and comparative analysis with A. fumigatus and A. oryzae. Nature 438 (7071):1105-1115.
Gniadkowski M, Fiett J, Borsuk P, Hoffmanzacharska D, Stepien PP and Bartnik E (1991). STRUCTURE AND EVOLUTION OF 5S RIBOSOMAL RNA GENES AND PSEUDOGENES IN THE GENUS ASPERGILLUS. Journal of Molecular Evolution 33 (2):175-178.
Goffeau A, Barrell BG, Bussey H, Davis RW, Dujon B, Feldmann H, Galibert F, Hoheisel JD, Jacq C, Johnston M, Louis EJ, Mewes HW, Murakami Y, Philippsen P, Tettelin H and Oliver SG (1996). Life with 6000 genes. Science 274 (5287):546-&.
Grinyer J, Hunt S, McKay M, Herbert BR and Nevalainen H (2005). Proteomic response of the biological control fungus Trichoderma atroviride to growth on the cell walls of Rhizoctonia solani. Current Genetics 47 (6):381-388.
Grinyer J, McKay M, Nevalainen H and Herbert BR (2004). Fungal proteomics: Initial mapping of biological control strain Trichoderma harzianum. Current Genetics 45 (3):163-169.
Harrison P, Kumar A, Lan N, Echols N, Snyder M and Gerstein M (2002). A small reservoir of disabled ORFs in the yeast genome and its implications for the dynamics of proteome evolution. Journal of Molecular Biology 316 (3):409-419.
Hubbard T, Andrews D, Caccamo M, Cameron G, Chen Y, Clamp M, Clarke L, Coates G, Cox T, Cunningham F, Curwen V, Cutts T, Down T, Durbin R, Fernandez-Suarez XM, Gilbert J, Hammond M, Herrero J, Hotz H, Howe K, Iyer V, Jekosch K, Kahari A, Kasprzyk A, Keefe D, Keenan S, Kokocinsci F, London D, Longden I, McVicker G, Melsopp C, Meidl P, Potter S, Proctor G, Rae M, Rios D, Schuster M, Searle S, Severin J, Slater G, Smedley D, Smith J, Spooner W, Stabenau A, Stalker J, Storey R, Trevanion S, Ureta-Vidal A, Vogel J, White S, Woodwark C and Birney E (2005). Ensembl 2005. Nucleic Acids Research 33 (January 1):D447-D453.
Jones T, Federspiel NA, Chibana H, Dungan J, Kalman S, Magee BB, Newport G, Thorstenson YR, Agabian N, Magee PT, Davis RW and Scherer S (2004). The diploid genome sequence of Candida albicans. Proceedings of the National Academy of Sciences of the United States of America 101 (19):7329-7334.
Kanehisa M, Goto S, Kawashima S, Okuno Y and Hattori M (2004). The KEGG resource for deciphering the genome. Nucleic Acids Research 32 (Database Issue):D277-D280.
Kellis M, Patterson N, Endrizzi M, Birren B and Lander ES (2003). Sequencing and comparison of yeast species to identify genes and regulatory elements. Nature 423 (6937):241-254.
Koonin EV (2005). Orthologs, Paralogs, and Evolutionary Genomics. Annual Review of Genetics 39 (0):309-338.
Koonin EV and Galperin MY (2002). Information Sources for Genomics. In: ed. Sequence - Evolution - Function. Norwell, Massachusetts. pp. 51-110
Korf I (2004). Gene finding in novel genomes. BMC Bioinformatics 5 59.
Korf I, Flicek P, Duan D and Brent MR (2001). Integrating genomic homology into gene structure prediction. Bioinformatics 17 (90001):140S-148.
Kupfer DM, Drabenstot SD, Buchanan KL, Lai HS, Zhu H, Dyer DW, Roe BA and Murphy JW (2004). Introns and splicing elements of five diverse fungi. Eukaryotic Cell 3 (5):1088-1100.
Lee Y, Sultana R, Pertea G, Cho J, Karamycheva S, Tsai J, Parvizi B, Cheung F, Antonescu V, White J, Holt I, Liang F and Quackenbush J (2002). Cross-referencing eukaryotic genomes: TIGR Orthologous Gene Alignments (TOGA). Genome Research 12 (3):493-502.
Lewis SE, Searle SMJ, Harris N, Gibson M, Iyer V, Richter J, Wiel C, Bayraktaroglu L, Birney E, Crosby MA, Kaminker JS, Matthews BB, Prochnik SE, Smith CD, Tupy JL, Rubin GM, Misra S, Mungall CJ and Clamp ME (2002). Apollo: a sequence annotation editor. Genome Biology 3 (12):research0082.0081 - 0082.0014.
Loftus B (2003). Genome sequencing, assembly and gene prediction in fungi. In: D. K. Arora and G. G. Khachatourians, ed. Applied Mycology and Biotechnology. v. 3:Fungal Genomics. Amsterdam. pp. 65-81
Loftus BJ, Fung E, Roncaglia P, Rowley D, Amedeo P, Bruno D, Vamathevan J, Miranda M, Anderson IJ, Fraser JA, Allen JE, Bosdet IE, Brent MR, Chiu R, Doering TL, Donlin MJ, D'Souza CA, Fox DS, Grinberg V, Fu J, Fukushima M, Haas BJ, Huang JC, Janbon G, Jones SJM, Koo HL, Krzywinski MI, Kwon-Chung JK, Lengeler KB, Maiti R, Marra MA, Marra RE, Mathewson CA, Mitchell TG, Pertea M, Riggs FR, Salzberg SL, Schein JE, Shvartsbeyn A, Shin H, Shumway M, Specht CA, Suh BB, Tenney A, Utterback TR, Wickes BL, Wortman JR, Wye NH, Kronstad JW, Lodge JK, Heitman

J, Davis RW, Fraser CM and Hyman RW (2005). The Genome of the Basidiomycetous Yeast and Human Pathogen Cryptococcus neoformans. Science 307 (5713):1321-1324.
Lorenz MC (2002). Genomic approaches to fungal pathogenicity. Current Opinion in Microbiology 5 (4):372-378.
Lukashin AV and Borodovsky M (1998). GeneMark.hmm: New solutions for gene finding. Nucleic Acids Research 26 (4):1107-1115.
Lynch M and Conery JS (2000). The evolutionary fate and consequences of duplicate genes. Science (Washington D C) 290 (5494):1151-1155.
Machida M, Asai K, Sano M, Tanaka T, Kumagai T, Terai G, Kusumoto K-I, Arima T, Akita O, Kashiwagi Y, Abe K, Gomi K, Horiuchi H, Kitamoto K, Kobayashi T, Takeuchi M, Denning DW, Galagan JE, Nierman WC, Yu J, Archer DB, Bennett JW, Bhatnagar D, Cleveland TE, Fedorova ND, Gotoh O, Horikawa H, Hosoyama A, Ichinomiya M, Igarashi R, Iwashita K, Juvvadi PR, Kato M, Kato Y, Kin T, Kokubun A, Maeda H, Maeyama N, Maruyama J-i, Nagasaki H, Nakajima T, Oda K, Okada K, Paulsen I, Sakamoto K, Sawano T, Takahashi M, Takase K, Terabayashi Y, Wortman JR, Yamada O, Yamagata Y, Anazawa H, Hata Y, Koide Y, Komori T, Koyama Y, Minetoki T, Suharnan S, Tanaka A, Isono K, Kuhara S, Ogasawara N and Kikuchi H (2005). Genome sequencing and analysis of Aspergillus oryzae. Nature 438 (7071):1157-1161.
Majoros WH, Pertea M and Salzberg SL (2004). TigrScan and GlimmerHMM: two open source ab initio eukaryotic gene-finders. Bioinformatics (Oxford) 20 (16):2878-2879.
Majoros WH, Pertea M and Salzberg SL (2005). Efficient implementation of a generalized pair hidden Markov model for comparative gene finding. Bioinformatics (Oxford) 21 (9):1782-1788.
Martinez D, Larrondo LF, Putnam N, Gelpke MDS, Huang K, Chapman J, Helfenbein KG, Ramaiya P, Detter JC, Larimer F, Coutinho PM, Henrissat B, Berka R, Cullen D and Rokhsar D (2004). Genome sequence of the lignocellulose degrading fungus Phanerochaete chrysosporium strain RP78. Nature Biotechnology 22 (6):695-700.
Mayor C, Brudno M, Schwartz JR, Poliakov A, Rubin EM, Frazer KA, Pachter LS and Dubchak I (2000). VISTA: Visualizing global DNA sequence alignments of arbitrary length. Bioinformatics (Oxford) 16 (11):1046-1047.
Medina ML, Haynes PA, Breci L and Francisco WA (2005). Analysis of secreted proteins from Aspergillus flavus. Proteomics 5 (12):3153-3161.
Medina ML, Kiernan UA and Francisco WA (2004). Proteomic analysis of rutin-induced secreted proteins from Aspergillus flavus. Fungal Genetics and Biology 41 (3):327-335.
Metzenberg RL, Stevens JN, Selker EU and Morzyckawroblewska E (1985). IDENTIFICATION AND CHROMOSOMAL DISTRIBUTION OF 5S RIBOSOMAL RNA GENES IN NEUROSPORA-CRASSA. Proceedings of the National Academy of Sciences of the United States of America 82 (7):2067-2071.
Mewes HW, Amid C, Arnold R, Frishman D, Gueldener U, Mannhaupt G, Muensterkoetter M, Pagel P, Strack N, Stuempflen V, Warfsmann J and Ruepp A (2004). MIPS: Analysis and annotation of proteins from whole genomes. Nucleic Acids Research 32 (Database Issue):D41-D44.
Mulder NJ, Apweiler R, Attwood TK, Bairoch A, Bateman A, Binns D, Bradley P, Bork P, Bucher P, Cerutti L, Copley R, Courcelle E, Das U, Durbin R, Fleischmann W, Gough J, Haft D, Harte N, Hulo N, Kahn D, Kanapin A, Krestyaninova M, Lonsdale D, Lopez R, Letunic I, Madera M, Maslen J, McDowall J, Mitchell A, Nikolskaya AN, Orchard S, Pagni M, Pointing CP, Quevillon E, Selengut J, Sigrist CJA, Silventoinen V, Studholme DJ, Vaughan R and Wu CH (2005). InterPro, progress and status in 2005. Nucleic Acids Research 33 (January 1):D201-D205.
Nierman WC, Pain A, Anderson MJ, Wortman JR, Kim HS, Arroyo J, Berriman M, Abe K, Archer DB, Bermejo C, Bennett J, Bowyer P, Chen D, Collins M, Coulsen R, Davies R, Dyer PS, Farman M, Fedorova N, Fedorova N, Feldblyum TV, Fischer R, Fosker N, Fraser A, Garcia JL, Garcia MJ, Goble A, Goldman GH, Gomi K, Griffith-Jones S, Gwilliam R, Haas B, Haas H, Harris D, Horiuchi H, Huang J, Humphray S, Jimenez J, Keller N, Khouri H, Kitamoto K, Kobayashi T, Konzack S, Kulkarni R, Kumagai T, Lafton A, Latge J-P, Li W, Lord A, Lu C, Majoros WH, May GS, Miller BL, Mohamoud Y, Molina M, Monod M, Mouyna I, Mulligan S, Murphy L, O'Neil S, Paulsen I, Penalva MA, Pertea M, Price C, Pritchard BL, Quail MA, Rabbinowitsch E, Rawlins N, Rajandream M-A, Reichard U, Renauld H, Robson GD, de Cordoba SR, Rodriguez-Pena JM, Ronning CM, Rutter S, Salzberg SL, Sanchez M, Sanchez-Ferrero JC, Saunders D, Seeger K, Squares R, Squares S, Takeuchi M, Tekaia F, Turner G, de Aldana CRV, Weidman J, White O,

Woodward J, Yu J-H, Fraser C, Galagan JE, Asai K, Machida M, Hall N, Barrell B and Denning DW (2005). Genomic sequence of the pathogenic and allergenic filamentous fungus Aspergillus fumigatus. Nature 438 (7071):1151-1156.

verbeek R, Fonstein M, D'Souza M, Pusch GD and Maltsev N (1999). The use of gene clusters to infer functional coupling. Proceedings of the National Academy of Sciences of the United States of America 96 (6):2896-2901.

ıvlovic V, Garg A and Kasif S (2002). A Bayesian framework for combining gene predictions. Bioinformatics (Oxford) 18 (1):19-27.

ɔnnisi E (2000). Ideas fly at gene-finding jamboree. Science 287 (5461):2182-+.

ɔmenteria A, Lopez-Molina N, Ludwig A, Vivanco AB, Bikandi J, Ponton J and Garaizar J (2005). Genes and molecules involved in Aspergillus fumigatus virulence. Revista Iberoamericana de Micologia 22 (1):1-23.

uepp A, Zollner A, Maier D, Albermann K, Hani J, Mokrejs M, Tetko I, Guldener U, Mannhaupt G, Munsterkotter M and Mewes HW (2004). The FunCat, a functional annotation scheme for systematic classification of proteins from whole genomes. Nucleic Acids Research 32 (18):5539-5545.

ılamov AA (2005). unpublished observations.

ılamov AA and Solovyev VV (2000). Ab initio gene finding in Drosophila genomic DNA. Genome Research 10 (4):516-522.

ıerman D, Durrens P, Beyne E, Nikolski M and Souciet J-L (2004). Genolevures: Comparative genomics and molecular evolution of hemiascomycetous yeasts. Nucleic Acids Research 32 (Database Issue):D315-D318.

.ms AH, Gent ME, Robson GD, Dunn-Coleman NS and Oliver SG (2004). Combining transcriptome data with genomic and cDNA sequence alignments to make confident functional assignments for Aspergillus nidulans genes. Mycological Research 108 (Part 8):853-857.

:ater GSC and Birney E (2005). Automated generation of heuristics for biological sequence comparison. BMC Bioinformatics 6 31.

ɔlovyev VV (2002). Structure, Properties and Computer Identification of Eukaryotic genes. In Bioinformatics from Genomes to Drugs. Germany:Wiley-VCH. pp

torm CEV and Sonnhammer ELL (2002). Automated ortholog inference from phylogenetic trees and calculation of orthology reliability. Bioinformatics (Oxford) 18 (1):92-99.

atusov RL, Fedorova ND, Jackson JD, Jacobs AR, Kiryutin B, Koonin EV, Krylov DM, Mazumder R, Mekhedov SL, Nikolskaya AN, Rao BS, Smirnov S, Sverdlov AV, Vasudevan S, Wolf YI, Yin JJ and Natale DA (2003). The COG database: An updated version includes eukaryotes. BMC Bioinformatics 4 (41):

atusov RL, Natale DA, Garkavtsev IV, Tatusova TA, Shankavaram UT, Rao BS, Kiryutin B, Galperin MY, Fedorova ND and Koonin EV (2001). The COG database: New developments in phylogenetic classification of proteins from complete genomes. Nucleic Acids Research 29 (1):22-28.

enney AE, Brown RH, Vaske C, Lodge JK, Doering TL and Brent MR (2004). Gene prediction and verification in a compact genome with numerous small introns. Genome Research 14 (11):2330-2335.

orrents D, Suyama M, Zdobnov E and Bork P (2003). A genome-wide survey of human pseudogenes. Genome Research 13 (12):2559-2567.

'anden Wymelenberg A, Sabat G, Martinez D, Rajangam AS, Teeri TT, Gaskell J, Kersten PJ and Cullen D (2005). The Phanerochaete chrysosporium secretome: Database predictions and initial mass spectrometry peptide identifications in cellulose-grown medium. Journal of Biotechnology 118 (1):17-34.

Vaugh M, Hraber P, Weller J, Wu YH, Chen GH, Inman J, Kiphart D and Sobral B (2000). The Phytophthora Genome Initiative database: Informatics and analysis for distributed pathogenomic research. Nucleic Acids Research 28 (1):87-90.

Vood V, Gwilliam R, Rajandream MA, Lyne M, Lyne R, Stewart A, Sgouros J, Peat N, Hayles J, Baker S, Basham D, Bowman S, Brooks K, Brown D, Brown S, Chillingworth T, Churcher C, Collins M, Connor R, Cronin A, Davis P, Feltwell T, Fraser A, Gentles S, Goble A, Hamlin N, Harris D, Hidalgo J, Hodgson G, Holroyd S, Hornsby T, Howarth S, Huckle EJ, Hunt S, Jagels K, James K, Jones L, Jones M, Leather S, McDonald S, McLean J, Mooney P, Moule S, Mungall K, Murphy L, Niblett D, Odell C, Oliver K, O'Neil S, Pearson D, Quail MA, Rabbinowitsch E, Rutherford K,

Rutter S, Saunders D, Seeger K, Sharp S, Skelton J, Simmonds M, Squares R, Squares S, Stevens K, Taylor K, Taylor RG, Tivey A, Walsh S, Warren T, Whitehead S, Woodward J, Volckaert G, Aert R, Robben J, Grymonprez B, Weltjens I, Vanstreels E, Rieger M, Schafer M, Muller-Auer S, Gabel C, Fuchs M, Fritzc C, Holzer E, Moestl D, Hilbert H, Borzym K, Langer I, Beck A, Lehrach H, Reinhardt R, Pohl TM, Eger P, Zimmermann W, Wedler H, Wambutt R, Purnelle B, Goffeau A, Cadieu E, Dreano S, Gloux S, Lelaure V, Mottier S, Galibert F, Aves SJ, Xiang Z, Hunt C, Moore K, Hurst SM, Lucas M, Rochet M, Gaillardin C, Tallada VA, Garzon A, Thode G, Daga RR, Cruzado L, Jimenez J, Sanchez M, del Rey F, Benito J, Dominguez A, Revuelta JL, Moreno S, Armstrong J, Forsburg SL, Cerrutti L, Lowe T, McCombie WR, Paulsen I, Potashkin J, Shpakovski GV, Ussery D, Barrell BG and Nurse P (2002). The genome sequence of Schizosaccharomyces pombe. Nature (London) 415 (6874):871-880.

Xu Y, Mural RJ and Uberbacher EC (1997). Inferring gene structures in genomic sequences using pattern recognition and expressed sequence tags. Fifth International Conference on Intelligent Systems for Molecular Biology. Halkidiki, Greece, p.344-353

Zhang ZL and Gerstein M (2004). Large-scale analysis of pseudogenes in the human genome. Current Opinion in Genetics & Development 14 (4):328-335.

Zhang ZL, Harrison PM, Liu Y and Gerstein M (2003). Millions of years of evolution preserved: A comprehensive catalog of the processed pseudogenes in the human genome. Genome Research 13 (12):2541-2558.

Zmasek CM and Eddy SR (2002). RIO: Analyzing proteomes by automated phylogenomics using resampled inference of orthologs. BMC Bioinformatics 3 (14):

ELSEVIER

Applied Mycology and Biotechnology
An International Series
Volume 6. Genes, Genomics & Bioinformatics

Bioinformatics Packages for Sequence Analysis

***Yeisoo Yu, †Leah A. Santat, and §¶Sangdun Choi**
*Arizona Genomics Institute, University of Arizona, Tucson, AZ 85721, USA;
†Division of Biology, California Institute of Technology, Pasadena, CA 91125, USA;
§Department of Biological Sciences, College of Natural Sciences, Ajou University, Suwon, 443-749, Korea;
¶Department of Neurobiology and Anatomy, The University of Texas Medical School at Houston, Houston, TX 77030, USA

Research paradigms in modern biology are shifting from a single gene to a genome-wide scale. Two major contributions toward this new trend are large-scale genome sequencing and bioinformatics. Recently, bioinformatics has emerged as a new science field that provides computational tools for collecting and maintaining complex biological data. Along with an exponential accumulation of sequence data, many bioinformatics software and algorithms have been developed to assist in genome scale analyses. A comprehensive knowledge of these tools can help not only to understand gene functions and genome organizations, but also to provide an opportunity to develop new tools that can answer many biological questions.

1. INTRODUCTION

The amount of sequence information available from the public database is exponentially increasing. By January 2006, over 100 gigabases of sequences, representing 55 million entries from at least 200,000 different organisms, were deposited into GenBank. The database is several hundred times larger than it was a decade ago. Advanced sequencing technologies and model organism genome projects were the major driving forces behind the explosion of new sequence information during this past decade. This genome data will provide fundamental information to biological and biomedical researchers that will enable them to better understand gene functions and regulations of different model organisms.

Today's biological research requires parallel strategies to simultaneously gather, examine and integrate the large amount of information. Biologists often face the need for genome-wide or cross-genome analysis of their genes of interest. Thus, without good data handling skills, researchers cannot achieve their ultimate research goals. Bioinformatics can provide biologists with powerful tools for collecting, maintaining, distributing, and analyzing huge amounts of genome data.

* Corresponding author: Sangdun Choi

Bioinformatics is a new science field that examines complex biological data on the basis of statistics and computer science. It can give biological meaning to the data by discovering structural and functional relationships that help to explain biological phenomena.

Many sequence analysis tools have been developed and successfully used for interpreting genome data. As biologists, we are using one or more programs on a daily basis without knowing which software is more suitable to analyze the data. In this chapter, we describe several bioinformatics programs which are commonly used for genome sequencing to make sense of sequence assembly, similarity search, repeat identification, and gene annotation.

2. NATIONAL CENTER FOR BIOTECHNOLOGY INFORMATION (NCBI)

NCBI was established in 1988 as a division of the National Library of Medicine (NLM) at the National Institutes of Health (NIH). Its mission is defined as the development, distribution, and maintenance of various molecular databases and computer software in order to support biological and biomedical studies at the molecular level. Regardless of the complicated NCBI structure, it is divided into two major categories in terms of data flow: sequence submission and retrieval.

2.1. Sequence Submission System

GenBank is the sequence depository site, which provides two programs to support sequence submission, BankIt and Sequin. BankIt (http://www.ncbi.nlm.nih.gov/BankIt) is a web based sequence submission tool that can be used for depositing a few sequences when the annotation is not complicated. Submission is accomplished in four steps: general submission information (contact information and release date), reference information (author, publication and citation information), source information (organism and source description), and input sequence (molecular type and sequence). BankIt does not require any special tools to submit sequences other than a web browser, and the submission directions are fairly easy to follow.

Sequin (http://www.ncbi.nlm.nih.gov/Sequin/index.html) is a stand-alone program used to submit and update long complex sequences and annotation information. It runs on Macintosh, PC and UNIX operating systems and is available from the NCBI Sequin ftp site (ftp://ftp.ncbi.nih.gov/sequin) with documentations and instructions. Sequin has a restriction on reading input files. Thus, submitters must prepare their sequences by following specific instructions (FASTA file is the standard format). Though more steps are involved in Sequin submission, it provides sophisticated tools to review and verify the sequence and annotation. Submission is finished by sending the Sequin output file (sqn file) via e-mail to GenBank. A Sequin quick guide is available from the Sequin web site at http://www.ncbi.nlm.nih.gov/Sequin/QuickGuide/sequin.htm.

2.2. GenBank Division for Submission

GenBank maintains databases according to the nature of the DNA sequence. Submitters have a choice of divisions to which they can deposit their sequences based on the source of sequences. It is categorized into 17 divisions listed in Table 1. Divisions of PRI, ROD, MAM, VRT, INV, PLN, BCT, VRL and PHG contain sequences from specific organisms whereas EST, HTG, STS and GSS contain sequences generated by specific technologies from various organisms.

Table 1. Sequence submission divisions in GenBank.

Division Abbreviation	Data Source
PRI	Primate sequences
ROD	Rodent sequences
MAM	Other mammalian sequences
VRT	Other vertebrate sequences
INV	Invertebrate sequences
PLN	Plant, fungal and algal sequences
BCT	Bacterial sequences
VRL	Viral sequences
PHG	Bacteriophage sequences
SYN	Synthetic sequences
UNA	Unannotated sequences
EST	Expressed sequence tags
PAT	Patent sequences
STS	Sequence tagged sites
GSS	Genome survey sequences
HTG	High-throughput genome sequences
HTC	Unfinished high-throughput cDNA sequences

dbEST: Expressed Sequence Tags (EST) are short and single pass sequences from mRNA via cDNA (complimentary DNA) cloning procedures (Adams et al. 1991). It represents gene expression profiles in a specific cell, tissue and organ, or in a specific developmental stage in a normal or stressed growth condition. Currently 32 million entries are available from GenBank (dbEST release 011306; http://www.ncbi.nlm.nih.gov/dbEST/index.html).

dbSTS: Sequence Tagged Sites (STS) contain short, unique sequences on chromosomes or genomes used to generate genetic maps (Olson et al. 1989). About 374,000 STSs are available in GenBank (release 073004; http://www.ncbi.nlm.nih.gov/dbSTS/index.html).

dbGSS: Short, single pass sequences from genomic DNA origin are deposited in the GSS (Genome Survey Sequence) division. Entries are comprised of genomic sequences from exon trapping, Alu PCR, and end sequences of large insert genomic clones such as BAC, cosmid, fosmid, and YAC (Venter et al. 1996; Mahairas et al. 1999; Siegel et al. 1999; Batzoglou et al. 1999). About 13 million entries are available from GenBank (release 011306; http://www.ncbi.nlm.nih.gov/dbGSS/index.html).

dbHTG: High-Throughput Genome sequences (usually called shotgun sequences) from large scale genome sequencing projects are deposited into the HTG division (Ouellette and Boguski 1997). Based on the degree of completion, the phase number is divided into 3 types: Phase1 submission means unfinished, and sequence contigs are not ordered. Phase2 sequences are also unfinished, but sequence contigs are ordered. Phase3 sequences are finished with achieved contiguity of less than 1 base error in 10,000 bases. Finished sequences are transferred to the organism specific databases (e.g., PRI, MAM, PLN, etc.)

WGS: Assembled contigs and annotation data from Whole Genome Shotgun (WGS) (Fleischmann et al. 1995; Venter et al. 1998) sequencing projects are submitted to the WGS division in GenBank. Nucleotide sequences are transferred to BLAST WGS, and protein sequences go to a BLAST non-redundant (nr) database. Scaffold or supercontig information can be submitted to GenBank with specific format (agp format) that contains contig orders and orientation information. Over a hundred WGS projects, including human and mouse, are listed in GenBank. Detailed information can be found at http://www.ncbi.nlm.nih.gov/Genbank/WGSprojectlist.html.

2.3. Sequence Retrieval System

NCBI's Entrez (http://www.ncbi.nlm.nih.gov/Entrez/index.html) is an integrated database retrieval system. Its cross-reference system allows researchers to

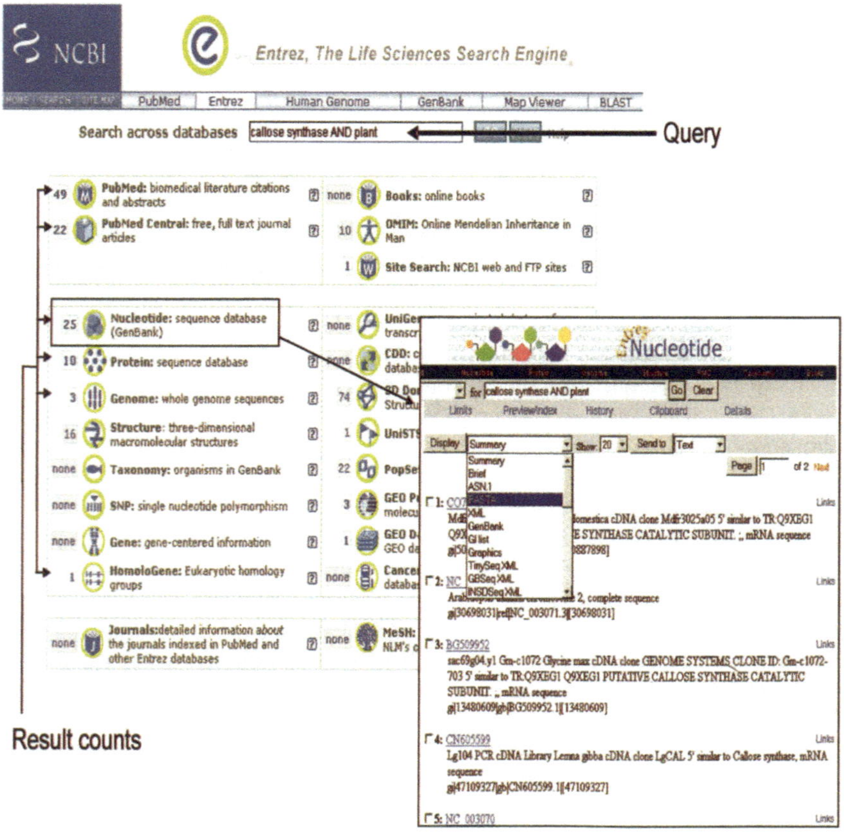

Fig. 1. An example of Entrez search.

access nucleotide, protein, or genome information as well as related research articles and relevant records from 31 databases using text based query. The query search can be refined using Boolean Operators: "AND", "OR", and "NOT". Figure 1 is an example of an Entrez search using the text query "callose synthase AND plant". The number of positive entries from the query is displayed, and 25 nucleotide sequences with a pull-down display menu are shown.

Batch Entrez (http://www.ncbi.nlm.nih.gov/entrez/batchentrez.cgi?) is another convenient tool to retrieve a large number of sequence data at once. Batch Entrez can read a text file containing either the GI (GenInfo Identifier) or accession number in one entry per line, and it can provide the sequences based on the user's preference of data format. Detailed help documentation is also available at http://www.ncbi.nlm.nih.gov/entrez/query/static/help/helpdoc.html.

3. SEQUENCE ANALYSIS TOOLS

3.1. Sequence Assembler

Two well known methods are applied to generate genome sequences. One is called the Clone By Clone (CBC) approach and the other is called the Whole Genome Shotgun (WGS) method. Figure 2 shows the general procedure for both CBC and WGS methods. Both methods create many small overlapping sequences or reads,

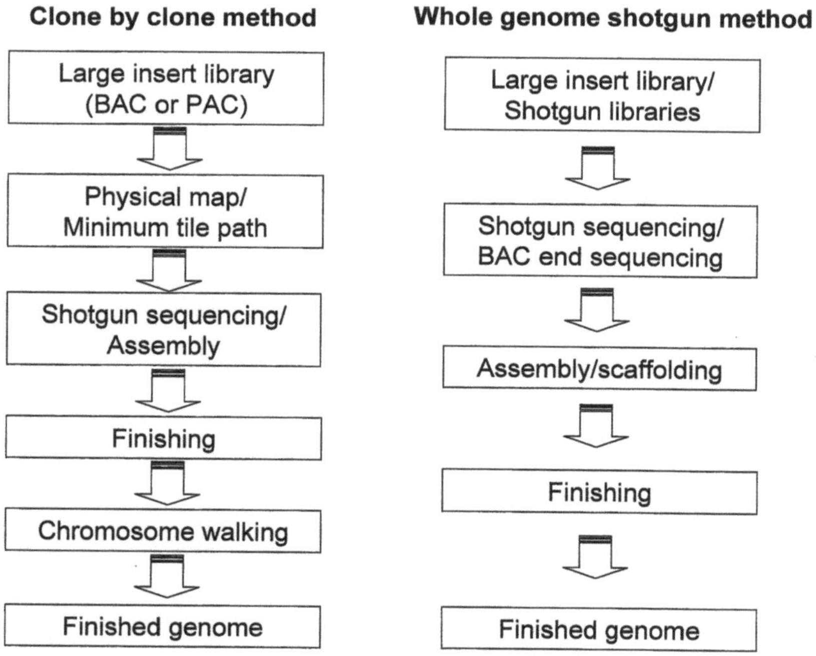

Fig. 2. Simplified procedures involved in Clone By Clone shotgun (CBC) and Whole Genome Shotgun (WGS) methods.

which are eventually assembled by computer software to build the sequence contigs. The CBC method sequences shotgun clones derived from minimum tiling BAC (bacterial artificial chromosome) or PAC (P1-derived artificial chromosome) clones. Although progress is slow using the CBC method, this method generates high quality data across the genome, even in highly repetitive regions. The WGS method uses sequences from shotgun clones derived directly from genomic DNA and from various insert sizes of genomic clones (10-150kb). Assembly is fast with WGS, but it permits low quality data and mis-assembly in repeat regions.

3.1.1 Phred/Phrap/Consed package

Phred, Phrap and Consed package is UNIX based software most widely used in CBC shotgun sequence assembly, and it provides the standardized quality assurance for many genome projects.

The software Phred (Ewing et al. 1998; Ewing and Green 1998) calls bases by reading electropherogram or trace files as raw data and assigns a quality value to each base. The quality value (QV) or phred value is a base call error probability, and it is calculated by the formula: $QV = -10\ log_{10}(p)$, where p is the probability that the base call is an error. A phred value of 20 represents one base call error in 100 bp (p=0.01), and a phred value of 30 represents one base call error in 1000 bp (p=0.001). Generally, a phred value greater than or equal to 20 is considered a high quality sequence. Phred generates FASTA formatted sequences, quality values, and phd (phred) files that can be used to assemble sequences using Phrap.

Fig. 3. Consed is a program for viewing and editing Phrap assembly. A: Snapshots of Consed assembly view, B: Aligned reads window, C: Trace window.

Phrap (http://www.genome.washington.edu/UWGC/analysistools/Phrap.cfm) is a program for assembling shotgun sequences based on sequence overlaps. Before assembly, vector sequences are removed from an individual read using the program cross_match, which comes with the package. Three input files are required to run Phrap: vector screened sequences, quality values, and phd files. Forward and reverse sequences, along with sequencing chemistry, are recognized by Phrap when a pre-defined naming convention is used. Manyreads and longreads versions of Phrap and cross_match are recommended when more than 64,000 reads are assembled or a sequence longer than 64,000 bp is included in a single assembly. The software Consed is required in order to display and edit the assembly by reading the Phrap output file (ace file).

Consed (Gordon et al. 1998) is a program for viewing and editing Phrap assembly in its finishing phase. It shows a global assembly view with forward-reverse pairs, read depth, and repeat match. Thus, this information can be easily used for finishing procedures. Independent from Phrap assembly, Consed also allows the breaking or joining of contigs by comparing them. *In silico* digestions can be generated and compared to real digestions to verify the overall assembly. For a primer walk in the finishing phase, Consed provides a built-in primer picking function. Figure 3 shows examples of Consed screen shots. Autofinish (Gordon et al. 2001) is a part of the Consed program. It generates an experiment list focused on filling gaps, improving the quality in low quality areas, and determining the orientation of contigs.

3.1.2 Whole genome assembler

Unlike CBC assembly, whole genome assembly is a challenging procedure. Whole genome assembly processes hundreds of thousands of reads at once, leading to possible mis-assembly due to repeat sequences. Development for WGS assembler thus focuses on reducing computation time and resolving mis-assembly problems caused by repeat sequences in the genome.

Arachne (Batzoglou et al. 2002; Jaffe et al. 2003) uses quality score associated sequences for whole genome shotgun sequence assembly. It utilizes the forward and reverse pair information within a similar insert size library and removes vector and other contaminated sequences in an initial assembly. It also identifies potential repeats by clustering reads, excludes the repeats from the assembly, and merges overlapping read pairs to make contigs. Read pairs from larger insert clones are used later to build supercontigs. Arachne-simulated assembly of the *Drosophila* genome showed about 98% coverage with an N50 contig length of 324 kb and an N50 supercontig length of 5,143 kb. The N50 length is the length x such that 50% of the sequence is contained in contigs of length x or greater. The mouse whole genome assembly was made possible with Arachne 2. The contigs have an N50 length of 24.8 kb, whereas the supercontigs have an N50 length that is approximately 700-fold larger at 16.9 Mb.

PCAP (Huang et al. 2003) is a contig assembly program using parallel computer processors. First, it removes vector and low quality areas from reads. Then, it uses BLAST2 to identify pairs of reads that contain potential overlaps. Identification of repetitive regions in reads is based on deep coverage by longer approximate matches. The score of every overlap is adjusted to reflect the depths of coverage for the two regions in the overlap. The consensus sequence of a contig is generated by constructing an alignment of reads in the contig. Human chromosome 20's assembly was simulated with PCAP, and it showed an N50 contig and a scaffold length of 41 kb and 2Mb, respectively.

Phusion (Mullikin and Ning 2003) first groups sequence reads by determining the number of times that sequences of length k (called k-mer word) occur in the data, and it eliminates reads representing highly redundant k-mer sequences. It generates a reads list and matrix based on reads showing a less repetitive or unique k-mer distribution. Phrap uses paired reads information, along with the above information, to assemble sequences. An iterative Phrap with read pairs from different size insert clones will merge contigs and make supercontigs.

3.1.3 EST clustering

Expressed Sequence Tag (EST) provides useful information because it is a profile of expressed gene sequences. It usually does not contain full length gene sequences

Table 2. List of GenBank databases for BLAST search.

	Database	Description
Nucleotide Search	Nr	Non-redundant nucleotide data of GenBank, [1]EMBL and [2]DDBJ not including EST, STS, GSS and HTGS database
	Month	Newly releases nucleotide sequences (within 30 days) in GenBank, EMBL and DDBJ
	dbEST	Non-redundant EST sequences in GenBank, EMBL and DDBJ
	dbSTS	Non-redundant STS sequences in GenBank, EMBL and DDBJ
	Mouse EST	Non-redundant mouse ESTs in GenBank, EMBL and DDBJ
	Human EST	Non-redundant human ESTs in GenBank, EMBL and DDBJ
	Other EST	Non-redundant ESTs in GenBank, EMBL and DDBJ excluding mouse and human ESTs
	Yeast	*Saccharomyces cerevisiae* genomic sequences
	E. coli	*Escherichia coli* genomic nucleotide sequences
	Vector	Vector sequence database
	Mito	Mitochondria sequence database
	Alu	Alu repeat sequence database
	GSS	Single pass genome survey sequences including BAC end sequences
	HTGS	Phase1 and 2 high throughput genomic sequence database
Protein Search	Nr	All non-redundant GenBank [3]CDS translations+[4]PDB+SwissProt+[5]PIR+[6]PRF
	Month	Newly released (within 30 days) or revised GenBank CDS translation, SwissProt and PIR
	SwissProt	The last major release of the Swiss-Prot protein sequence database
	Yeast	*Saccharomyces cerevisiae* protein sequences derived from yeast genome sequence
	E. coli	*Escherichia coli* translated coding sequences
	PDB	Protein data bank archive of macromolecular structural data

[1]EMBL: European Molecular Biology Laboratory, [2]DDBJ: DNA Data Bank of Japan, [3]CDS: Coding Sequence, [4]PDB: Protein Data Bank, [5]PIR: Protein Information Resource, [6]PRF: Protein Research Foundation.

since about 600 bp sequences are generated from the 5′ and 3′ end of cDNA clones. EST also permits low quality bases due to single pass sequencing, and some sequences are highly redundant in certain genes. In order to overcome these disadvantages and to collect more unique sequences (UniGenes), clustering ESTs is necessary. Phrap, TIGR assembler (Pop and Kosack 2004) and CAP3 (Huang and Madan 1999) are used to cluster or assemble EST data.

CAP3 uses FASTA sequence files as an input for assembly, and two additional files (FASTA quality, and forward and reverse pair information) can be used to correct the assembly. CAP3 removes low quality regions from the 5′ and 3′ end of sequences, and it detects overlaps between input reads. These overlaps are used to join the sequences and thus make contigs. Forward and reverse information (constraints) is used to correct the assembly. Once corrected, it writes consensus sequences and the quality value of each base. The output or assembly file (ace file) can be viewed using the Consed program. Lucy is a program used to prepare raw DNA sequences for EST or shotgun assembly. It removes low quality or vector sequences in raw data, and it provides high quality sequences and quality files for assembly (Chou and Holmes 2001).

3.2. Pairwise and Multiple Sequence Alignments

Sequence alignment is a daily task of most biologists in order to find the relationship or similarity between biological sequences. Pairwise sequence alignment tries to find the optimal alignment in parts of sequences (local alignment) or in entire sequences (global alignment). In a global alignment, all of the nucleotides or amino acids in both sequences participate in the alignment. Thus it is useful for aligning closely-related sequences. Local alignment finds and aligns related regions within sequences. It is more flexible than global alignment and is useful in identifying related regions that appear in a different order in two sequences.

Multiple sequence alignment is an extension of pairwise alignment that allows the identification of common regions within several sequences. This tool is mostly used for building phylogenetic trees and creating sequence profiles which can be used to search distantly related sequences in the database.

3.2.1 *BLAST*

BLAST (Basic Local Alignment Search Tool) (Altschul et al. 1990; Altschul et al. 1997) is the most popular local alignment program for similarity search and sequence alignment. The BLAST algorithm generates a list of short word matches (default word size is 3 for protein and 11 for nucleotide) in query sequences. The database is then searched for the occurrences of these words. The matching words are extended to the local alignment between two sequences and these extensions are continued until the score is below the threshold. A BLAST search can be performed at NCBI's BLAST server using the web browser or at any local computer by installing the BLAST software (stand-alone BLAST). Stand-alone BLAST reduces the search time significantly by avoiding on-line communication and allowing a batch blast (submit multiple queries at once) against local databases downloaded from

GenBank or created by the user. The setup information for standalone BLAST can be found at NCBI's ftp site (ftp://ftp.ncbi.nlm.nih.gov/blast/documents/blast.html).

Several BLAST programs are available based on the search purpose and on the query and database relation. The FASTA file format is a standard input requirement for all BLAST programs and database format. The formatdb program is necessary in order to do local BLAST searches.

Nucleotide search: BLASTN is used to compare nucleotide query sequences against nucleotide databases. NCBI provides several databases to compare query sequences. Table 2 shows the NCBI databases for a BLAST search. MegaBLAST uses a greedy algorithm (Zhang et al. 2000) to perform a nucleotide search using a word size of 28 as a default (BLASTN uses a word size of 11), which makes the search 10 times faster for closely related sequences. Web MegaBLAST allows multiple query searches with FASTA format sequences or accession numbers.

Protein search: The BLASTP program is used to compare an amino acid sequence query against amino acid sequence databases using BLOSUM62 (Henikoff and Henikoff 1992) as a default substitution matrix. More specialized protein searches can be done with PHI-BLAST and PSI-BLAST (Altschul et al. 1997). Pattern-Hit Initiated (PHI)-BLAST is designed to search for proteins that contain a pattern in the query specified by the user. Position-Specific Iterated (PSI)-BLAST is designed to find distantly related protein sequences using a PSSM (Position-Specific Scoring Matrix) generated from each search. PSI-BLAST is the most sensitive BLAST program.

3.2.2 Translated query or database search

BLASTX: Compares a nucleotide query sequence translated in all reading frames against a protein sequence database. This program can be used to find potential translation products of an unknown nucleotide sequence.

TBLASTN: Compares a protein query sequence against a nucleotide sequence database dynamically translated in all reading frames.

TBLASTX: Compares the six-frame translations of a nucleotide query sequence against the six-frame translations of a nucleotide sequence database.

Blast 2 sequences (bl2seq): bl2seq (Tatusova and Madden 1999) is designed to directly compare two sequences. Most programs mentioned above, such as BLASTN (1st query- nucleotide and 2nd query- nucleotide), BLASTP (protein and protein), BLASTX (nucleotide and protein), and TBLASTN (protein and nucleotide), are used to align two query sequences.

3.2.3 *FASTA*

Another local alignment tool is the FASTA program developed by Pearson and Lipman (1988), which is available over the web or by download. FASTA searches short sequences called k-tuples (which are similar to words in BLAST) to identify ungapped alignments. The alignments are tested and merged into a local alignment in order to find the optimal local alignment based on the threshold and score. FASTA provides tools similar to BLAST. However, it also performs global alignments which are not provided by BLAST. The FASTA program is used to compare a DNA sequence to a DNA database or a protein sequence to a protein database (equivalent to BLASTN and BLASTP). FASTX/FASTY is the same as BLASTX in that it compares translated DNA against a protein database. TFASTX/TFASTY is used to compare a protein sequence against a translated nucleotide database (similar to TBLASTN). ALIGN performs global alignment between two protein or nucleotide queries. FASTA programs can be found at URL http://fasta.bioch.virginia.edu.

3.2.4 *BLAT*

BLAT stands for BLAST-like alignment tool developed by Jim Kent (2002) and is designed to effectively align EST sequences to genomic sequences. Its algorithm is similar to BLAST in that it finds word matches and extends hits to form high scoring pairs (HSPs). However, BLAT builds an index of the database first and then performs queries. Moreover, it extends alignments on any number of perfect or nearly perfect hits and provides a large alignment. BLAT also gives unspliced alignment between mRNA and DNA sequences so that splicing sites are easily detected in the alignment. Web search is available at http://genome.ucsc.edu/cgi-bin/hgBlat and stand-alone is also available at http://www.soe.ucsc.edu/~kent/exe. Standard output is in tabular format (tab delimited pls file), and BLAST pairwise output is also available.

3.2.5 *CLUSTALW*

Multiple sequence alignment is a tool used to study closely related genes or proteins in order to find the evolutionary relationships between genes and to identify shared patterns among functionally or structurally related genes. A popular program for multiple sequence alignment is ClustalW (Higgins et al. 1994; http://www.ebi.ac.uk/clustalw). Both progressive global and local alignments can be done in ClustalW. The user has the option to control parameters to make the best alignments (e.g., word size, matrix, gap open, extension, etc.). It also provides two phylogenetic trees, a cladogram (equal length of branched tree showing common ancestry) or a phylogram (unequal length of branched tree showing evolutionary distances). Alignment can be further edited using the Jalview program (Clamp et al. 2004; http://www.ebi.ac.uk/jalview).

3.3. Gene Prediction

Gene prediction is an important aspect of genome projects. In eukaryotes, gene prediction and annotation is not a simple process due to the various sizes of introns (noncoding sequences) located between exons (coding sequences). In addition, many genes have alternative splice variants which make eukaryotic gene structure and length difficult to predict.

Many gene prediction programs have been developed for genome wide annotation. They are generally categorized into three groups. The first group uses an *ab initio* approach to predict genes directly from nucleotide sequences. Prediction programs in this group utilize statistical models to differentiate the promoter, coding or noncoding regions, as well as intron-exon junctions in genomic sequences. Hidden Marcov Models (HMMs) is a popular model used to make gene prediction programs, such as Grail (Xu et al. 1994), FGENESH (Solovyev et al. 1995; Salamov and Solovyev 2000), HMMgene (Krogh 1997), MZEF (Zhang 1997), and GENSCAN (Burge and Karlin 1997; Burge and Karlin 1998). The second group uses a similarity based approach to identify gene structure using a sequence alignment between genomic sequence and transcript (EST and cDNA) or protein databases. This approach has recently been expanded to genomic sequence comparison (comparative approach) between evolutionarily related species in order to identify functional regulatory elements which tend to be conserved through evolution. AAT (Huang et al. 1997), CRASA (Chuang et al. 2003), and AGenDA (Taher et al. 2003) belong to this group. The third group combines the *ab initio* method and similarity based approach. Procrustes (Gelfand et al. 1996; Mironov et al. 1998), FGENESH+ (Salamov and Solovyev, 2000), GenomeScan (Yeh et al. 2001), and GeneWise (Birney et al. 2004) are available for this approach. Table 3 shows the prediction programs listed above.

Splicing site prediction is important in choosing the correct gene models on the basis of accurate intron-exon boundaries. Many programs use computational models based on consensus dimer sequences in donor sites, acceptor sites, and branch points (about 30bp upstream of acceptor site). They also use sequence alignments between transcripts and genomic sequences to predict splicing sites in genomic sequences. NetGene2 (Hebsgaard et al. 1996; http://www.cbs.dtu.dk/services/NetGene2), SplicePredictor (Brendel and Kleffe 1998; http://bioinformatics.iastate.edu/cgi-bin/sp.cgi), or GeneSplicer (Pertea et al. 2001; http://www.tigr.org/tdb/GeneSplicer/gene_spl.html) is used for splicing site prediction.

Table 3. List of gene prediction programs. *Ab initio* program: FGENESH, Genscan, Grail, MZEF and HMMgene. Similarity based program: CRASA, AAT and AgenDa. Combined program: GenomeScan, Procrustes and FGENESH+.

Program	URL and Training Set
FGENESH	http://www.softberry.com/berry.phtml Human, Mouse, Fruit fly, Monocot, Dicot, *S. pombe*, *Neurospora*, Fish, Algae, *Aspergillus*
Genscan/ Genscan+	http://genes.mit.edu/GENSCAN.html Vertebrates, *Arabidopsis*, Maize
Grail	http://compbio.ornl.gov/Grail-1.3 Human, Mouse, *Arabidopsis*, *Drosophila*, *E. coli*
MZEF	http://rulai.cshl.org/tools/genefinder Human, Mouse, *Arabidopsis*, Yeast
HMMGene	http://www.cbs.dtu.dk/services/HMMgene Human and other vertebrates, *C. elegans*
CRASA	http://crasa.sinica.edu.tw/bioinformatics/bioinformatics.html
AAT	http://genome.cs.mtu.edu/aat/aat.html
AGenDa	http://bibiserv.techfak.uni-bielefeld.de/agenda
GenomeScan	http://genes.mit.edu/genomescan.html Vertebrates, *Arabidopsis*, Maize
Procrustes	http://www-hto.usc.edu/software/procrustes/wwwserv.html
FGENESH+	http://www.softberry.com/berry.phtml

tRNAScan-SE (Lowe and Eddy 1997) identifies transfer RNA genes in genomic sequences by searching for conserved A & B box promoter sequences and progressively identifying various stem-loop structures. It provides tabular and secondary structure as the standard output. tRNA analysis is performed on-line at http://www.genetics.wustl.edu/eddy/tRNAscan-SE.

Although some areas of the genome rely only on *ab initio* or similarity based approaches due to prediction failure or lack of experimental data, a combined approach generally increases the accuracy of gene annotation.

3.4. Repeat Identification

Repetitive sequences occupy a large portion of most eukaryotic genomes and are divided into tandem (including simple sequence repeats, SSRs) and interspersed repeats. Transposable elements (TEs), one type of interspersed repeats, are the most abundant in the repeat family. Transposable elements are further classified into two groups: class I and class II. A class I TE is called a retrotransposon, and it transposes using a RNA intermediate. This element encodes reverse transcriptase and other viral proteins (*gag* and *pol*). It is subdivided into LTR type (*gypsy* and *copia* group) and non-LTR type [long interspersed repetitive element (LINE) and short interspersed repetitive element (SINE)] based on long terminal repeats (LTRs). A class II TE is a DNA element that moves from one site in the genome to another. It has terminal inverted repeats (TIRs) and encodes a transposase that moves from one position to another within a genome. *P* elements in the fruit fly and *Ac, Spm,* and *Mu* elements in maize are well studied transposons (Girard and Freeling 1999; Wessler 2001).

Recently, another type of repeat element called miniature inverted-repeat transposable element (MITE) has been identified. MITEs are usually less than 600 bases in size and have short terminal inverted repeats. They can be divided into two groups based on TIRs and target site duplication: *Tourist* and *Stowaway* (Bureau and Wessler 1992; Bureau and Wessler, 1994; Jiang et al. 2003).

Repeat finding programs often use a similarity based approach to find and annotate the repeat regions in genomic sequences. However, Juretic and co-workers (2004) made some efforts to annotate TEs in the rice genome with HMMER (Eddy 1998), the hidden Marcov model software package.

3.4.1 *RepeatMasker*

RepeatMasker (http://repeatmasker.org) is a wildly used program to find interspersed repeats (LINEs, SINEs, LTRs and DNA elements), simple sequence repeats (SSRs), and low complexity regions in the sequences using similarity search against a well defined repeat database. RepeatMasker uses the cross_match program as a default search engine which results in a high sensitivity but slow speed when long sequences are searched. WU-BLAST is added as an optional search engine to increase the search speed. A user defined repeat database can be used to search against in stand-alone RepeatMasker.

3.4.2 *RECON*

RECON (Bao and Eddy 2002) allows the *de novo* detection and classification of repeat families in genomic sequences. The RECON algorithm detects and groups repeats from genomic sequences by performing a blast itself. This approach helps repeat annotation by determining repeat boundaries in genomic sequences and by enabling identification of new repeat elements. RECON is available from http://www.genetics.wustl.edu/eddy/recon.

3.5. Other Programs

3.5.1 *PipMaker*

PipMaker (Schwartz et al. 2000) is a tool used to align two sequences, and it generates a percent identity plot (PIP) and a dot plot as output. It requires two FASTA format sequences and repeat information from RepeatMasker. Exon information may be optionally submitted. MultiPipMaker allows the submission of multiple FASTA sequences (up to 20) and performs multiple sequence alignments in order to analyze relationships among input sequences. PipMaker analysis can be performed at http://pipmaker.bx.psu.edu/pipmaker. The zPicture program (Ovcharenko et al. 2004; http://zpicture.dcode.org) also provides similar dynamic alignment and visualization of two sequences.

3.5.2 *rVISTA*

Regulatory Vista (rVISTA; Loots and Ovcharenko 2004) is a computational tool used to identify evolutionarily conserved transcription factor binding sites (TFBSs) by performing multiple alignments of orthologous sequences followed by the prediction of TFBSs using the TRANSFAC database (Matys et al. 2003) collected from eukaryotic transcription factors. The output from zPicture can be transferred to the rVISTA program without further modification. This program can assist in understanding the function of conserved non-coding sequences and can identify the potential *cis*-regulatory elements in genomic sequences. rVIST2.0 is available at http://rvista.dcode.org.

3.5.3 *MUMmer*

MUMmer (Delcher et al. 2002; Kurtz et al. 2004) is a tool that allows the rapid alignment of two large nucleotide or protein sequences as well as the alignment of two genomes. It uses the suffix tree algorithm to find a minimum of 20bp exact match as an alignment anchor and then extends the alignment to generate a pair-wise alignment like BLAST. MUMmer is comprised of several programs, such as MUMmer for finding the maximal exact match, NUCmer for aligning closely related sequences, and PROmer for aligning far related sequences. Programs are selected on the basis of related sequences participating in alignments. Alignment outputs are converted to dot and percent identity plots using mummerplot and gnuplot. The software is available at ftp://ftp.tigr.org/pub/software/MUMmer.

3.5.4 *EMBOSS*

EMBOSS (Rice et al. 2000) stands for The European Molecular Biology Open Software Suite. Currently, more than 100 programs are available in the EMBOSS package. The programs are grouped according to the varying analyses they perform, such as alignment, display, edit, enzyme kinetics, nucleotide analysis, protein analysis, and phylogeny. Many more applications will be added in the near future. It is difficult to describe all of the programs here, but EMBOSS is mainly used for sequence alignment, restriction map creation, CpG island analysis, primer design, sequence extraction, sequence retrieval from the database, codon usage analysis, protein motif analysis, and much more. It runs on a UNIX environment with the command line mode (http://www.hgmp.mrc.ac.uk/Software/EMBOSS/download.html). The java graphical user interface version (JEMBOSS; Carver and Bleasby, 2003) is also available (http://www.rfcgr.mrc.ac.uk/Software/EMBOSS/Jemboss/download).

4. MOLECULAR DATABASE

Each year, Nucleic Acids Research (http://nar.oxfordjournals.org) provides collective database information covering various areas in biological research. In 2004, 548 biological databases were updated in 11 hierarchical classifications, which helped users easily find the database that they needed (Galperin 2004).

Pfam database: Pfam (Beatman et al. 2004) is a comprehensive collection of protein domains and families represented by multiple sequence alignment and profile-HMMs of SwissProt and TrEMBL protein data. It is divided into two groups: pfam-A and pfam-B. pfam-A is a collection of protein domains from manual multiple alignments. pfam-B is automatic collection of conserved domains. More than 7,400 families are currently listed (June 2004). Protein, nucleotide, and keyword searches are carried out using the web service at the pfam site (http://pfam.wustl.edu).

SCOP database: The Structural Classification of Proteins (SCOP; Andreeva et al. 2004) is a collection of proteins classified on the basis of known protein structure. Proteins are classified as alpha, beta, alpha and beta, multi-domain, membrane, cell surface, or small proteins. 70,859 domains were classified as 2,845 families, 1,539 superfamilies, and 945 folds (Release 1.69; July 2005) (http://scop.mrc-lmb.cam.ac.uk/scop).

RefSeq: Reference Sequence (RefSeq; Pruitt and Maglott 2001) is a non-redundant collection of DNA, RNA, and protein sequences for major research organisms. Each RefSeq entry represents a stable reference for gene identification, mutation analysis, polymorphism discovery, and comparative analysis. It is manually curated and periodically updated. The RefSeq Release 15 (January 1, 2006) includes 2,273,764 proteins from 3,244 organisms (http://www.ncbi.nlm.nih.gov/RefSeq).

TIGR Gene Indices: TIGR Gene Indices (Quackenbush et al. 2001) is an EST database based on public EST sequences. EST sequences provide expression gene profiles in a genome. However, due to the nature of the short, single pass sequences, usage is limited. To overcome this problem, assembly strategy is applied on public ESTs. 92 Gene Indices covering animals, plants, and fungi were available at http://www.tigr.org/tdb/tgi/index.shtml in January, 2006. EST clustering methods are described at http://www.tigr.org/tdb/tgi/software.

MetaCyc: MetaCyc (Krieger et al. 2004) is a multi-organism metabolic pathway and enzyme database primarily for microorganisms and plants. It provides information on metabolic pathways along with compounds, enzymes, and genes. 700 pathways from more than 600 different organisms are collected in MetaCyc, representing about 4,900 enzymatic reactions. One of MetaCyc's applications is to serve as a reference database for predicting the metabolic network from its annotated genome, such as by the software Pathway Tools (Karp et al. 2002). The database can be browsed by submitting enzyme, pathway, or gene name as a query (http://metacyc.org).

5. CONCLUSION

Knowing and utilizing the appropriate program is important to many biologists in order to investigate biological questions. In this chapter, we have described a handful of bioinformatics tools used for analyzing sequence data. Two important tools that we did not discuss are the UNIX operating system and Perl scripting. Many sequence analysis programs are written and operated in the UNIX environment. Therefore, understanding UNIX gives more power to extensively utilize analysis programs as well as efficiently handle input and output files. Perl is the most popular computer programming language used in genomics. Perl script allows the identification of patterns in huge data sets or in text inputs and outputs. Independent from existing programs, Perl script can add more flexible ways to organize and analyze data.

We are currently immersed in a flood of genomic information. However, it is more likely that we are merely snorkeling in a small pool etched into the floor of a vast ocean. As many organisms are sequenced, we realize that we need more and more sequence information in order to explain similarities and differences within genomes and between genomes. Consequently, biology and bioinformatics must work together to design new algorithms and programs for analyzing genomes. These collective efforts will give us profound knowledge to better understand the diversity of living organisms.

REFERENCES

Adams MD, Kelley JM, Gocayne JD, Dubnick M, Polymeropoulos MH, Xiao H, Merril CR, Wu A, Olde B, Moreno RF, Kerlavage AR, McCombie WR and Venter JC (1991). Complementary DNA sequencing: expressed sequence tags and human genome project. Science 252:1651-1656.

Altschul SF, Gish W, Miller W, Myers EW and Lipman DJ (1990). Basic local alignment search tool. J Mol Biol 215:403-410.

Altschul SF, Madden TL, Schaffer AA, Zhang J, Zhang Z, Miller W and Lipman DJ (1997). Gapped BLAST and PSI-BLAST: a new generation of protein database search programs. Nucleic Acids Res 25:3389-3402.

Andreeva A, Howorth D, Brenner SE, Hubbard TJ, Chothia C and Murzin AG (2004). SCOP database in 2004: refinements integrate structure and sequence family data. Nucleic Acids Res 32 Database issue:D226-229.

Bao Z and Eddy SR (2002). Automated de novo identification of repeat sequence families in sequenced genomes. Genome Res 12:1269-1276.

Bateman A, Coin L, Durbin R, Finn RD, Hollich V, Griffiths-Jones S, Khanna A, Marshall M, Moxon S, Sonnhammer EL, Studholme DJ, Yeats C and Eddy SR (2004). The Pfam protein families database. Nucleic Acids Res 32 Database issue:D138-141.

Batzoglou S, Berger B, Mesirov J and Lander ES (1999). Sequencing a genome by walking with clone-end sequences: a mathematical analysis. Genome Res 9:1163-1174.

Batzoglou S, Jaffe DB, Stanley K, Butler J, Gnerre S, Mauceli E, Berger B, Mesirov JP and Lander ES (2002). ARACHNE: a whole-genome shotgun assembler. Genome Res 12:177-189.

Birney E, Clamp M and Durbin R (2004). GeneWise and Genomewise. Genome Res 14:988-995.

Brendel V and Kleffe J (1998). Prediction of locally optimal splice sites in plant pre-mRNA with applications to gene identification in *Arabidopsis thaliana* genomic DNA. Nucleic Acids Res 26:4748-4757.

Bureau TE and Wessler SR (1994). Stowaway: a new family of inverted repeat elements associated with the genes of both monocotyledonous and dicotyledonous plants. Plant Cell 6:907-916.

Bureau TE and Wessler SR (1992). Tourist: a large family of small inverted repeat elements frequently associated with maize genes. Plant Cell 4:1283-1294.
Burge C and Karlin S (1997). Prediction of complete gene structures in human genomic DNA. J Mol Biol 268:78-94.
Burge CB and Karlin S (1998). Finding the genes in genomic DNA. Curr Opin Struct Biol 8:346-354.
Carver T and Bleasby A (2003). The design of Jemboss: a graphical user interface to EMBOSS. Bioinformatics 19:1837-1843.
Chou HH and Holmes MH (2001). DNA sequence quality trimming and vector removal. Bioinformatics 17:1093-1104.
Chuang TJ, Lin WC, Lee HC, Wang CW, Hsiao KL, Wang ZH, Shieh D, Lin SC and Ch'ang LY (2003). A complexity reduction algorithm for analysis and annotation of large genomic sequences. Genome Res 13:313-322.
Clamp M, Cuff J, Searle SM and Barton GJ (2004). The Jalview Java alignment editor. Bioinformatics 20:426-427.
Delcher AL, Phillippy A, Carlton J and Salzberg SL (2002). Fast algorithms for large-scale genome alignment and comparison. Nucleic Acids Res 30:2478-2483.
Eddy SR (1998). Profile hidden Markov models. Bioinformatics 14:755-763.
Ewing B and Green P (1998). Base-calling of automated sequencer traces using phred. II. Error probabilities. Genome Res 8:186-194.
Ewing B, Hillier L, Wendl MC and Green P (1998). Base-calling of automated sequencer traces using phred. I. Accuracy assessment. Genome Res 8:175-185.
Fleischmann RD, Adams MD, White O, Clayton RA, Kirkness EF, Kerlavage AR, Bult CJ, Tomb JF, Dougherty BA, Merrick JM and et al. (1995). Whole-genome random sequencing and assembly of *Haemophilus influenzae* Rd. Science 269:496-512.
Galperin MY (2004). The Molecular Biology Database Collection: 2004 update. Nucleic Acids Res 32 Database issue:D3-22.
Gelfand MS, Mironov AA and Pevzner PA (1996). Gene recognition via spliced sequence alignment. Proc Natl Acad Sci U S A 93:9061-9066.
Girard L and Freeling M (1999). Regulatory changes as a consequence of transposon insertion. Dev Genet 25:291-296.
Gordon D, Abajian C and Green P (1998). Consed: a graphical tool for sequence finishing. Genome Res 8:195-202.
Gordon D, Desmarais C and Green P (2001). Automated finishing with autofinish. Genome Res 11:614-625.
Hebsgaard SM, Korning PG, Tolstrup N, Engelbrecht J, Rouze P and Brunak S (1996). Splice site prediction in *Arabidopsis thaliana* pre-mRNA by combining local and global sequence information. Nucleic Acids Res 24:3439-3452.
Henikoff S and Henikoff JG (1992). Amino acid substitution matrices from protein blocks. Proc Natl Acad Sci U S A 89:10915-10919.
Higgins DG (1994). CLUSTAL V: multiple alignment of DNA and protein sequences. Methods Mol Biol 25:307-318.
Huang X, Adams MD, Zhou H and Kerlavage AR (1997). A tool for analyzing and annotating genomic sequences. Genomics 46:37-45.
Huang X and Madan A (1999). CAP3: A DNA sequence assembly program. Genome Res 9:868-877.
Huang X, Wang J, Aluru S, Yang SP and Hillier L (2003). PCAP: a whole-genome assembly program. Genome Res 13:2164-2170.
Jaffe DB, Butler J, Gnerre S, Mauceli E, Lindblad-Toh K, Mesirov JP, Zody MC and Lander ES (2003). Whole-genome sequence assembly for mammalian genomes: Arachne 2. Genome Res 13:91-96.
Jiang N, Bao Z, Zhang X, Hirochika H, Eddy SR, McCouch SR and Wessler SR (2003). An active DNA transposon family in rice. Nature 421:163-167.
Juretic N, Bureau TE and Bruskiewich RM (2004). Transposable element annotation of the rice genome. Bioinformatics 20:155-160.
Karp PD, Paley S and Romero P (2002). The Pathway Tools software. Bioinformatics 18 Suppl 1:S225-232.
Kent WJ (2002). BLAT--the BLAST-like alignment tool. Genome Res 12:656-664.
Krieger CJ, Zhang P, Mueller LA, Wang A, Paley S, Arnaud M, Pick J, Rhee SY and Karp PD (2004). MetaCyc: a multiorganism database of metabolic pathways and enzymes. Nucleic Acids Res 32 Database issue:D438-442.
Krogh A (1997). Two methods for improving performance of an HMM and their application for gene finding. Proc Int Conf Intell Syst Mol Biol 5:179-186.
Kurtz S, Phillippy A, Delcher AL, Smoot M, Shumway M, Antonescu C and Salzberg SL (2004). Versatile and open software for comparing large genomes. Genome Biol 5:R12.

Loots GG and Ovcharenko I (2004). rVISTA 2.0: evolutionary analysis of transcription factor binding sites. Nucleic Acids Res 32:W217-221.

Lowe TM and Eddy SR (1997). tRNAscan-SE: a program for improved detection of transfer RNA genes in genomic sequence. Nucleic Acids Res 25:955-964.

Mahairas GG, Wallace JC, Smith K, Swartzell S, Holzman T, Keller A, Shaker R, Furlong J, Young J, Zhao S, Adams MD and Hood L (1999). Sequence-tagged connectors: a sequence approach to mapping and scanning the human genome. Proc Natl Acad Sci U S A 96:9739-9744.

Matys V, Fricke E, Geffers R, Gossling E, Haubrock M, Hehl R, Hornischer K, Karas D, Kel AE, Kel-Margoulis OV, Kloos DU, Land S, Lewicki-Potapov B, Michael H, Munch R, Reuter I, Rotert S, Saxel H, Scheer M, Thiele S and Wingender E (2003). TRANSFAC: transcriptional regulation, from patterns to profiles. Nucleic Acids Res 31:374-378.

Mironov AA, Roytberg MA, Pevzner PA and Gelfand MS (1998). Performance-guarantee gene predictions via spliced alignment. Genomics 51:332-339.

Mullikin JC and Ning Z (2003). The phusion assembler. Genome Res 13:81-90.

Olson M, Hood L, Cantor C and Botstein D (1989). A common language for physical mapping of the human genome. Science 245:1434-1435.

Ouellette BF and Boguski MS (1997). Database divisions and homology search files: a guide for the perplexed. Genome Res 7:952-955.

Ovcharenko I, Loots GG, Hardison RC, Miller W and Stubbs L (2004). zPicture: dynamic alignment and visualization tool for analyzing conservation profiles. Genome Res 14:472-477.

Pearson WR and Lipman DJ (1988). Improved tools for biological sequence comparison. Proc Natl Acad Sci U S A 85:2444-2448.

Pertea M, Lin X and Salzberg SL (2001). GeneSplicer: a new computational method for splice site prediction. Nucleic Acids Res 29:1185-1190.

Pop M and Kosack D (2004). Using the TIGR assembler in shotgun sequencing projects. Methods Mol Biol 255:279-294.

Pruitt KD and Maglott DR (2001). RefSeq and LocusLink: NCBI gene-centered resources. Nucleic Acids Res 29:137-140.

Quackenbush J, Cho J, Lee D, Liang F, Holt I, Karamycheva S, Parvizi B, Pertea G, Sultana R and White J (2001). The TIGR Gene Indices: analysis of gene transcript sequences in highly sampled eukaryotic species. Nucleic Acids Res 29:159-164.

Rice P, Longden I and Bleasby A (2000). EMBOSS: the European Molecular Biology Open Software Suite. Trends Genet 16:276-277.

Salamov AA and Solovyev VV (2000). Ab initio gene finding in Drosophila genomic DNA. Genome Res 10:516-522.

Schwartz S, Zhang Z, Frazer KA, Smit A, Riemer C, Bouck J, Gibbs R, Hardison R and Miller W (2000). PipMaker--a web server for aligning two genomic DNA sequences. Genome Res 10:577-586.

Siegel AF, Trask B, Roach JC, Mahairas GG, Hood L and van den Engh G (1999). Analysis of sequence-tagged-connector strategies for DNA sequencing. Genome Res 9:297-307.

Solovyev VV, Salamov AA and Lawrence CB (1995). Identification of human gene structure using linear discriminant functions and dynamic programming. Proc Int Conf Intell Syst Mol Biol 3:367-375.

Taher L, Rinner O, Garg S, Sczyrba A, Brudno M, Batzoglou S and Morgenstern B (2003). AGenDA: homology-based gene prediction. Bioinformatics 19:1575-1577.

Tatusova TA and Madden TL (1999). BLAST 2 Sequences, a new tool for comparing protein and nucleotide sequences. FEMS Microbiol Lett 174:247-250.

Venter JC, Adams MD, Sutton GG, Kerlavage AR, Smith HO and Hunkapiller M (1998). Shotgun sequencing of the human genome. Science 280:1540-1542.

Venter JC, Smith HO and Hood L (1996). A new strategy for genome sequencing. Nature 381:364-366.

Wessler SR (2001). Plant transposable elements. A hard act to follow. Plant Physiol 125:149-151.

Wessler SR, Bureau TE and White SE (1995). LTR-retrotransposons and MITEs: important players in the evolution of plant genomes. Curr Opin Genet Dev 5:814-821.

Xu Y, Mural RJ and Uberbacher EC (1994). Constructing gene models from accurately predicted exons: an application of dynamic programming. Comput Appl Biosci 10:613-623.

Yeh RF, Lim LP and Burge CB (2001). Computational inference of homologous gene structures in the human genome. Genome Res 11:803-816.

Zhang MQ (1997). Identification of protein coding regions in the human genome by quadratic discriminant analysis. Proc Natl Acad Sci U S A 94:565-568.

Zhang Z, Schwartz S, Wagner L and Miller W (2000). A greedy algorithm for aligning DNA sequences. J Comput Biol 7:203-214.

Applied Mycology
and Biotechnology
An International Series
Volume 6. Bioinformatics

A Survey of Computational Methods Used in Microarray Data Interpretation

Brian Tjaden[1] and Jacques Cohen[2]
[1]Computer Science Department, Wellesley College, Wellesley, MA 02481, USA (btjaden@wellesley.edu); [2]Volen Center for Complex Systems, Brandeis University, Waltham, MA 02454, USA (jc@cs.brandeis.edu).

In a companion chapter in this volume, Wilson et al. (this volume, chapter by Wilson et al.) provide a detailed account of the experimental design and statistical analysis of microarray data. Their chapter is of interest to researchers planning microarray experiments capable of yielding data that can be statistically analyzed to insure reliable levels of confidence. In contrast, the present chapter emphasizes what can be done with the gathered data so as to simplify the huge task of interpreting the expression levels of tens of thousands of genes. In the companion chapter the authors assume the availability of statistical programs that are often used in the design of experiments. In this chapter we explore in greater detail the algorithms that process the collected data to obtain further information about cell behavior. Many of the algorithms described here aim at grouping similar data. We also explore microarray usage that is not addressed in the companion chapter.

1. INTRODUCTION

Considering the storage capabilities of present-day computers, the size of the data generated by a *single* microarray experiment is relatively modest. An array with a capacity for measuring the transcription expression of 30,000 genes and involving two variants of cells identifiable by two colours can be internally represented in a computer using 60,000 storage units (possibly words). Half of them represent the relative degree of hybridization, the other half the percentage of each individual colour as measured for each array position. However, in a time-series microarray experiment, dozens if not hundreds of single experiments may be required to estimate how gene product concentrations vary with time. In the pharmaceutical industry the effect of drugs is often estimated by microarray experiments. It is natural then to infer that, with the decreasing costs of microarrays and scanners, the

Corresponding author: Brian Tjaden

available data will grow at a very fast rate, very likely exponentially as in the case of genomic data.

As explained in the accompanying chapter, the present quality of microarray data is relatively poor (noisy) compared to that of genomic sequences. In ranking order of accuracy of laboratory measurements, structural protein data (in the PDB) has the highest quality, followed by genomic data; unfortunately, microarray data takes a distant third position. Yet, storage capacity is not the most significant problem facing bioinformaticians dealing with microarrays. A major bottleneck is the processing time needed to properly *interpret* their data. By interpretation we mean obtaining quantitative information about a variety of biological facts that increase our knowledge about cell behavior. That interpretation has to take into account the noisy nature of microarray data and insure that the obtained results attain good levels of confidence. When dealing with tens of thousands of gene expression data it becomes necessary to group together the genes whose expression behavior is similar. The class of algorithms that perform this task is called *clustering*. From a computer science perspective the determination of the *best* among various possible clusters for a large set of data is a very time consuming task. By *best* it is meant the one that groups together objects that are most similar, using a mathematical measure of similarity. Thus, inevitably, we have to conciliate accuracy and efficiency by designing *approximate* microarray clustering algorithms that run in a reasonable time, say of the order of minutes, using presently available hardware.

The objective of this chapter is to present the algorithmic aspects of the interpretation of microarray data. The companion chapter by Wilson et al. contains valuable information about the biological facets of obtaining microarray data and using specific software to obtain results that are useful to biologists. In contrast, we are concerned about conveying to the reader the various algorithmic options that are available for interpreting microarray data, their advantages and drawbacks (Quackenbush 2001). In addition, we also describe types of microarray usage other than those for determining gene expression properties, for example, in finding single nucleotide polymorphisms (SNPs). In explaining the material in this chapter, we purposely avoid complex mathematical notation. It is assumed that the reader is savvy in elementary algebra, probability, and is familiar with some computer programming and web usage. This work is directed to bioinformaticians and biologists who wish to have a deeper understanding of the capabilities and drawbacks of the various types of microarray data analyses (Jones and Pevzner 2004). Section 2 covers the issues of adequately defining similarity in expression among several genes. Foremost in this definition is establishing how closely two or more gene expression data can be grouped into clusters having the same characteristics, for example being activated or repressed in tandem. Section 3 describes various clustering algorithms; these can be classified according to whether or not human intervention guides the grouping of similar gene data. In supervised learning a user establishes an initial training set in which she defines subsets of objects as belonging to a desired grouping. When confronted with a new object the computer selects the most likely subset exhibiting similar characteristics; the new object is then added to that subset. These supervised methods are also known as *classification*.

In unsupervised learning no such training set is used. The unsupervised algorithms repeatedly group and regroup the various objects using a similarity measure; the regrouping is done using heuristics that, hopefully, converge to a near-optimal solution. These unsupervised methods are known as *clustering*. The term "clustering" is also used informally to indicate the ensemble of supervised and unsupervised methods. The reader already familiar with topics in bioinformatics will certainly notice that clustering algorithms are closely related to those that construct phylogenetic trees. The latter kind groups species according to genomic similarity, whereas the former groups data according to similarity in gene expression. In both cases the notion of *distance between two objects* guides the clustering, or the tree construction. Following the descriptions of the clustering algorithms, we present a summary of the ongoing work in attempting to deduce genetic networks from microarray time-series data. The three final sections of the paper deal respectively with: (1) open-source and commercial software available to perform clustering and how they relate to the two preceding sections, (2) the current status and applications of microarray usage, and (3) final general remarks concerning the material presented in this work and future directions in microarray analyses.

2. MICROARRAY FORMALISM

2.1 Microarray Data Notation and Formalism

From the conceptual and algorithmic point of view a microarray is a one-dimensional array M, indexed by an integer i identifying a known gene G_i. The content of M_i is initially a large set S_i of theoretically "identical" sequences s_i of nucleotides (A, C, G, U). Each sequence $s_i \in S_i$ represents the complement of a sequence s_i' that, when hybridized with the sequences s_i, can identify the expression of the given gene G_i. The relative number of hybridized sequences for each set S_i provides a relative measure of how much G_i is expressed for a given experiment. Let A_i be an array of real numbers expressing the relative amounts of hybridization for each G_i in a given experiment.

When multiple experiments are conducted, M can be viewed as a two-dimensional array, indexed by an integer i identifying a known gene G_i, and an integer j identifying a particular experiment trial E_j. Then A_{ij} is the relative amount of hybridization for each G_i in E_j. It is assumed throughout this chapter that the rows of the two-dimensional matrix A correspond to genes and the columns of A correspond to experiments. Thus, the ith row of the matrix, denoted A_i, is a vector representing the expression pattern (or profile) of gene G_i across the set of experiments. In typical microarray applications, dozens of experiments are performed, often in replicate, for thousands of genes per array, so that the matrix M may contain thousands of rows and dozens of columns (Fig. 1). As described in the accompanying chapter, in the case of two-colour microarray experiments, each entry in the matrix reflects the log ratio of the relative amounts of hybridization in the two channels. The algorithms that we are interested in deal with extracting valuable biological information from A. While the previous chapter by Wilson et al. discusses

	Experiment 1	Experiment 2	Experiment 3	Experiment 4	...	Experiment *m*-1	Experiment *m*
G_1	0.6	4.4	1.3	1.0	...	3.1	2.2
G_2	1.5	2.6	5.2	0.8	...	2.8	2.9
G_3	0.7	3.7	2.4	1.9	...	1.5	1.6
G_4	0.3	0.7	0.2	1.3	...	4.9	3.0
G_5	3.1	3.0	2.1	1.4	...	4.2	0.9
...	...	...	...	...	...	...	...
G_{n-1}	1.8	2.5	1.8	0.7	...	2.7	3.1
G_n	0.5	3.4	3.0	0.5	...	1.8	2.5

Fig. 1: An example of the hybridization matrix *A* for *n* genes assayed by *m* microarray experiments. Each entry represents the relative amount of hybridization for each gene in each experiment.

experimental procedures for obtaining the matrix *A*, including microarray platforms, expression measurement protocols, image spot quantitation, and background correction, we use the expression matrix *A* as our starting point for downstream analyses of microarray data.

2.2 Data Transformation and Normalization

Often one of the first steps, once the matrix *A* is obtained, is *log* transforming the expression measurements of *A*. Taking the logarithm of the expression values provides several benefits. Measurement variation may be stabilized since variations of *log* expression values are less dependent on the absolute magnitude of the expression measurements. Logarithms can also reduce the skewness of highly variable gene expression distributions. Furthermore, when replicate expression measurements are performed for a single gene, the measurements follow a *log*normal distribution in many cases, justifying the *log* transformation. In addition to the *log* transformation, normalization of expression data is important to balance the expression measurement values so that meaningful biological comparisons can be made. The most common approach to normalizing expression levels is based on total hybridization normalization. It is assumed that for two or more microarray experiments, the RNA hybridized in each experiment is equivalent. With this assumption, the expression measurements of all genes in an experiment are normalized so that the mean or median expression values of all experiments are equal. In terms of the matrix *A*, this normalization corresponds to dividing each entry in a particular column of the matrix *A* by the mean value of the given column. In addition to the total hybridization normalization mentioned above, numerous advanced techniques exist such as locally weighted linear regression, rank invariant methods, and ratio statistics (Hogg and Craig 1994). Normalization helps address many of the biases inherent in microarray experiments, including unequal quantities of starting RNA, variations in labeling or detection efficiencies, and differences in hybridization levels (Quackenbush 2002). The chapter by Wilson et al. in this volume provides an in depth description of the problems involved in normalization.

2.3 Distance Measures

When interpreting a matrix A of microarray data, one of the fundamental operations, which serves as a basis for sophisticated analyses such as clustering, is determining the similarity of two genes' expression patterns. For any two genes G_x and G_y, it is useful to determine the similarity (or dissimilarity) of the vectors A_x and A_y, representing the expression profiles for the two genes. Since similarity measures are inversely related to distance measures, one measure can readily be transformed to the other as appropriate to the application. Examples are provided below for both similarity and distance measures.

The most common distance measure for analyzing microarray data is the Euclidean distance, also called the L_2 norm. The Euclidean distance between two gene expression patterns, A_x and A_y, is given by:

$$d_{x,y} = \sqrt{\sum_{j=1}^{m} \left(A_{xj} - A_{yj}\right)^2},$$

where m is the number of experiments conducted, i.e., the length of the vectors A_x and A_y. This measure has the property of obeying the triangle inequality, namely that the sum of the distances $d_{x,y}$ and $d_{y,z}$ is at least $d_{x,z}$, for any x, y, and z. This property is useful in insuring the appropriateness of certain clustering methods. The Euclidean distance incorporates information on the magnitude of difference between expression patterns, which may not be the best measure of dissimilar expression for two genes. Figure 2 illustrates gene expression profiles for four genes, G_w, G_x, G_y, and G_z. While the expression profiles for genes G_w and G_y have the smallest Euclidean distance, these two genes do not appear co-regulated. Alternatively, the expression profiles for genes G_w and G_x have the same pattern of regulation even though their profiles have a larger distance. Rather than use the magnitude of difference between expression profiles, the shape of gene expression patterns is more commonly used to compare gene expression across various experimental conditions. To capture the shape of expression patterns, the rows of the matrix A can be normalized appropriately, such as subtracting from each entry in a row the mean of the row and dividing each entry by the standard deviation of the row. As an alternative to the Euclidean distance measure, correlation can be used as a measure of similarity for gene expression patterns.

The most common similarity measure for analyzing microarray data is the correlation coefficient, or Pearson correlation (Hogg and Craig 1994). The correlation coefficient for two gene expression patterns, A_x and A_y, is given by:

$$r_{x,y} = \frac{m\sum_{j=1}^{m}\left(A_{xj}A_{yj}\right) - \sum_{j=1}^{m}A_{xj}\sum_{j=1}^{m}A_{yj}}{\sqrt{\left[m\sum_{j=1}^{m}\left(A_{xj}\right)^2 - \left(\sum_{j=1}^{m}A_{xj}\right)^2\right]\left[m\sum_{j=1}^{m}\left(A_{yj}\right)^2 - \left(\sum_{j=1}^{m}A_{yj}\right)^2\right]}},$$

where m is the number of experiments conducted, i.e., the length of the vectors A_x and A_y. This measure assesses the correlation of two gene expression patterns. The value of the correlation coefficient ranges between 1.0 and -1.0, where 1.0 corresponds to perfectly correlated expression patterns, 0.0 corresponds to entirely uncorrelated expression patterns, and -1.0 corresponds to perfectly anti-correlated expression patterns. As an example, genes G_w and G_z in Figure 2 have perfectly anti-correlated expression patterns.

The correlation coefficient does not obey the triangle inequality. However, the correlation measure has the advantage of being invariant under linear transformations of gene expression patterns. In other words, if two expression patterns have the same relative shape but different magnitudes, then they will be perfectly correlated, such as the expression patterns for genes G_w and G_x in Figure 2. The correlation coefficient is based on the assumption that each of the gene's expression values follows a Gaussian distribution.

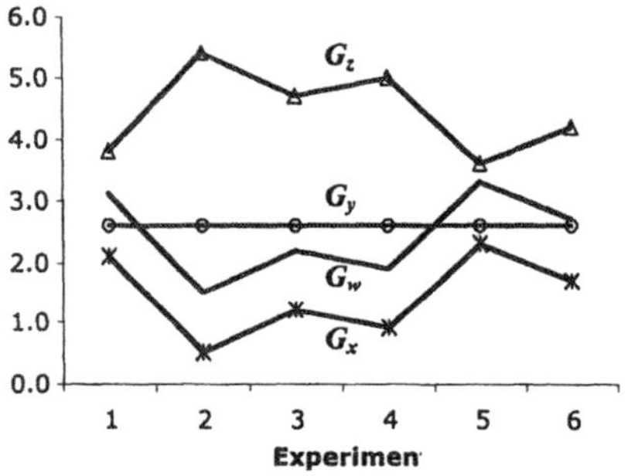

Fig. 2. Four gene expression patterns across 6 experiments are depicted. In terms of Euclidean distance, G_w and G_y have the closest expression patterns. However, since the shape of the expression patterns of G_w and G_x are identical, these two patterns are the most similar in terms of correlation. $r_{w,x} = 1.0$; $r_{w,y} = 0.0$; $r_{w,z} = -1.0$

Two popular variations of the correlation coefficient are the Spearman rank correlation and Kendall's τ (Hogg and Craig 1994). In contrast to the Pearson correlation coefficient, these two similarity measures do not assume that gene expression values approximate a Gaussian distribution. Rather, they are non-parametric and they have the advantage of being robust with respect to outlier expression data values. The Spearman rank correlation represents the correlation between the rank of the magnitude of the expression values in the two gene expression patterns, and Kendall's τ represents correlation based on the relative ordering of the expression value ranks.

In the remainder of this chapter, the terms "distance" and "similarity," when relating to two genes' expression patterns, will refer to the abovementioned measures of Euclidean distance and correlation coefficient, respectively. Further, for a set of expression profiles, the term "mean" or "centroid" will be used to denote the average value in each coordinate (i.e., experiment) of all profiles in the set.

3. CLUSTERING

Cluster analysis is the art of finding groups in a given data set such that objects in the same group are as similar to each other as possible and as dissimilar to objects in other groups as possible (Jiang et al. 2004). Dozens of clustering algorithms have been proposed and applied to microarray data sets, each algorithm yielding different results. One of the challenges for researchers using exploratory techniques such as clustering is choosing among the various approaches. A few of the more common clustering methods are detailed in the following pages. In general, there is no one best clustering method, but rather each offers different advantages and disadvantages making it appropriate for particular sets of gene expression data.

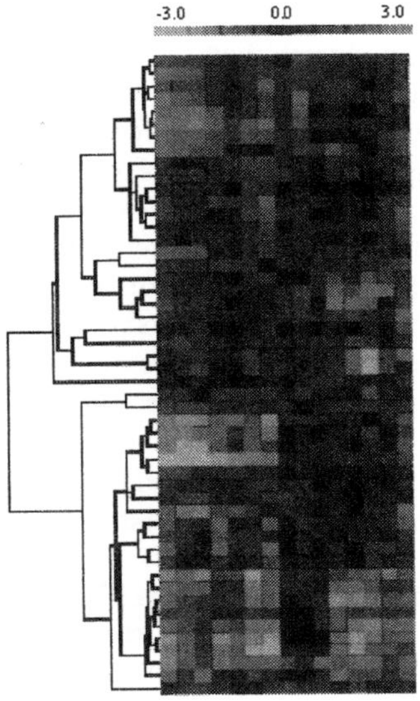

Fig. 3. An example of gene expression data from the organism *Saccharomyces cerevisiae* clustered using hierarchical clustering. The resulting hierarchy is depicted on the left-hand side of the figure. The stripe on the top indicates the spectrum of shades corresponding to the contents of the wells.

One of the reasons for the variety of clustering methods is that there is rarely an unbiased criteria which can be used to evaluate whether one method better partitions data objects than another method. Unsupervised clustering is appropriate when there is no information about which data objects should be grouped together. In contrast, supervised clustering is applicable when cluster information is available for a subset of the data objects; this information is used in determining the needed criteria (training) to subsequently group data objects for which no cluster information is known. Both unsupervised and supervised clustering methods are described below.

Since most formulations of clustering approaches are computationally prohibitive for large data sets, such as those generated by microarray experiments, clustering methods tend to rely on heuristics and approximations. In addition, clustering data objects which are unrelated will still produce clusters, even though the clusters may not be meaningful. Thus, while clustering can be a useful approach for exploring large amounts of gene expression data, researchers must exercise care to ensure they are applying the clustering techniques appropriately.

Most commonly with microarray data (e.g., Figure 3), the objects to be clustered are genes, i.e., the rows of the expression matrix A. It is worth noting, however, that experiments can be clustered similarly, simply by employing the same clustering approaches on the transpose of the matrix A, thereby clustering the columns of the matrix, i.e., the experiments. Indeed, finding phenotypically similar groups can provide useful insights into the data; this analysis is known as *experiment-based* clustering as opposed to *gene-based* clustering.

Clustering gene expression data has numerous useful applications. Genes with related functions often have similar expression patterns, so identifying groups of genes with similar expression patterns may suggest possible roles for genes with unknown function based on the known functions of genes that are placed in the same cluster. For instance, clustering can be applied to identify groups of genes which are expressed at different phases of the cell cycle, such as sporulation. Clustering of genes can also be utilized as a preprocessing step in inferring regulatory networks. For example, sets of clustered genes can be used to reduce the size of regulatory networks to be inferred. Also, clustering can be employed, in connection with sequence data, to identify DNA sequence patterns specific to each expression cluster. For instance, given a set of co-expressed genes as determined from clustering analysis, regulatory motifs such as transcription factor binding sites can be identified by searching for patterns common to the upstream DNA sequences of the clustered genes. Each of the abovementioned applications has been employed successfully in the model organism *Saccharomyces cerevisiae* and has led to new genomic insights. In the following two sections, we review a few of the unsupervised (i.e., clustering) and supervised (i.e., classification) methods.

3.1 Unsupervised Methods

3.1.1 Hierarchical clustering

Many clustering methods are hierarchical in that they produce a set of nested clusters resembling a dendrogram or phylogenetic tree (Eisen et al. 1998). Each leaf of the hierarchy corresponds to a gene and levels of the hierarchy correspond to partitions of genes into different numbers of clusters (Fig. 4). Hierarchical clustering is a greedy clustering approach, meaning that the best choices are made at each step in the process without regard for future choices. This approach has the advantage of being simple and fast; it produces a final clustering result in the form of a hierarchical tree that is easy to visualize. As a result, hierarchical clustering is the most popular clustering technique. The hierarchical clustering approach proceeds as

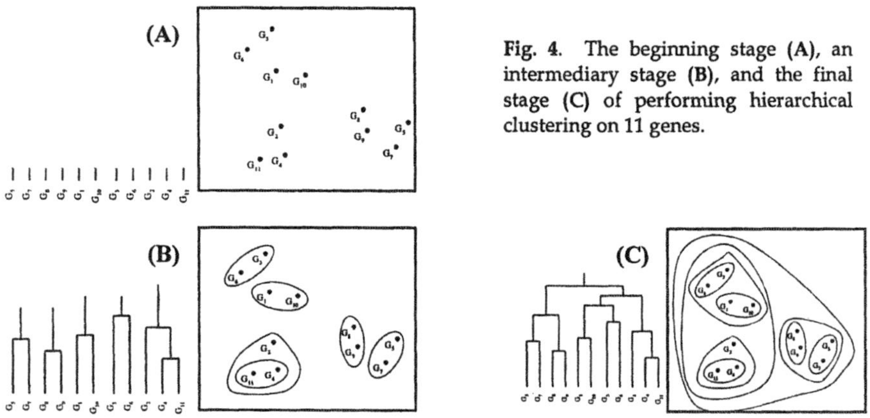

Fig. 4. The beginning stage (**A**), an intermediary stage (**B**), and the final stage (**C**) of performing hierarchical clustering on 11 genes.

follows:

- Let each gene expression pattern be a cluster containing one data point

- Repeat the following step until all clusters are merged into a single cluster (the root of the hierarchical tree): Find the two clusters with the smallest distance between them and merge them into a single cluster

The results of hierarchical clustering can vary depending on how the distance between two clusters of points is determined. In *single-link clustering,* the distance between two clusters is the shortest distance between any point in one cluster and any point in the other cluster. In *average-link clustering,* the distance between clusters is the distance between cluster centroids. In *complete-link clustering,* the distance between clusters is the farthest distance between any point in one cluster and any point in the other cluster. Different inter-cluster distance measures require different amounts of computation and affect the final topography of the clustering hierarchy.

3.1.2 K-Means

The *k*-means clustering method is a common and relatively simple heuristic method for partitioning data points into *k* clusters (Tavazoie et al. 1999). Unlike hierarchical clustering, no nested clusters are constructed. Instead, for a pre-determined number of clusters *k*, the *k*-means algorithm partitions genes into one of *k* groups such that the sum of distances from each gene expression profile to its cluster center is minimized. As mentioned in section 2, cluster centers (centroids) are generally calculated as the mean of all expression profiles in a given cluster (Figure 5). *k*-means is actually a special case of a maximum-likelihood algorithm applied to a

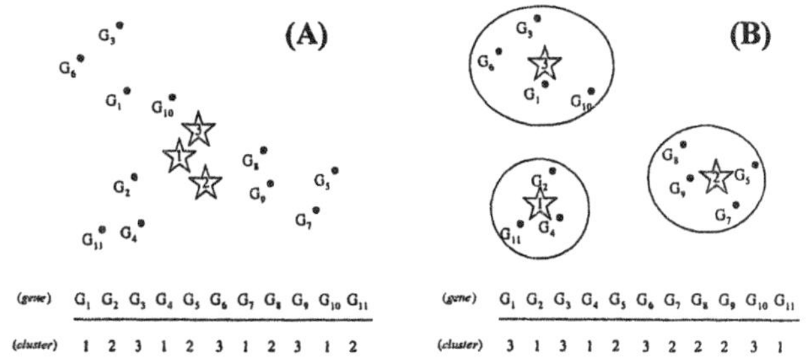

Fig. 5. The start **(A)** and end **(B)** stages of *k*-means clustering. Stars represent the mean of all points assigned to the same cluster. In this example, $k = 3$.

mixture density in which the mixture components are Gaussian distributions with equal variance (Bishop 1995).

- Randomly assign each gene a cluster number between 1 and *k*
- Repeat until no gene is assigned to a different cluster from the one obtained in the previous iteration (i) For each of the *k* clusters, calculate the mean of all points which are assigned the same cluster number and (ii) For each gene, determine the distance from the gene's expression pattern to each of the *k* means, and assign the gene to the cluster number of the closest of these means

k-means is a fast method for clustering expression data, but has the limitation of requiring prior knowledge of the number of clusters, *k*. If *k* is unknown, one can try different values of k and choose the value which yields the most plausible clustering result.

3.1.3 Self-Organizing Maps

Like k-means clustering, self-organizing maps (SOMs) create a virtual expression profile for each cluster which serves as a surrogate for the genes belonging to the cluster (Tamayo et al. 1999). Virtual expression profiles do not correspond to actual gene expression profiles as determined from the microarray data, but rather are computer generated expression profiles that are meant to represent a group of actual expression profiles. In k-means clustering, for example, the mean of a group of profiles is a virtual profile. SOMs also assign each gene to the cluster whose virtual expression profile is most similar to the gene's expression profile, just as in the case of k-means. However, SOMs differ from k-means clustering in how they determine the virtual expression profiles for each cluster.

A meta-structure, such as a grid or a lattice, is imposed on the clusters, so that each cluster is connected to a set of neighboring clusters as defined by the meta-structure. The virtual expression profiles for each cluster are determined by sampling genes and determining the cluster which is most similar to the gene's expression profile. The virtual expression profile of the most similar cluster is then updated, along with the virtual expression profile of the neighboring clusters.

- Create random expression profiles for each cluster in the meta-structure
- Repeat until the virtual expression profiles stabilize
  - Sample genes at random and determine the virtual expression profile which is most similar to the gene's expression profile
  - Update the virtual expression profile, as well as that of its neighbors in the meta-structure, to reflect its similarity to the gene's expression profile
- Assign each gene to the cluster whose virtual expression profile is most similar to the gene's expression profile

SOMs are implemented as neural networks, where each neuron in the network corresponds to a cluster or a virtual expression pattern (Kohonen 1997). The neural network is utilized to adjust the meta-structure to better represent the clusters. In the above algorithm, neural networks perform the step of updating virtual profiles. SOMs have the advantage of constructing clusters that conform to a meta-structure, which is often a two-dimensional grid, and thus, is easily represented visually. However, one of the main drawbacks of utilizing SOMs for clustering is that they require the user to specify the meta-structure, including the number of clusters. Such structure in microarray data is rarely known prior to clustering.

3.1.4 Graph-theoretic approaches

Graph-theoretic clustering approaches typically model gene expression patterns as a graph with nodes and edges. Each node in the graph corresponds to a gene and each edge between two nodes in the graph is weighted based on the similarity of the two genes' expression profiles. CAST (Cluster Affinity Search Technique) is a classic heuristic algorithm for clustering which operates by searching for cliques (groups of closely connected nodes) in the graph (Hartuv et al. 1999). Like k-means, CAST has a user-defined parameter called a threshold which effectively dictates the number of clusters into which the algorithm will group points.

- Repeat until all points are clustered
  - Choose a nonclustered point and place it into its own cluster
  - Repeat until no points are added/removed from the cluster
    - If the average distance from any nonclustered point to the points within the cluster is less than some threshold, then add the point to the cluster
    - If the average distance from any point in the cluster to the other points in the cluster is greater than the threshold, then remove the point from the cluster
  - Mark all points in the cluster as clustered

3.1.5 Model based clustering

Model based clustering operates on the assumption that gene expression data originates from a finite mixture of underlying probability distributions (Ramoni et al. 2001). Each cluster corresponds to a different distribution, and generally, the distributions are assumed to be Gaussians. The parameters of each distribution (i.e., cluster) are estimated by maximizing the likelihood of the expression data (Hogg and Craig 1994). The *k*-means approach is a special case of model based clustering, where *all* the distributions are assumed to be Gaussians with equal variance.

- Randomly generate parameters (in the case of Gaussians, the parameters would be the mean and standard deviation or covariance matrix) describing each probability distribution (i.e., cluster)
- Repeat until the parameters of each distribution converge
  - For each gene, estimate the probability that the gene's expression pattern was generated from each of the distributions
  - For each distribution, estimate the parameters of the distribution so as to maximize the likelihood of the expression data given the probability that each gene was generated from the distribution
- Assign each gene to the distribution which generates the gene's expression profile with maximum probability

Model based clustering has the advantage of providing the probability that each gene belongs in each cluster. However, model based clustering operates under the assumption that expression data comes from particular probability distributions, which may not be a reasonable assumption for many microarray data sets.

3.1.6 Principal component analysis

Principal component analysis (PCA) is a linear algebra technique that is akin to singular value decomposition (Raychaudhuri et al. 2000). Although PCA can be used to cluster expression data, PCA is more commonly used as a preprocessing step before applying other clustering algorithms. Its goal is to reduce the dimensionality of the expression data. Since some experiments may be more informative than others

Fig. 6. The expression profiles of 11 genes are plotted along the standard X and Y coordinate axes (solid arrows). The principal components are shown (dotted arrows) and represent vectors which best distinguish the variance in the plotted data points.

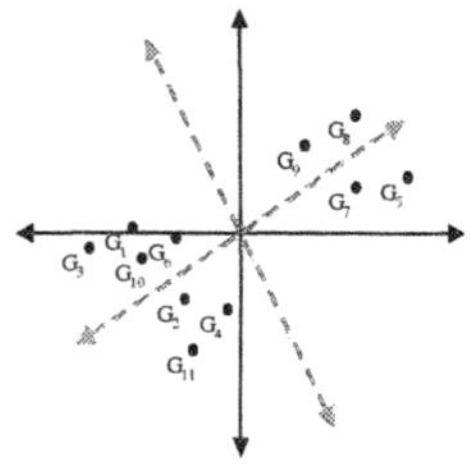

in a given set of microarray experiments, PCA offers a method for identifying a small set of virtual experiments which explains most of the variance in the data (Figure 6). The small set of virtual experiments is not necessarily a subset of the given microarray experiments, but rather each virtual experiment in the small set is a linear combination of the entire given set of microarray experiments.

- Calculate the covariance matrix of A, the matrix of gene expression data
- Compute the eigenvalues and eigenvectors of the covariance matrix (Hogg and Craig 1994)
- Choose the eigenvectors with the largest eigenvalues, and cluster the expression data (using any appropriate clustering method) in the reduced dimensional space defined by the chosen eigenvectors

Each eigenvector corresponds to a principal component. Intuitively, each component is a linear combination of microarray experiments. The eigenvectors with large eigenvalues are the ones that contain the most information. Those with small eigenvalues are assumed to capture only residual noise in the expression data. PCA has the advantage of reducing the dimensionality of the data by summarizing the microarray experiments. However, the summarization may come at the price of making the expression data less biologically interpretable by researchers.

3.2 Supervised Methods

In contrast to unsupervised clustering where gene clusters are *a priori* unknown, supervised classification is often used when clusters are known for a subset of genes. Those with known clusters, i.e., the training set, can then be used to guide the clustering of genes with unknown clusters. For instance, since genes with related functions often have similar expression patterns, supervised classification may be used to suggest possible roles for genes with unknown function based on the similarity of their expression patterns to those genes with known functions. In this case, the previously annotated genes correspond to genes with known clusters and hypothetical genes correspond to those with unknown clusters. In supervised methods it is very important to select the training set carefully. A good training set should exhibit all of the different patterns which we hope to discern in the data. One of the simplest supervised classification methods is called k-nearest neighbors. To classify a target gene, the k closest genes in the training set are found, i.e., the k genes whose expression profiles are most similar to that of the target gene. The target gene is then assigned to the cluster containing the highest number of the k nearest neighbors. The *k*-nearest neighbors' method is a straightforward classification approach that works well when the clusters are compact and the number of clusters is small. However, for genes on the border between clusters, which usually occurs as the number of clusters increases, *k*-nearest neighbors performs unreliably.

Support vector machines (SVMs) are a sophisticated supervised classification method (Brown et al. 2000). Assuming that genes from the training set fall into one of two classes or clusters, an SVM will first map gene expression profiles into a higher dimensional space and then attempt to find a hyperplane which effectively separates the expression profiles of genes from the two classes. Once a separating hyperplane has been established from the training data, genes with unknown classes can be classified based on where their expression profiles fall relative to the hyperplane. SVMs are generally employed as binary classifiers, e.g., when genes belong to one of two classes. While choosing appropriate functions and parameters for an SVM may be more of an art than a science, SVMs are often easier to implement and use than other machine learning techniques such as neural networks (Mitchell 1997).

4. BEYOND CLUSTERING

One of the most challenging applications of microarrays is to infer from their data the properties of the underlying genetic networks. In this section we first review the concept of genetic networks (GNs) in a context often used in bioinformatics; we then show how time-series microarray data can be employed to attempt to deduce GNs representing the dynamic behavior of a cell being studied.

A genetic network can be abstractly represented by a directed graph. Each of its nodes characterizes a gene, and the edges between two nodes *i* and *j* stand for the interaction between those genes. The edges are labeled by the signs + or – to denote activation or repression. For example, a directed edge between *i* and *j* labeled by a minus sign indicates that gene *i* represses the output of the product of gene *j*. A GN graph may also contain Boolean connectors (AND, OR) to indicate the conjunction or the disjunction of the effects of repressions and activations acting upon a given node. Note that the graph may contain loops, including self-loops (genes that affect themselves). The reader will notice that the clustering of microarray data indicating the behavior of thousands of genes allows researchers to replace many of the nodes of a GN graph by their centroid representatives; this in turn implies a significant decrease in processing times needed to analyze the reduced graphs.

It can be shown (Bower and Bolouri 2001) that for each graph representing a GN there is a corresponding set of non-linear differential equations. The variables in those equations reflect the concentrations of gene products at a given time. The non-linearity stems from the fact that gene product concentrations are not linear (they can be approximated by sigmoids).

Given initial conditions for the concentrations of gene products, a numerical solution to the system of differential equations yields a number of curves - each corresponding to a gene - showing how those concentrations vary with time. The curves may also indicate the convergence to stationary states in which all concentrations attain a constant level. The reader might already surmise that GNs are coarse representations of metabolic and signaling pathways. Nevertheless, by omitting details of individual reactions, GNs are convenient abstract tools for studying gene interactions. GNs are also used to develop simulation models of gene behavior (Tomita et al. 1999).

The present challenge of using microarray data to infer GNs amounts to solving a *reverse-engineering* problem: Given the set of curves representing how gene product concentrations vary with time, attempt to generate the corresponding GN. From a mathematical perspective the problem amounts to deriving a GN or a system of differential equations that best describes the set of curves obtained by microarray time series experiments.

In the previous section we have shown the role of clustering in determining the genes that exhibit analogous or opposite actions. Referring back to Figure 2, the correlation factors -1, 0, and +1 correspond to pairs of genes that suppress, remain unaltered, or activate each other. There is a relationship between these factors and the labeling of the edges of a GN. The actual direction of the arrows in a GN can be inferred, for example, by performing microarray experiments with cells having certain genes knocked out. James Collins and his group have had significant success in reconstructing GNs from microarray data obtained by perturbations of individual genes (Gardner et al. 2003).

Other attempts to solve the reverse-engineering problem include using Bayesian approaches (Nachman et al. 2004), and information theory (Liang et al. 1998) . The on-line Proceedings of the Pacific Symposium in Biocomputing (http://psb.stanford.edu/psb-online) contain a wealth of information about the computational aspects of genetic networks.

5. TOOLS FOR MICROARRAY ANALYSIS

Numerous computer packages exist for microarray data analysis. Both commercial products as well as freely available packages can be found on the World Wide Web. Most products contain tools for data transformation, normalization, and clustering, as described in this chapter. A few of the more popular applications for microarray data analysis are listed below. The website for each application is listed along with the application's availability, either as freely available software or as a commercial product for purchase. The table below differs from the one in the companion chapter since it is focused on clustering packages.

Tool	Free	Website
ArrayVision		http://www1.amershambiosciences.com/
Bioconductor	*	http://www.bioconductor.org/
Cluster and TreeView	*	http://rana.lbl.gov/EisenSoftware.htm
ExpressionProfiler	*	http://www.ebi.ac.uk/expressionprofiler/
GeneCluster	*	http://www.broad.mit.edu/cancer/software/software.html
GeneDirector		http://www.biodiscovery.com/
Genes@Work	*	http://www.research.ibm.com/FunGen/FGDownloads.htm
GeneSifter		http://www.genesifter.net
GeneSpring		http://www.agilent.com/chem/genespring
J-Express Pro		http://www.molmine.com/
MAANOVA	*	http://www.jax.org/staff/churchill/labsite/software/anova/
MIDAS and MEV	*	http://www.tigr.org/software/
Resolver		http://www.rosettabio.com/products/resolver/default.htm
Spotfire		http://www.spotfire.com/

Most of these packages allow the user to select various clustering techniques such as those discussed throughout this chapter. By examining the results of several different clustering techniques, a user will be better prepared to assess the plausibility of the results. Gene expression data is available in a number of public repositories. The table below lists some of the publicly available microarray databases which contain mycology related expression information.

6. CURRENT STATUS AND FUTURE APPLICATIONS OF MICROARRAY USAGE

In addition to more traditional applications of microarray experiments, such as assaying gene expression levels, microarrays have been used for a variety of other purposes. For example, one of the first steps for scientists after newly sequencing a genome is to annotate the genes by computational analysis. Gene prediction programs are fairly accurate for long, well-conserved genes. However, these programs are less reliable for (1) short genes (less than 100 nucleotides in length), (2) genes which do not code for proteins, and (3) regions of genes that are transcribed but not translated.

Database	Website
ArrayExpress	http://www.ebi.ac.uk/arrayexpress/
ChipDB	http://staffa.wi.mit.edu/chipdb/public/
CIBEX	http://cibex.nig.ac.jp/index.jsp
ExpressDB	http://salt2.med.harvard.edu/ExpressDB/
Gene Expression Omnibus	http://www.ncbi.nlm.nih.gov/geo/
MAD	http://mad.jax.org/
MUSC Database	http://proteogenomics.musc.edu/
PUMAdb	http://puma.princeton.edu/
Stanford Database	http://genome-www5.stanford.edu/
UNC Database	https://genome.unc.edu/
Yale Database	http://info.med.yale.edu/microarray/
yMGV	http://www.transcriptome.ens.fr/ymgv/

Microarrays can be used to assay transcript expression, not only of annotated coding regions of genes, but of the entire genome including intergenic regions (Kapranov et al. 2002). Transcripts which are detected in intergenic regions by microarray experiments may suggest previously undiscovered genes or untranslated regions (UTRs) of annotated genes. This is an example of the great enabling power of microarrays. A single microarray experiment can provide a snapshot of the entire transcript expression of an organism under given conditions. By assaying expression of known genes as well as providing predictive guides for identifying new genes, microarrays facilitate the analysis of rapidly growing genomic data thus increasing our understanding of the cellular machinery of microorganisms. Comparative genome hybridization and the study of genetic variability have become increasingly important in biology for understanding fungal diseases and the mechanisms by which fungi provide drugs such as antibiotics. Microarrays enable rapid comparative genome hybridizations, resequencing, and subsequent genotyping of any organism.

Resequencing applications have become increasingly important in ascertaining the genotypic differences between strains with different phenotypic characteristics. To study the variation of single nucleotide polymorphisms (SNPs), one may consider a small genomic sequence *S*, containing that nucleotide *N*. In designing a microarray to detect SNPs, one constructs 4 probes, each specifying variants of *S* containing a nucleotide that can replace *N*.

For applications that do not require a specific nucleotide sequence, a global view of genome content can be achieved through comparative genome hybridizations. Such comparisons can help us better understand the evolution of fungi strains, differentiate fungal pathogens from nonpathogens, and identify differences in gene content between fungi strains or their subtypes. This information, coupled with whole genome expression analysis provides genotypic and phenotypic information that can be used to increase our knowledge of the underlying differences in pathogenesis between two related strains; this information also facilitates the characterization of genes and their functions. Combining *in silico* analyses with comparative genome hybridization assists in determining the genetic heterogeneity of an organism and provides a valuable tool for the mycology laboratory.

Another useful application of microarrays is genome-wide location analysis (Ren et al. 2000). It enables researchers to identify targets of DNA or RNA binding proteins throughout an entire genome via chromatin immunoprecipitation (ChIP). Genome-wide location analysis proceeds as follows. Cells are harvested and disrupted, and DNA fragments cross-linked to a protein of interest are enriched by immunoprecipitation with a specific antibody. Following reversal of the cross-links, enriched DNA is amplified, labeled, and hybridized to a microarray. Microarray probes which evince hybridization correspond to regions of the genome which interact with the protein of interest. Genome-wide location analysis has proven to be a useful technique for elucidating protein-DNA or protein-RNA interactions. For instance, using such an approach, the binding sites of a transcription factor can be identified throughout the genome.

Like DNA microarrays, protein arrays are emerging as a high-throughput platform for identifying protein-protein, protein-antibody, protein-ligand, or protein-drug interactions (Zhu and Snyder 2001). Traditional proteomics techniques, such as 2D gel electrophoresis or chromatography combined with mass spectrometry, are relatively expensive and they may miss proteins expressed at lower levels. Protein arrays are rapidly becoming a valuable tool both to detect protein expression and to investigate protein interactions and function. As with their DNA microarray cousins, protein arrays aim to provide high-throughput experiments, assaying the expression or interaction of many proteins in parallel (Emili and Cagney 2000).

The most common type of protein arrays contain a large number of probes consisting of either proteins or their ligands. The protein array platform is usually either a glass slide or a membrane. Analysis of protein arrays is similar to that of DNA microarrays; this nascent technology has already been used for diagnostics, prognostics, and drug discovery and development.

7. CONCLUSION

It is likely that in a decade or so we will enter the era of personalized medicine in which medication will be prescribed according to the genomic makeup of individual patients. To achieve this objective we will first have to increase the accuracy of microarray experiments and reduce their cost. Once this is done, we will be confronted with a most unusual situation. We will be generating an exponentially increasing number of microarray experiments each of which will have to be analyzed within reasonable times. As we mentioned in the introduction, the exponential nature of the exact clustering algorithms will force us to design increasingly efficient and effective approximate algorithms that, to be widely used, will have to be almost linear, pretty much like BLAST performs in huge genomic data bases. That will be a great challenge for bioinformaticians in the years to come.

REFERENCES

Bishop CM (1995). Neural Networks for Pattern Recognition. Oxford University Press.

Bower JM and Bolouri H (2001). Computational Modeling of Genetic and Biochemical Networks. MIT Press, Cambridge, MA.

Brown MPS, Grundy WN, Lin D, Cristianini N, Sugnet C, Furey TS, Ares M and Haussler D (2000). Knowledge-based analysis of microarray gene expression data using support vector machines. Proceedings of the National Academy of Science USA 97: 262-267.

Eisen MB, Spellman PT, Brown PO and Botstein D (1998). Cluster analysis and display of genome-wide expression patterns. Proceedings of the National Academy of Sciences USA 95(25): 14863-14868.

Emili AQ and Cagney G (2000). Large-scale functional analysis using peptide or protein arrays. Nature Biotechnology 18: 393-397.

Gardner T, Bernardo D, Lorenz D and Collins J (2003). Inferring Genetic Networks and Identifying Compound Mode of Action Via Expression Profiling. Science 301: 102-105.

Hartuv E, Schmitt A, Lange J, Meirer-Ewert S, Lehrach H and Shamir R (1999). An algorithm for clustering cDNAs for gene expression analysis. Proceedings for the Third Annual International Conference on Research in Computational Molecular Biology: 188-197.

Hogg RV and Craig A (1994). Introduction to Mathematical Statistics. 5th edition. Prentice Hall.

Jiang D, Tang C and Zhang A (2004). Cluster analysis for gene expression data: A survey. IEEE Transactions on Knowledge and Data Engineering: 1370-1386.

Jones N and Pevzner P (2004). An Introduction to Bioinformatics Algorithms. MIT Press. http://www.bioalgorithms.info/

Kapranov P, Cawley SE, Drenkow J, Bekiranov S, Strausberg RL, Fodor SPA and Gingeras TR (2002). Large-scale transcriptional activity in chromosomes 21 and 22. Science 296(5569): 916-919.

Kohonen T (1997). Self-Organizing Maps. Springer-Verlag.

Liang S, Fuhrman S and Somogyi R (1998). REVEAL: A General Reverse Engineering Algorithm for Inference of Genetic Network Architectures. Pacific Symposium on Biocomputing 3:18-29.

Mitchell T (1997). Machine Learning. McGraw Hill.

Nachman I, Regev A and Friedman N (2004). Inferring Quantitative Models of Regulatory Networks from Expression Data. Bioinformatics 20 Suppl. 1:S248-S256.

Quackenbush J (2001). Computational Analysis of Microarray Data. Nature Review Genetics 2: 418-427.

Quackenbush J (2002). Microarray data normalization and transformation. Nature Review Genetics 32: 496-501.

Ramoni MF, Sebastiani P and Kohane IS (2001). Cluster analysis of gene expression dynamics. Proceedings of the National Academy of Sciences USA 99: 9121-9126.

Raychaudhuri S, Stuart JM and Altman RB (2000). Principal components analysis to summarize microarray experiments: application to sporulation time series. Pacific Symposium on Biocomputing: 455-466.

Ren B, Robert F, Wyrick JJ, Aparicio O, Jennings EG, Simon I, Zeitlinger J, Schreiber J, Hannett N, Kanin E, Volkert TL, Wilson CJ, Bell SP and Young RA (2000). Genome-wide location and function of DNA binding proteins. Science 290(5500): 2306-2309.

Tamayo P, Slonim D, Mesirov J, Zhu Q, Kitareewan S, Dmitrovsky E, Lander ES and Golub TR (1999). Interpreting patterns of gene expression with self-organizing maps: methods and application to hematopoietic differentiation. Proceedings of the National Academy of Science USA 96(6): 2907-2912.

Tavazoie S, Hughes JD, Campbell MJ, Cho RJ and Church GM (1999). Systematic determination of genetic network architecture. Nature Genetics 22(3): 281-285.
Tomita M, Hashimoto K, Takahashi K, Shimizu T, Matsuzaki Y, Miyoshi F, Saito K, Tanida S, Yugi K, Venter JC and Hutchison C (1999). E-CELL: Software environment for whole cell simulation. Bioinformatics 15(1):72-84.
Zhu H and Snyder M (2001). Protein arrays and microarrays. Current Opinion in Chemical Biology 5: 40-45.

Applied Mycology
and Biotechnology
An International Series
Volume 6. Bioinformatics

Computational Methods in Genome Research

Manoj Bhasin and G. P. S. Raghava
Institute of Microbial Technology, Sector 39 A, Chandigarh, India (raghava@imtech.res.in)

Computational biology has revolutionized biological and medical research. In the last two decades, a large number of computer methods have been developed to analyze DNA, RNA and protein sequences. These computer methods are playing a vital role in extracting useful information from sequences of genomes. These computational methods have been developed by different academic groups all over the world to serve the biological community. The methods are available as stand alone programs or on-line web servers. Most of these software packages are available free for academicians (freeware). In this chapter, we have described the major computational methods available for biologists to extract information from sequences. This chapter covers computational methods f r i) genome annotation; ii) comparative genomics; iii) protein structure prediction; iv) functional classification of proteins; and v) identification of potential vaccine candidates. These software packages are not available from a single source so it is not easy for users to obtain the software of their interest. In order to overcome this problem, attempts have been made to collect and compile a list of free biological software programs that includes software at EMBL and Indiana University. A catalog of biological software (Biocatalog) is also available on the internet (Rodriguez-Tome 1998). Recently, a repository of free software in biology has been created at Institute of Microbial Technology, Chandigarh, India which contains more than 800 free software packages.

1. INTRODUCION

The development of nucleotide sequencing technologies (Sanger et al. 1977; Maxam and Gilbert 1977). has allowed biologists to look into the primary sequences of genes and study their genetic implications. This yielded sequence data has resulted in creation of nucleotide databases like the EMBL nucleotide database, Genbank and DNA Databank of Japan (DDBJ) (Benson et al. 2004; Miyazaki et al. 2004). In the last two decades, sequencing technology has undergone tremendous improvement that lead to the development of fully automatic sequencing from

Corresponding author: G. P. S. Raghava

manual sequencing, due to the advent of shotgun sequencing technology. This development in technology has shifted the focus of biologists from sequencing of individual genes to the complete genome of an organism. In 1995, the first whole-genome sequence (1830 kb) of thebacterium *Haemophilus influenzae* Rd was announced (Fleischmann et al. 1995). After 1995, the number of organisms whose genomes have been sequenced is increasing at an exponential rate. The latest estimates show that around 47 archaeal, 600 bacterial and 450 eukaryote sequencing projects are in existence or have been completed. High throughput sequencing has produced a large amount of data that increases the demand of databases to maintain these sequences. The databases allow for the quick retrieval of sequences to promote knowledge-based studies. In addition to databases of raw data, a large number of specific databases are also available over the internet. The experimental analysis of large-scale genomic data is nearly impossible as it is a labor and time-intensive task. So, there is a need for computational tools to process the information rapidly. Bioinformatics tools can provide useful information to facilitate the experimental analysis of genomic data.

At present, thousands of computer software packages, web servers and databases have been developed to solve biological problems. In the 1980s, most software and databases were developed for stand-alone systems and one needs to install the programs on an individual workstation. After 1990, most of the software and databases were developed for web servers, where multiple users can run these algorithms via the internet without installing them locally on their desktop machines. These software programs are scattered across numerous web sites and are not arranged in any systematic manner. Most of the material is available for free but not utilized intensively because of the difficulty in finding the appropriate packages for specific needs. Biological software can be divided in two categories, i) commercial software and ii) free software (freeware). Commercial software packages have been developed by various companies for commercial purposes depending on the demand of the market. These software packages are user-friendly and provide multiple functions from a single point. In addition to usability, integration and user-interface, they also provide documentation and customer support. The major limitations of these commercial software packages are high cost, non-availability of source code, and up-to-date algorithms. In most of the cases costs of commercial software are too high, and it becomes nearly impossible for the biological community of developing countries to afford them. In contrast, freeware programs are usually developed by academicians, researchers or scientists for the benefit of public. Freeware is essential for promoting science. There are a few disadvantages of freeware, which include i) non-availability of proper documentation; and ii) implementation and distribution of the software. In the literature, only a few articles are available on freeware repositories. There is no comprehensive review on biological software repositories. By recognizing the importance of freeware in biology, the European Molecular Biology Laboratory (EMBL) at Heidelberg and Indiana University have created software repositories to collect and distribute free biological software. In this chapter we summarize the existing bioinformatics tools including resources from where one can download them.

2. ANALYSIS OF NUCLEOTIDE AND PROTEIN SEQUENCES

In this era of genomics, genetic sequences are being generated at an exponential rate due to the sequencing of human and other microbial genomes. In order to analyze these sequences one needs to know where the data is publicly available and how to use it. This information will be very useful for biologists as they can design and conduct experiments based on the available data. At present, a large number of DNA, RNA and protein databases are available online as shown in Figure 1. Nucleotide sequence data is first deposited to GenBank, EMBL and DDBJ after automated annotation and analysis of the data (Miyazaki et al. 2004; Benson et al. 2004). In addition to the general nucleotide databases that contain genome information, there are specialized databases that contain DNA sequences from viruses, mitochondria and chloroplasts as illustrated in Table 1. All of the databases listed in Table 1 are available via the internet at the URLs noted. These databases have numerous tools for the retrieval of data. There are number of tools for the analysis of genomic sequences to assign probable function, family and structure.

This type of analysis is useful for the molecular biologist to design experiments. A summary of bioinformatics tools available for the analysis of DNA sequences is provided in Table 2. The tools for analysis of features are available online (Majoros et al. 2003; Rozen and Skaletsky 2000; Schneider et al. 1982; Rice et al. 2000,

Table 1. A compilation of databases related to DNA sequences.

Name	Description	Available
EMBL	Collection of a large number of databases related to genomic sequences	http://www.ebi.ac.uk
GenBank	Comprehensive sequence database that contains publicly available DNA sequences for more than 119 000 different organisms.	http://www.ncbi.nlm.nih.gov. (Benson et al. 2004).
DDBJ	Data submitted mainly by Japanese genome projects and sequencing teams.	http://www.ddbj.nig.ac.jp (Miyazaki et al. 2004).
GOLD	A comprehensive resource for accessing information related to completed and ongoing genome projects worldwide. The database currently provides information on 350 genome projects, of which 48 have been completed and their analyses published.	http://igweb.integratedgenomics.com/GOLD/ (Bernal et al. 2004).
TIGR	A database of Gene Indices	http://www.tigr.org/tdb/tdb.html (Quakenbush et al. 2001).
HIVDB	Contain data on HIV genetic sequences	http://www.hiv.lanl.gov/content/hiv-db/mainpage.html (Kuiken et al. 2001).
GOBASE	Contains integrated sequence, RNA secondary structure and biochemical and taxonomic information about mitochondria and chloroplast.	http://megasun.bch.umontreal.ca/gobase/gobase.html (Shimko et al. 2001).
SNP	Contains information about human SNP.	http://www.broad.mit.edu/snp/human

Fig. 1. A diagrammatic view of important bioinformatics databases and tools for DNA, RNA and protein analysis. In figure Db# and Ptm$ refer to database and post-translational modification respectively.

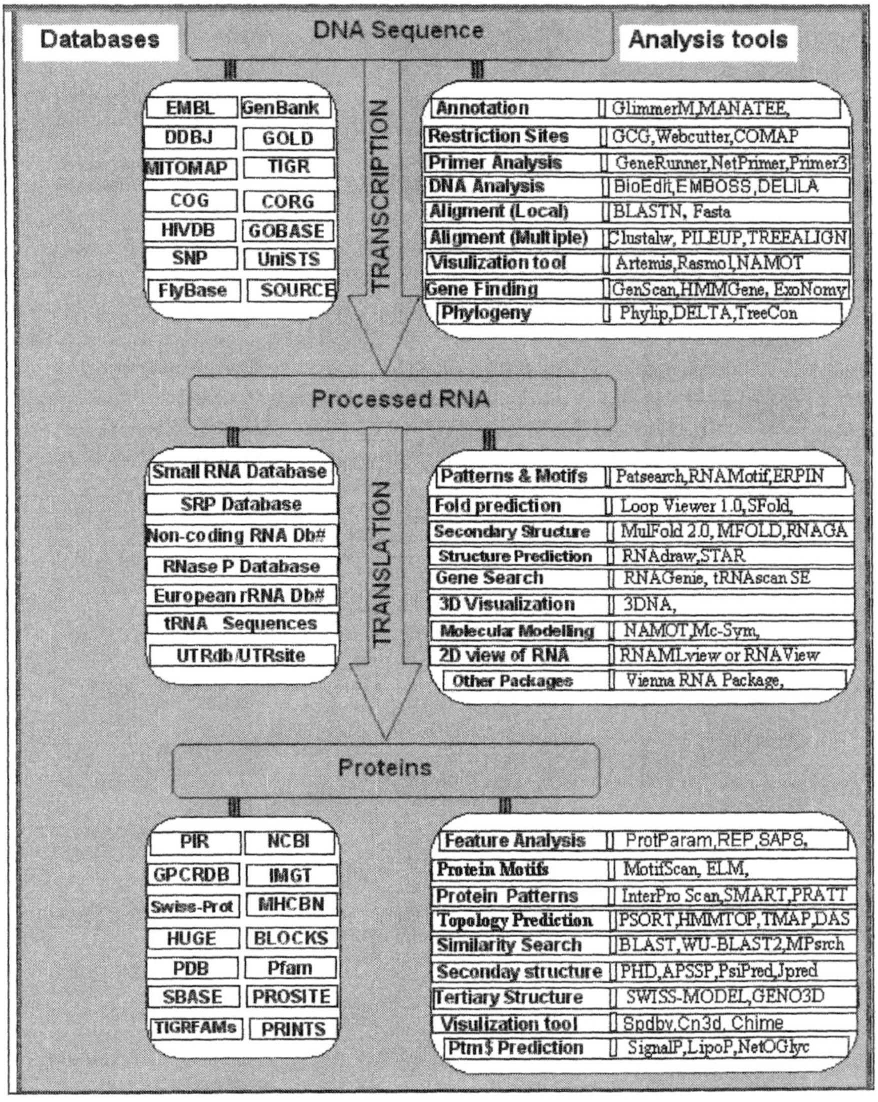

Hall, unpublished, Raghava, 1994; Raghava, 1995; Raghava and Sahni 1994). Pprobable gene functions and families of novel sequences can be assigned by performing similarity searches against annotated sequences in the major databases using software like BLAST and FASTA (Altschul et al. 1990; Pearson and Lipman 1988; Issac and Raghava 2002). Conserved patterns in sequences can be identified using software tools for multiple alignments like PILEUP and CLUSTALX (Thompson et al. 1994). The evolutionary history of a sequence can be traced through PHYLIP and TREECON programs as shown in Table 2 (J. Felsenstein, unpublished; Van de Peer 1994).

Table 2. A list of major bioinformatic tools for the analysis of DNA sequences.

Major Area	Program	Application or Function	URL (Reference)
Restriction sites	GMAP	A program for locating potential restriction sites.	http://www.imtech.res.in/raghava/progs/gmap (Raghava and Sahni 1994)
	WEBCUTTER	Online software for restriction mapping of nucleotide sequences.	http://www.firstmarket.com/cutter/cut2.html
Primer Design	GENERUNNER	Allow the designing of primer from nucleotide sequence.	http://www.generunner.com/
	NETPRIMER	Software for analysis of primers.	http://www.premierbiosoft.com/netprimer/
	PRIMER3	Picks primers for PCR reactions depending on oligonucleotide melting temperature, size, GC content, and primer-dimer possibilities and PCR product size.	http://www.broad.mit.edu/genome_software/other/primer3.html (Rozen and Skaletsky 2000)
DNA Analysis	BIOEDIT	Contains a large number of software programs which can be used for DNA or protein analysis.	http://www.mbio.ncsu.edu/BioEdit/bioedit.html (Hall, unpublished)
	EMBOSS	A suite of software programs that is useful for the analysis of DNA or protein sequences.	http://www.hgmp.mrc.ac.uk/Software/EMBOSS/ (Rice et al. 2000).
	DNAOPT	A program for optimizing gel conditions.	http://www.imtech.res.in/raghava/progs/danopt (Raghava 1995)
	DNASIZE	Computation of DNA fragment sizes.	http://www.imtech.res.in/raghava/progs/dnasize/ (Raghava 1994).
Visualization Tools	ARTEMIS	Genome viewer and annotation tool that allows visualization of sequence features and the results of analyses within the context of the sequence, and its six-frame translation	http://www.sanger.ac.uk/Software/Artemis/ (Rutherford et al. 2000).
	RASMOL	Software for looking at macromolecular structure and its relation to function.	http://www.umass.edu/microbio/rasmol/
Phylogeny	PHYLIP	A program for inferring phylogenies	http://evolution.genetics.washington.edu/phylip.html
	DELTA	A flexible method for encoding taxonomic descriptions for computer processing	http://biodiversity.uno.edu/delta/ (Askevold et al. 1994).
	TREECON	A package for the construction of phylogenetic trees	http://iubio.bio.indiana.edu:7780/archive/00000138/ (Van de Peer 1994).

The analysis of genomic sequences to identify regions that code for RNA is also very useful for molecular biologists. There are three major classes of RNA (mRNA, rRNA and tRNA) generated during the process of transcription. Information related to RNA is available from many repositories or databases as shown in Table 3. A brief description about each database and the corresponding URL are noted in Table 3. Since RNA is a very important molecule involved in transcription, splicing and other biochemical processes, many bioinformatics tools have been devised to analyze RNA sequences as depicted in Table 4. These tools are mostly available for prediction of RNA secondary structure (folded form) from the nucleotide sequence (Chen et al. 2000). (See Table 4). The prediction of structure can be performed using the tools based on energy minimization (e.g. MULFOLD) or based on conserved patterns in the sequence (RNADRAW) (Mtazura and Wenborg et al. 1996). Tools for finding conserved motifs and patterns from RNA sequences are also available from online resources (e.g. PATSEARCH, RNA MOTIF, and ERPIN) (Grillo et al. 2003).

In addition to the analysis of genomic and RNA sequences, analysis of amino acid sequences is also important for biologists. The analysis of primary amino acid sequence is useful to predict the secondary and tertiary structure of a protein and thereby its function. There are many databases, which contain information about proteins, as shown in Table 4. The primary sequences of proteins can be retrieved from databases like SWISS-PROT, PIR and NCBI (Boeckmann et al. 2003; Barker et al. 2000; Pruitt et al. 2003). Databases containing three-dimensional structures of proteins are also available online (e.g. PDB) (Robinson et al. 2003). There are many other protein databases which have specific functions such as MHCBN, IMGT (Robinson et al. 2003; Bhasin and Raghava 2003). The analysis of protein sequences in terms of structure and function is important from the perspective of a molecular biologist. In the arena of bioinformatics, a large number of tools are available for the analysis of protein sequences as shown in Table 5. Computational programs for the analysis of protein properties, specifically immunological properties, are also available on the web. Assignment of post-translational modification sites and protein topologies are also important from the biologist's point of view. In the last decade, highly accurate tools have been developed for predicting the post-transnational modification sites (NETOGLYC, LIPOP and SIGNALP) and topology or sub cellular localization (Nielsen et al. 1997; Juncker et al. 2003; Hansen et al. 1998; Nakai and Kanehisa 1991; Tusnady and Simon 2001). These tools can provide important insights about the biological functions of proteins. A summary of various bioinformatics tools or programs related to proteins is shown in Table 5.

3. GENOME ANNOTATION

The first step in genome annotation involves the integration of features revealed by the DNA and protein sequences into a systematic view of the organism's molecular machinery. Annotation can be divided into two processes; i) direct analysis of DNA sequences to locate coding regions and repeated elements, and ii) prediction of function and structure of the proteins encoded in the genome. In all organisms, coding regions are differentiated from neighboring non-coding regions by specific features. Detecting these features is essential to transforming sequence data into a fully annotated genome. Once a gene is identified or predicted, the next step is to assign a putative function, identify possible homologs in other organisms

Table 3. Databases related to RNA and protein sequences.

Database	Description	Availability (Reference)
Small RNA Database	Compilation of small RNA sequences including nuclear, nucleolar, cytoplasmic and mitochondrial small RNAs from eukaryotic organisms and small RNAs from prokaryotic cells and viruses.	http://mbcr.bcm.tmc.edu/smallRNA/smallrna.html (Perumal et al. 1999).
SRPDB	Signal recognition particle database. Provides aligned, annotated and phylogenetically ordered sequences related to the structure and function of SRPs.	http://psyche.uthct.edu/dbs/SRPDB/SRPDB.html (Rosenbald et al, 2003).
European rRNA Database	Curated database that contains complete or nearly complete LSU rRNA sequences in aligned form. Incorporates secondary structure information for each rRNA sequence.	http://rrna.uia.ac.be/lsu/ (Wuyts et al. 2004).
tRNA Sequence Database	Contains 3279 sequences of tRNA genes and tRNAs.	http://www.uni-bayreuth.de/departments/biochemie/trna/ (Sprinzl et al. 1998).
SWISS-PROT	A curated database of protein sequences.	http://www.ebi.ac.uk/swissprot/ (Boeckmann et al. 2003).
PROSITE	Consists of biologically significant patterns and profiles designed in such a way that with appropriate computational tools it can rapidly and reliably help to determine to which known family of proteins (if any) a new sequence belongs, or which known domain(s) it contains	http://www.expasy.org/prosite/ (Hulo et al. 2004).
MHCBN	MHC-binding, non-binding, and TAP-binding peptides.	http://www.imtech.res.in/raghava/mhcbn (Bhasin et al. 2003).
PIR	A comprehensive, quality controlled and well-organized protein sequence information resource.	http://pir.georgetown.edu/ (Barker et al. 2000).
PRINTS	Collection of protein fingerprints.	http://www.bioinf.man.ac.uk/dbbrowser/PRINTS/ (Attwood et al. 2003).

gene and within the ome, and to postulate its role in the biology of the organism. By comparing the genetic complement and genome organization of related organisms, novel insights may be realized regarding their evolutionary relationships.

Table 4: A compilation of important computational tools related to RNA

Major Area	Software	Description	Availability (Reference)
Patterns & Motifs	PATSEARCH	Pattern-matching tool that can find a well-defined pattern in a given sequence(s) or database (primary or specialized) divisions.	http://bighost.area.ba.cnr.it /BIG/PatSearch/ (Grillo et al. 2003).
	RNA MOTIF	A program for finding RNA motifs.	ftp://ftp.scripps.edu/pub/ macke/
Fold Prediction	LOOP VIEWER 1.0	Graphical representation of RNA folding.	http://softwareseek.progen ote.net/downloads/loopvie wer.hqx
	SFOLD	Predicts RNA secondary structures, assesses target accessibility, and provides tools for the rational design of RNA-targeting nucleic acids.	http://sfold.wadsworth.org /index.pl
Secondary Structure Prediction	MULFOLD 2.0	Software for prediction of RNA secondary structure by free energy minimization.	http://softwareseek.progen ote.net/downloads/mulfold. hqx
	MFOLD	RNA/DNA secondary structure prediction.	http://bioweb.pasteur.fr/se qanal/interfaces/mfold-simple.html
	RNAGA	Prediction of common secondary structures of RNAs.	http://bioweb.pasteur.fr/se qanal/interfaces/rnaga.html (Chen et al. 2000).
Structure Prediction	RNADRAW	An integrated program for RNA secondary structure calculation and analysis.	http://iubio.bio.indiana.edu /soft/molbio/ibmpc/rnadra w-readme.html (Matzura and Wenborg et al. 1996).
	STAR	Structure analysis of RNA using three different algorithms.	http://wwwbio.leidenuniv. nl/~Batenburg/STAR.html
3D Visualizatio n	3DNA	A package for analyzing and rebuilding 3-dimensional nucleic acid structures.	http://rutchem.rutgers.edu /%7Exiangjun/3DNA/ (Lu and Olson 2003)
Molecular Modeling	NAMOT	A useful nucleic acid modeling tool.	http://namot.lanl.gov/
2D view of RNA	RNAMLVIEW	Generates 2-dimensional displays of RNA/DNA secondary structures with tertiary interactions.	http://ndbserver.rutgers.ed u/servlet/RNAView.Frame2 DMgr (Waugh et al. 2000).

estimating complete annotation of a genome includes information regarding gene location and organization, transcripts and products of those genes, as well as regulation and control of expression, translation and degradation. This process included boundaries between coding and non-coding sequence, identification of DNA features associated with gene structures, and translation of protein coding

Table 5. A list of computational tools for analysis of protein sequences.

Major Area	Software	Description	Availability
Feature Analysis	PROTPARAM	Allows the computation of various physical and chemical parameters for a given protein.	http://us.expasy.org/tools/protparam.html (Gasteiger et al. 2003).
	REP	Searches a protein sequence for repeats.	http://www.embl-heidelberg.de/~andrade/papers/rep/search.html (Andrade et al. 2000).
	PSA	Analysis of protein properties.	http://www.imtech.res.in/raghava/psa
Protein Motifs	MOTIFSCAN	Scans a sequence against protein profile databases.	http://hits.isb-sib.ch/cgi-bin/PFSCAN?
	ELM	Eukaryotic linear motif resource for functional sites in proteins.	http://elm.eu.org/ (Puntervoll et al. 2003).
Protein Patterns	INTERPRO	Assists in finding domains and family assignment by performing an integrated search in PROSITE, PFAM, PRINTS databases.	http://www.ebi.ac.uk/interpro/scan.html (Mulder et al. 2003).
	SMART	Allows the identification and annotation of genetically mobile domains and the analysis of domain architectures.	http://smart.embl-heidelberg.de/ (Schultz et al. 1998).
	PRATT	Allows the user to search for conserved patterns in sets of unaligned protein sequences.	http://www.ebi.ac.uk/pratt/ (Jonassen et al. 1995).
Visualization tools	SPDBV	A user-friendly program that allows visualization and analysis of 3D structures of proteins.	http://us.expasy.org/spdbv/ (Guex and Peitsch 1997).
	Cn3D	A macromolecular structure viewer.	http://www.biosino.org/mirror/www.ncbi.nlm.nih.gov/Structure/cn3d/
	CHIME	A free program to show molecular structure in three dimensions.	http://www.umass.edu/microbio/chime/
Post-transnational Modifications	SIGNALP	Prediction of signal peptide cleavage sites.	http://www.cbs.dtu.dk/services/SignalP/ (Nielsen et al. 1997).
	LIPOP	Prediction of lipoproteins and signal peptides in Gram negative bacteria.	http://www.cbs.dtu.dk/services/LipoP/ (Juncker et al. 2003).
	NETOGLYC	Prediction of N-glycosylation sites in human proteins.	http://www.cbs.dtu.dk/services/NetNGlyc / (Hansen et al. 1998).

genes into protein sequence. The following subsections describe two of the major challenges in genome annotation; repeat prediction and gene prediction.

3.1. Repeat Prediction

The genomes of all organisms, particularly eukaryotic organisms, contain repetitive elements of varying lengths that can occupy a significant fraction of the total DNA content. For example, the human genome consists of more than 50% repeated sequences of various types (Lander et al. 2001). Repeats play a vital role in a number of regulatory functions and are responsible for instability of genomes. Many tandem repeats like the trinucleotide motifs, (e.g. CCG; CAG; AAG; CTG; GCG etc.) are associated with diseases such as fragile X, myotonic dystrophy, Huntington's, ataxia and others. Thus, identification of repeat elements is an important task in annotating a genome. Genomic repeat elements can be divided in two categories; i) tandem repeats which are usually confined to specific chromosomal regions, and ii) interspersed repeats mainly represented by inactive (pseudogenes) copies of historically or contemporarily active transposable elements (Strachan and Read 1999). Tandem repeats are grouped into three major subclasses; satellites, minisatellites and microsatellites (Strachan and Read 1999). Satellite repeats are composed of very long tandem arrays of short units usually present at centromeres. Minisatellites consists of tandem repeats of short units with lengths of about 7 to 64 bp located near telomeres, while microsatellite repeats are highly repetitive sequences consisting of 1 to 6 bp segments that are repeated up to 5 times the unit length as tandem arrays dispersed throughout all the chromosomes. Similarly, interspersed repeats can also be sub grouped into 5 types: SINEs (Short Interspersed Nuclear Elements) of 80-300 bp long units, LINEs (Long Interspersed Nuclear Elements) that are 6000-8000 bp long, LTRs (Long Terminal Repeats) that are 300 - 1000 bp long, and DNA transposons of variable lengths with two short inverted repeats flanking the element (Smit, 1996). Several repeat-finding algorithms have been developed to detect repeats, and these programs can be divided into two groups based on the type of repetitive DNA they identify; i) Tandem repeat finders and ii) interspersed repeat finders. Table 6 lists the major repeat finder programs available.

3.2. Gene Prediction

Correct predictions of gene location and structure are major challenges in the post genomic era, particularly for eukaryotic genomes. In the last decade a large number of computer programs have been developed for scanning genomic sequences to locate DNA segments that encode proteins. Prokaryotic genes may be predicted with considerable accuracy if one knows the codon usage pattern of the organism in question. A simple, long ORF (open reading frame) in a prokaryotic DNA sequence can be predicted as protein coding. The problem with gene prediction in prokaryotes lies in identifying the promoter and regulatory region.

Unlike prokaryotic genes, the eukaryotic genes are neither continuous nor contiguous. They are separated by long stretches of intergenic DNA and their coding sequences are interrupted by non-coding introns. Coding sequences occupy just a small fraction of a typical higher eukaryotic genome. Additionally, some eukaryotic genes are alternatively spliced -- i.e. they have more than one possible exon assembly. The arrangement of genes in genomes is also prone to exceptions. Some genes are nested (overlapping) within each other (Dunham et al. 1999). The presence of pseudogenes further complicates the identification of protein coding regions. Regulatory sequences usually located upstream of coding sequences can sometimes be found downstream and within the introns of genes. In prokaryotic systems, genes are simple in structure where introns do not split protein-coding regions and they are comparatively easy to identify. However, finding genes in eukaryotic genomic sequences is far from being a trivial problem. Unlike prokaryotic genomes, the coding regions in eukaryotes represent only a small proportion of the eukaryotic genome and are mostly found to lie in non-repetitive regions of the genome. The major existing methods used for gene prediction are listed in Table 6.

Table 6. List of major gene finders and repeat finder software.

Name	Description	URL/Reference
DOTTER	Finds repeats without prior knowledge using dot plot.	Sonnhammer and Durbin 1995
SPUTNIK	Finds small repeats using recursive algorithm	abajian.net/sputnik
TANDYMAN	Finds all exact repeats in an entire genome sequence	www.stdgen.lanl.gov/tandyman/index.html
TROLL	Tandem Repeat occurrence locator based on slight modification of the Aho-Corasick algorithm.	Castelo et al. 2002
FORREPEAT	FORRepeats: detects repeats on entire chromosomes and between genomes using novel data structure called factor oracle	Lefebyre et al. 2003
REPUTER	Applications of repeat analysis on a genomic scale	Kurtz et al. 2001
SRF	Identification of repeat sequences using Fourier transformation.	Sharma et al. 2004
GLIMMER	Primary microbial gene finder at TIGR, and has been used to annotate the complete genomes	www.tigr.org/software/glimmer/ (Majoros et al. 2003).
EGPRED	Similarity aided ab initio method for gene prediction	http://www.imtech.res.in/ragahava/egpred (Issac and Raghava 2004).
GENSCAN	Identification of complete gene structures in genomic DNA.	http://genes.mit.edu/GENSCAN.html (Burg and Karlin 1997).
HMMGENE	Prediction of vertebrate and *C. elegans* genes	http://www.cbs.dtu.dk/services/HMMgene/ (Krog 1997).
FTG	Prediction of protein coding regions using Fourier transform	http://www.imtech.res.in/raghava/ftg (Issac et al. 2002).

4. COMPARATIVE GENOMICS

Comparative genomics is playing major role in extracting useful information from biological sequences. One important aspect of comparative genomics is the comparison of proteomes (the complete protein set) of two or more organisms. In addition, it involves the comparison of gene locations, relative gene order, and regulation. It also involves an examination of such events such as gene loss, duplications, and horizontal gene transfer. Such analyses aim to go beyond mere descriptions of similarities and differences, and they are directed toward the development of models and rules that might explain such events (Tatusov et al. 1997). What can we expect comparative genomics to reveal? One of the major goals of comparative genomics is to attempt prediction of gene function. Even for well studied bacteria such as *E. coli* (~ 4600 genes) and the well studied yeast, *S. cerevisiae* (~ 6500 genes), only 60-70% of the genes have known or predicted functions. An important goal is to understand the role of the remaining 30-40% of the genes. The field of comparative genomics has led to the development of novel tools and resources as well as new terminologies and vocabularies. A few important terminologies are defined here: Homology is the relationship of any two characters (such as two proteins that have similar sequences) that have descended, usually through divergence, from a common ancestral character. Homologs are genes/proteins with similar sequences that can be attributed to a common ancestor of the two organisms during evolution. Homologs can either be orthologs,paralogs, or xenologs. Orthologs are homologs that have evolved from a common ancestral gene by speciation. They usually have similar functions. Paralogs are homologous genes/proteins that are related or produced by duplication within a genome followed by subsequent divergence. They often have different functions. Xenologs are homologs that are related by an interspecies (horizontal transfer) of the genetic material for one of the homologs. The functions of the xenologs are quite often similar. Analogues are non-homologous genes/proteins that have descended convergently from unrelated ancestors. They have similar functions although they are unrelated.

Comparative genomics is a powerful approach for deciphering function through sequence comparisons, gene order, and regulation. These studies can also reveal insights into the recruitment of enzymes in a pathway. Specialized software tools can help to reveal how enzymes and domains are recruited and how enzymes are specifically lost in some lineages. In other words, comparative genomics may be useful to help us understand the genetic basis of diversity in organisms, both speciation and variation, events that are important aspects of evolutionary biology (Snel et al. 2000). Comparative genomic studies will also shed important light on the pathogenesis of organisms, as well as help in understanding and identifying human disease genes. Another important benefit of such analyses is the identification and development of novel drug targets (Frishman et al. 2003). These may be either virulence genes, uncharacterized essential genes, or species-specific genes. There are number of programs and databases which allow comparative analysis, and they are listed in Table 7.

Table 7. Software used for comparative genomics.

Software	Description	URL (Reference)
BLASTN	Method for rapid searching of nucleotide and protein databases. Since the BLAST algorithm detects local as well as global alignments, regions of similarity embedded in otherwise unrelated proteins can be detected.	http://www.ncbi.nih.gov/BLAST/ (Altschul et al. 1990).
GWFASTA	Compares a DNA sequence to another DNA sequence.	http://www.imtech.res.in/raghava/gwfasta (Issac and Raghava 2002).
BLAST	Sequence alignments provide a powerful way to compare novel sequences with previously characterized genes/proteins. Both functional and evolutionary information can be inferred from well designed queries and alignments.	http://www.ncbi.nlm.nih.gov/BLAST/ (Altschul et al. 1990).
GWBLAST	A genome wide BLAST server.	http://www.imtech.res.in/raghava/gwblast
MPSRCH	Smith/Waterman sequence comparison at EBI.	http://www.ebi.ac.uk/MPsrch/
TREEALIGN	Phylogenetic alignment of homologous sequences.	http://bioweb.pasteur.fr/seqanal/interfaces/treealign-simple.html (Hein, 1990).
MBGD	Facilitates comparative genomics from various points of view such as ortholog identification, paralog clustering, motif analysis and gene order comparison.	http://mbgd.genome.ad.jp/
STRING	A tool for the retrieval of interacting genes/proteins.	http://string.embl.de/ (Snel et al. 2000).
PEDANT	It allows protein extraction, description and tools for analysis.	http://pedant.gsf.de/ (Frishman et al. 2003).
GENECENSUS	Tools for analysis of genomic data.	http://bioinfo.mbb.yale.edu/genome/

5. PROTEIN STRUCTURE PREDICTION

Knowledge of protein three-dimensional structure or tertiary structure (3D) is a basic prerequisite for understanding the function of a protein. Currently, the main techniques used to determine protein 3D structure are X-ray crystallography and nuclear magnetic resonance (NMR). In X-ray crystallography the protein is crystallized and then using X-ray diffraction the structure of protein is determined. Determination of 3D structure by X-ray crystallography is not always straightforward and sometimes takes as much as three to five years. NMR is another useful technique to determine the protein structure. The advantage of NMR over X-ray crystallography is that the protein can be studied in an aqueous environment that may resemble its actual physiological state more closely. The main limitation of NMR is that it is only suitable for small proteins that have less than 150 amino acids. The gap between known protein sequences and the known protein structure is increasing exponentially. Thus, there is a need to develop the computational techniques to predict protein structures. Computer-aided protein conformation/tertiary structure prediction could facilitate i) the prediction of tertiary structures for proteins with known sequences and unknown structures, ii) understanding of protein folding, iii) engineering of proteins so that new functions may be incorporated, and iv) drug designing.

The problem of protein structure prediction has been approached through three main routes: 1) computer simulation based on empirical energy calculations, 2) knowledge based approaches using information derived from structure-sequence relationships from experimentally determined protein 3-D structures; and iii) hierarchical methods. Each approach has its merits and limitations.

5.1. Energy Minimization Based Methods

Protein structure predictions based on energy minimization methods are rooted in observations that native protein structures correspond to a system at thermodynamic equilibrium with a minimum free energy. Energy-based methods do not make *a priori* assumptions about the coding properties of amino acids. Rather attempts to locate the global minimum in surface free energy of the protein molecule is assumed to correspond with the native conformation of the molecule. Methods based on the principle of energy minimization can be classified broadly in two categories; i) static minimization methods and ii) dynamical minimization methods. The major software packages based on energy minimizations are AMBER; CHARMS; ECEPP; and GROMOS (Pearlman et al. 1995; van Gunsteren and Berendsen 1990; Brooks et al. 1990). Energy calculations offer the advantage of being based on physicochemical principles but are hampered by the large number of degrees of freedom to be considered and the limited performance of energy functions. There are essentially two major problems with methods based on energy calculations. First, the computations required for assigning protein structure based on energy minimization are beyond the reach of presently available computers. Secondly, the interaction potentials used for such calculations are not good enough to model the native structure of a protein at atomic detail (Somorjai 1990).

5.2. Knowledge Based Approaches

5.2.1. Homology modeling

Presently, homology modeling is the most powerful method for predicting the tertiary structure of proteins in cases where a query protein has sequence similarity to a protein with known atomic structure. (Blundell et al. 1987; Sali et al. 1990; Sutcliffe et al. 1987). These methods are based on the observation that structures are more conserved than sequences. Therefore, an accurate molecular model of a protein may be constructed by assigning a conformation that is based on sequence alignment, followed by model building and energy minimization. Due to the availability of plentiful genome sequence data, the number of protein sequences is increasing at an exponential rate, and the gap between the number of sequences and their corresponding structures is widening. Therefore, construction of protein models is becoming an increasingly important technique (Orengo et al. 1992). The first crucial step in homology modeling involves generation of a structure-based alignment between the query protein and the sequence with known three-dimensional structure (Pascarella and Argos 1992). For cases of low homology (less than 20 % identity) the quality of the optimal alignments produced by automatic methods is often poor. A conceptually different approach to homology modeling is based on distance geometry. In this prospective, the tertiary template restrictions are translated into distance restraints that are used as input for distance geometry programs (Havel and Snow 1991; Sali and Blundell 1993). Homology-based modeling approaches fail in the absence of homologous structures.

5.2.2. Threading Approach

The concept of threading protein sequences through alternative folding motifs involves the construction of misfolded model structures, where an incorrect sequence is deliberately built onto the backbone of another protein. Threading a sequence through a fold requires a specific alignment between the amino acid sequence of the protein under consideration and the corresponding amino acid residue positions of the folding motif. The known structure establishes a set of possible amino acid positions in three-dimensional space. The query sequence is made similar to the known structure by placing its amino acids into their aligned positions. The primary aim of these methods is to select the most probable fold for a given sequence or to recognize suitable sequences that might fold into a given structure. The threading method is normally applied only on proteins whose amino acid sequences accept one of the protein folds previously studied by experimental techniques. The success of threading depends on the number of available folds whose structures are known at a level of atomic detail. In cases the atomic structure of folds are known then a query protein sequence can fitted with the known fold.

5.3. Hierarchical Approach

An alternate strategy for prediction of protein structures from their amino acid sequences uses the hierarchy of protein structure from primary to secondary and secondary to tertiary. An intermediate step in understanding the relationship between amino acid sequence and tertiary structure is to predict an intermediate state such as the secondary structure of a protein. This procedure involves constructing a model for the secondary structure from amino acid sequence data and use of the secondary structure model to build a tertiary structure prediction. There are a number of algorithms that have been developed for secondary modeling of proteins. Presently available methods can be classified into i) statistical methods, ii) physiochemical methods, (iii) artificial intelligence (AI) based methods, vi) evolutionary information based methods, and v) combinatorial methods (Rost 1996; Mcguffin et al. 2000; Cuff et al. 1998). Unfortunately, the prediction accuracy of secondary structures from sequence information is only about 80%. In using secondary structure models to predict tertiary structures attempts have been made to predict tight-turns and super secondary structures in addition to helices, turns, sheets and strands (Kaur and Raghava 2003a; Kaur and Raghava 2003b; Kaur and Raghava 2004).

Table 8: A list of major software packages for protein structure prediction.

Software Program	*Use or Function*	*URL (Reference)*
PHD	A method for sequence analysis and structure prediction	http://www.embl-heidelberg.de/predictprotein/predictprotein.html (Rost 1996).
APSSP2	Advanced protein secondary structure prediction server.	http://www.imtech.res.in/raghava/apssp2/
PSIPRED	Allows prediction of protein secondary structure, topology of transmembrane domains and fold prediction.	http://bioinf.cs.ucl.ac.uk/psipred/ (Mcguffin et al. 2000).
JPRED	A consensus method for predicting protein secondary structure.	http://www.compbio.dundee.ac.uk/~www-jpred/ (Cuff et al. 1998)
BETATPRED2	Predicts beta turns in proteins from multiple alignments using neural networks.	http://www.imtech.res.in/raghava/betatpred2 (Kaur and Raghva 2003a).
GAMMAPRED	Predicts gamma turns in proteins from multiple alignments using neural networks.	http://www.imtech.res.in/raghava/gammmapred (Kaur and Raghava 2003b).
ALPHAPRED	Predicts alpha turns in proteins from multiple alignments using neural networks.	http://www.imtech.res.in/raghava/alphapred (Kaur and Raghava 2004).
SWISS-MODEL	An automated comparative protein modeling server.	http://www.expasy.org/swissmod/SWISS-MODEL.html (Peitsch et al. 1995).
GENO3D	Automatic modeling of protein three-dimensional structures.	http://geno3d-pbil.ibcp.fr/ (Combet et al. 2002).
CPHMODELS	Fold recognition/homology modeling.	http://www.cbs.dtu.dk/services/CPHmodels/
Meta Fold Recognition Server	Allows submission to multiple servers.	http://bioinfo.pl/Meta/ (Ginalski et al. 2003).
HMMSTR	Predicts the secondary, local, super secondary, and tertiary structures of proteins from sequences.	http://www.bioinfo.rpi.edu/~bystrc/hmmstr/server.php (Bystroff and Shao 2002).
AMBER	A set of molecular mechanics force fields for the simulation of biomolecules.	http://amber.scripps.edu/ (Pearlman et al. 1995).
CHARMS	A set of programs for molecular simulation.	(Gunsteren and Berendsen 1990).

5.4. Benchmarking of Structure Prediction Methods

A major problem in the field of protein structure prediction is to assess the performance of existing methods. Methods have been developed using different sets of proteins and using different criteria for evaluation. In order to assist the developers and users, an open world wide experiment was initiated in 1994 called the Critical

Assessment of Techniques for Protein Structure Prediction (CASP). CASP experiments aim to establish the current state of the art in protein structure prediction by identifying what progress has been made and highlighting where future efforts may be most productively focused. These activities are held in alternate years, and the sixth CASP was initiated in December 2004 (http://PredictionCenter.llnl.gov/casp6). In addition to CASP, a number of other experiments were initiated to assess the performance of structure prediction methods such as the Critical Assessment of Fully Automatic Structure Prediction Servers (CAFASP), and the Evaluation of Automatic protein structure predictions (EVA). These experiments allow evaluation of online web servers for protein structure prediction. Table 8 lists major software and web servers for protein structure prediction.

6. FUNCTIONAL ANNOTATION & CLASSIFICATION OF PROTEINS

6.1. Subcellular Localization

Information concerning the subcellular localization of a protein may provide an important clue to elucidate its function, because it must be in the proper subcellular compartment to perform its biological function (Eisenhaber and Bork 1998). Knowledge about subcellular localization is sometimes useful in understanding disease mechanisms and for developing novel drugs. Therefore, the experimental determination of the subcellular localization of a protein constitutes one step on the long way to determine its biological function (Chou 2001). A number of methods have been developed for the prediction of subcellular localization in prokaryotes as well as eukaryotes. Such predictions for prokaryotic proteins are easy in comparison to eukaryotes due to the complex organization of eukaryotic cells. Similarity searches using BLAST and FASTA are commonly used to obtain evidence that a protein may be localized to a specific cellular compartment, however, these methods often fail in the absence of sequence similarity between query and target proteins (Eisenhaber and Bork 1998). Another way to predict subcellular localization is to identify local sequence motifs such as signal peptides or nuclear localization signals. Proteins designated for the secretory pathway, the mitochondria and the chloroplast contain N-terminal targeting peptides that are recognized by translocation machinery. Thus, these prediction methods will only analyze the N-terminus of the peptide. The reliability of methods based on sorting signals is strongly dependent on the quality of the gene sequence in the 5'-region or in the protein N-terminal sequence (Hua and Sun 2001). The major problem for methods that detect N-terminal sorting signals is that start codons are predicted with less than 70% accuracy by various genome projects and gene prediction methods. Prediction methods based on N-terminal sorting signals will be inaccurate when the signals are missing or only partially included (Reinhardt and Hubbard 1998). In addition, known signals are not general enough to cover the resident proteins of each compartment. To overcome these limitations, a number of methods based on amino acid and dipeptide composition have been developed (Table 9).

on various approaches such as hidden Markov models (Jaakkota et al. 2000), hierarchical assignments (Attwood et al. 2002), amino acid composition (Karchin et al. 2002), and dipeptide composition (Bhasin and Raghava 2004c).

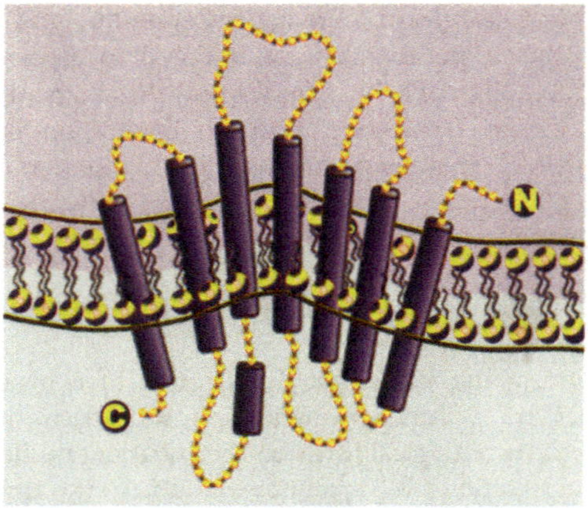

Fig. 2. GPCRs structure and topology in the cell membrane.

6.3. Nuclear Receptors

Nuclear receptors are key transcription factors that regulate crucial gene networks responsible for cell growth, differentiation and homeostasis. Recognition of nuclear receptors is crucial, because many of them are potential targets for developing therapeutic strategies for diseases like breast cancer and diabetes (Robinson-Rechavi and Laude 2003). All nuclear receptors consist of six distinct regions or domains (Figure 3). The N-terminal region (A/B) is highly variable, and contains one constitutionally active transactivation region (AF-1) and several autonomous transactivation domains. The A and B domains are variable in length from less than 50 to more than 500 amino acids. Recently, Bhasin and Raghava (2004) developed a method for predicting nuclear receptors (Bhasin and Raghava 2004a).

7. IDENTIFICATION OF VACCINE TARGETS

Traditionally vaccinations are achieved by injecting patients with a preparation of killed or seriously weakened (attenuated) virus or pathogen. Vaccines based on this approach can lead to potentially catastrophic results if, for some reason, the virus "catched" and the patient actually developed disease (Goldsby et al. 2000). Nevertheless, this approach has achieved limited success; however, the immunity raised is usually sufficient to provide only protection against individual isolates of a virus, and not for all isolates obtained. This is primarily due to the changing nature of viruses. In order to overcome the limitations of traditional vaccine design, there has been a significant change in the strategy for vaccine development in last few years. Presently, subunit vaccines are now employed in which vaccine candidates are derived from immunogenic peptides/regions in proteins instead of the complete antigenic protein. Most subunit vaccines are based on T cell epitopes. Therefore, identification of immunologically active regions/

7. IDENTIFICATION OF VACCINE TARGETS

Traditionally vaccinations are achieved by injecting patients with a preparation of killed or seriously weakened (attenuated) virus or pathogen. Vaccines based on this approach can lead to potentially catastrophic results if, for some reason, the virus "catched" and the patient actually developed disease (Goldsby et al. 2000).

Table 9: Prediction methods for subcellular localization of proteins and classification of proteins.

Program and Reference	Description	URL
PSORT (Nakai and Horton 1999)	For proteins of gram-negative bacteria. Based on rules derived from experimental data.	http://psort.nibb.ac.jp/form.html
iPSORT (Nakai and Kanehisa 1991)	For eukaryotic proteins. Based on amino acid sequences and features such as hydrophobicity and hydrophilicity.	http://www.hypothesiscreator.net/iPSORT
PSORT-B (Gardy et al. 2003).	For improved prediction of subcellular localization of proteins in gram-negative bacteria. Based on amino acid composition, presence of signal peptide, transmembrane alpha helices, motifs and similarity searches.	http://www.psort.org/psortb/
NNPSL (Reinhardt and Hubbard 1998)	For prokaryotic and eukaryotic proteins. Based on amino acid composition using ANN.	http://www.doe-mbi.ucla.edu/%7Eastrid/astrid.html
SUBLOC (Hua and Sun 2001)	For prokaryotic and eukaryotic proteins. Based on amino acid composition using SVM.	http://www.bioinfo.tsinghua.edu.cn/SubLoc
ESLPRED (Bhasin and Raghava 2004b)	For eukaryotic proteins. Based on amino acid, dipeptide composition, physiochemical properties using SVM.	http://www.imtech.res.in/raghava/eslpred/
GPCRPRED (Bhasin and Raghava 2004c)	For classification of GPCRs using SVM	http://www.imtech.res.in/raghava/gpcrpred
NRPred (Bhasin and Raghava, 2004a)	For classification of Nuclear receptors.	http://www.imtech.res.in/raghava/nrpred/

Nevertheless, this approach has achieved limited success; however, the immunity raised is usually sufficient to provide only protection against individual isolates of a virus, and not for all isolates obtained. This is primarily due to the changing nature of viruses. In order to overcome the limitations of traditional vaccine design, there has been a significant change in the strategy for vaccine development in last few years. Presently, subunit vaccines are now employed in which vaccine candidates are derived from immunogenic peptides/regions in proteins instead of the complete antigenic protein. Most subunit vaccines are based on T cell epitopes. Therefore, identification of immunologically active regions/epitopes recognized by T cells plays a crucial role in subunit vaccine design (Masignani et al. 2002; Singh and Raghava 2001, Rappuoli 2000). Experimental methods for the identification of such regions are costly and time-consuming. Therefore, computational methods for prediction of such sites are of great value. In the last decade, a large number of computational methods were developed to predict T/B cell epitopes for potential vaccine candidates as described in the following sections.

7.1. B-cell Epitopes

The antigenic regions of proteins that are recognized by binding sites or paratopes of immunoglobulin molecules are called B-cell epitopes. These epitopes may be linear (continuous) or conformational (discontinuous). These epitopes play an important role in peptide-based vaccines design, disease diagnosis and allergy research. The development of computational methods for prediction of B-cell epitopes remains a vital and challenging task due to the inherent complexity of antigen recognition. It is nearly impossible to predict conformational epitopes as it require knowledge of the tertiary structure for both antigens and antibodies. In the past, a number of algorithms were developed for predicting continuous/linear B-cell epitopes based on physicochemical properties such as hydrophilicity, flexibility, accessibility, and turns (Hopp and woods 1983; Kyte and Doolittle 1982; Karplus and Schulz 1985; Kolaskar and Tongaonkar 1990; Alix 2000; Odorico and Pellequer 2003). Recently, Saha and Raghava (2004). (http://www.imtech.res.in/raghava/bcepred/) studied the performance of several computational methods on a clean and large data set of B-cell epitopes (Saha and Raghava unpublished; http://www.imtech.res.in/raghava/bcipred/). The performance of algorithms based on physicochemical properties varied from 52.9% to 57.5%, whereas, combined methods showed 58.7% accuracy. Recently our group has used artificial neural networks to predict linear B cell epitopes (http://www.imtech.res.in/raghava/abcpred/).

7.2. T-cell Epitopes

Extracellular antigens are processed via an exogenous pathway and recognized by helper T (T_h) cells, whereas, intracellular antigens processed via the endogenous path are recognized by cytotoxic T-lymphocytes (Watts and Powis 1999). Earlier methods for prediction of T cell epitopes were based on analysis of experimentally determined T cell epitopes and are known as direct methods of T-cell epitope prediction (Table 10). These methods are based on the assumption that the conformation of a peptide is responsible for its recognition by T cells. These methods were superseded after the analysis of MHC peptide complexes by X- ray crystallography, which demonstrated that a peptide bound in the MHC groove has an extended conformation (Stern et al. 1994). It was also observed that binding of a peptide to an MHC allele requires more specificity than its recognition by T-cells. This started a new era of predictive methods called indirect methods of T-cell epitope prediction where the predictor identifies the MHC binding regions in an antigen rather than T-cell epitopes.

linearity in the data. Therefore, machine learning techniques like artificial neural networks (ANN) and support vector machines (SVM) have been introduced for MHC class II binder prediction. Machine learning

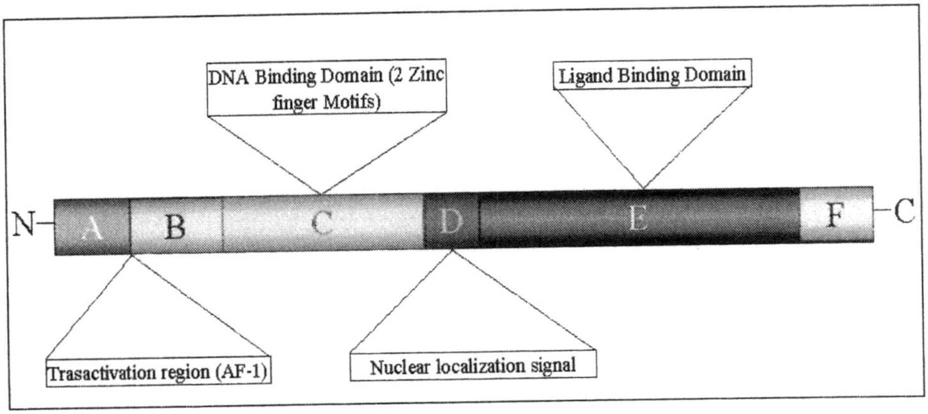

Fig. 3. Schematic representation of nuclear receptors.

based methods have achieved better accuracy compared to matrices and motif-based methods (Brusic et al. 1998; Bhasin and Raghava 2004d). The major existing methods for MHC class II binder prediction are summarized in Table 10.

7.2.2. Prediction of MHC class I binders

Several methods have been developed for prediction of MHC class I binding peptides from antigenic sequences (Table 10). Designing methods for MHC class I binders prediction is easy compared to predicting class II binders since the length of binding peptides is nearly fixed. Preliminary methods were based on motifs (Rammensee et al. 1995). SYFPEITHI is the most successful method based on refined motifs derived from pooled sequences and single peptide analysis exclusively of natural ligands. The motif-based methods have low accuracy, because all MHC binders do not contain exact motifs. Quantitative matrix-based methods consider all positions and residues in peptides to determine its MHC binding potential. These methods fail in handling non-linear data. In order to handle non-linearity in data and to adapt self-learning, machine learning techniques like artificial neural networks (ANN) and support vector machines have been introduced for prediction of MHC class I binders. All of the above approaches are knowledge-based where rules are derived from known binders and non-binders. Another alternative to the knowledge-based approach is a structure-based prediction in which the conformations of peptides to fit in the MHC groove are studied. Hanan Marglit and co-workers (2000). devised a method for prediction of MHC binding peptides on the basis of structural information (Schueler-Furman et al. 2000). These methods are quite slow and yet not fully developed due to limited information from the MHC and peptides.

ligands. The motif-based methods have low accuracy, because all MHC binders do not contain exact motifs. Quantitative matrix-based methods consider all positions and residues in peptides to determine its MHC binding potential. These methods fail in handling non-linear data. In order to handle non-linearity in data and to adapt self-learning, machine learning techniques like artificial neural networks (ANN) and support vector machines have been introduced for prediction of MHC class I binders. All of the above approaches are knowledge-based where rules are derived from known binders and non-binders. Another alternative to the knowledge-based approach is a structure-based prediction in which the conformations of peptides to fit in the MHC groove are studied. Hanan Marglit and co-workers (2000). devised a method for prediction of MHC binding peptides on the basis of structural information (Schueler-Furman et al. 2000). These methods are quite slow and yet not fully developed due to limited information from the MHC and peptides.

8. RESOURCES

8.1. Software at EMBL

EMBL is a major source of free biological software, and more than 300 software packages are available (Stoehr and Omond 1989). These software packages are also available from the European Bioinformatics Institute (http://www.ebi.ac.uk/), an outstation of EMBL. These programs are divided in four categories based on operating systems: i) MS-DOS or Windows, ii) Apple Macintosh, iii) UNIX, and iv) VAX-VMS (Fuchs 1990). The software available at EBI is kindly provided by its authors. A major advantage of this repository is that software can be obtained by email, ftp or http.

Email Server: To obtain information about software, users should send email to netserv@ebi.ac.uk with command "help software" in the body of the message.

- **FTP Server:** Software is obtained by anonymous ftp from EBI (ftp.ebi.ac.uk).
- **Web Server:** Software may be downloaded via the internet from http://www.ebi.ac.uk/

All of the files at EBI are converted into printable ASCII format, so they can be distributed via standard email. This repository encourages authors to make their software available to the scientific community.

8.2. Freeware at Indiana University

Indiana University offers a large collection of software packages for biology (Gilbert 2000). This repository allows users to browse, search and download available software packages from http://iubio.bio.indiana.edu/.

8.3. BioCatalog

The BioCatalog is a database that contains information about biology and genetics software (Rodriguez-Tome, 1998). It is different from other software repositories, because it maintains information about software rather than software itself. It can be accessed at http://www.ebi.ac.uk/biocat/. The catalog is freely distributed as an ASCII file. It categorizes software packages based on their functions. Users can download the catalog from EBI at ftp://ftp.ebi.ac.uk/databases/biocat/.

Table 10: Summary of methods used for prediction of potential vaccine candidates.

Program	Description	URL
MHC Class II Binder Prediction methods		
SYFPEITHI (Rammensee et al. 1999)	Motif based prediction of large numbers of MHC class I and class II alleles.	http://www.syfpeithi.de/
PROPRED (Singh and Raghava 2001)	Prediction of promiscuous binders for 51 HLA class II alleles using virtual matrices.	http://www.imtech.res.in/raghava/propred/ or http://bioinformatics.uams.edu/mirror/propred/
TEPITOPE (Sturniolo et al. 1999)	Prediction of promiscuous binders for 25 HLA class II alleles using virtual matrices.	PC program can be downloaded from (http://www.vaccinome.com/)
HLADR4PRED (Bhasin and Raghava 2004d)	Prediction of binders for HLA-DRB1*0401 using SVM and ANN.	http://www.imtech.res.in/raghava/hladr4pred/
T cell epitope Prediction		
CTLPRED (Bhasin and Raghava 2004e)	A direct method for CTL epitope prediction.	http://www.imtech.res.in/raghava/ctlpred/
EPIMER & OPTIMER (Meister et al. 1995)	Based on density of MHC binding motifs and their conformation.	---
AMPHI (Spouge et al. 1987; Margalit et al. 1987)	Based on the assumption that T cell epitopes have amphipathic alpha-helices.	---
MHC Class I Binder Prediction		
PROPRED1 (Singh and Raghava 2003)	Matrix based prediction of promiscuous binders of 47 MHC class I alleles.	http://www.imtech.res.in/raghava/propred1/ or http://bioinformatics.uams.edu/mirror/propred1/
nHLAPRED	Prediction of promiscuous binders for 67 MHC class I alleles using ANN and QM techniques.	http://www.imtech.res.in/raghava/nhlapred/ or http://bioinformatics.uams.edu/mirror/nhlapred/
BIMAS	Ranks potential peptides based on predicted half-time of dissociation to HLA class I molecules.	http://www-bimas.cit.nih.gov/molbio/hla_bind/

8.4. RFSB: Repository of Free Software in Biology

In order to promote free software resources in biology, the Bioinformatics Centre (BIC) at the Institute of Microbial Technology (IMTECH) in India has initiated a project called Public Domain Resources in Biology (PDRB). The major goal of this project is to collect, manage and distribute free biological software to the academic community via the World Wide Web. Under this project, a repository of free software has been developed (Raghava 2001a), and the latest version of the database contains more than 800 biological software packages. This is largest repository of free software in biology. The RFSB was created using POSTGRESQL, a free RDBMS program. The database stores the following information about each software package: i) program name; ii) category of software based on function; iii) operating system requirements; iv) main function; v) a brief description of the software; vi) reference (if published); vii) authors; viii) hardware requirements; ix) software requirements; and x) original ftp/http site for obtaining information and downloading the software. In addition to offering users the ability to download software packages, the database also allows the submission of new software via the internet.

9. CONCLUSION

Improvements in sequencing technologies has lead to the elucidation of complete genomes for a large number of organisms. Therefore, annotation of these genomes and assignment of functions for the corresponding genes and proteins is a major challenge in the field of genome research. Fortunately, a number of software packages and web servers have been developed which facilitate genome analysis and functional annotation. Within the scope of this chapter it is not possible to describe all the computer programs that have been developed for genomics, so we have focused primarily on those that are most popular. We have made an attempt to provide an overview of various computational tools that can assist in biological research.

REFERENCES

Alix AJP (2000). Predictive estimation of protein linear epitopes by using the program PEOPLE. Vaccine 18: 311-314.

Altschul S F, Gish W, Miller W, Myers EW and Lipman DJ (1990). Basic local alignment search tool. J Mol Biol 215: 403-410.

Andrade MA, Ponting C, Gibson T and Bork P (2000). Identification of protein repeats and statistical significance of sequence comparisons. J Mol Biol 298: 521-537.

Askevold IS and O'Brien CW (1994). DELTA, an invaluable computer program for generation of taxonomic monographs. Ann Entomol Soc Am 87: 1-16.

Attwood TK, Croning MD and Gaulton A (2002). Deriving structural and functional insights from a ligand-based hierarchical classification of G protein-coupled receptors. Protein Eng 15: 7-12.

Attwood TK, Bradley P, Flower DR, Gaulton A, Maudling N, Mitchell AL,Moulton G, Nordle A, Paine K, Taylor P, Uddin A and Zygouri C (2003). PRINT and its automatic supplement, prePRINTS. Nucleic Acids Res 31: 400-402.

Barker WC, Garavelli JS, Huang H, McGarvey PB, Orcutt BC, Srinivasarao GY, Xiao C, Yeh LS, Ledley RS, Janda JF, Pfeiffer F, Mewes HW, Tsugita A and Wu C (2000). The protein information resource (PIR). Nucleic Acids Res 28: 41-44.

Baxevanis AD (2003). The Molecular Biology Database Collection: 2003 update. Nucleic Acids Res 31:1-12.

Benson DA, Karsch-Mizrachi I, Lipman DJ, Ostell J and Wheeler DL (2004). GenBank: update. Nucleic Acids Res 32: D23-D26.

Bernal A, Ear U and Kyrpides N (2001). Genomes Online Database (GOLD): a monitor of genome projects world-wide. Nucleic Acids Res 29: 126-127.

Bhasin M and Raghava GPS (2004a). Classification of nuclear receptors based on amino acid composition and dipeptide composition. J Biol Chem 279:23262-23266.

Bhasin M and Raghava GPS (2004b). ESLpred: SVM based method for subcellular localization of eukaryotic proteins using dipeptide composition and PSI-BLAST. Nucleic Acids Res 32:W414-W419.
Bhasin M and Raghava GPS (2004c). GPCRpred: An SVM based method for prediction of families and subfamilies of G-protein coupled receptors. Nucleic Acids Res 32:W383-W389.
Bhasin M and Raghava GPS (2004d). SVM based method for predicting HLA-DRB1*0401 binding peptides in an antigen sequence. Bioinf 20: 421-423.
Bhasin M and Raghava GPS (2004e). Prediction of CTL epitopes using QM, SVM and ANN techniques. Vaccine (in press).
Bhasin M, Singh H and Raghava GPS (2003). MHCBN: a comprehensive database of MHC binding and non-binding peptides. Bioinf 19: 665-666.
Blundell TL, Sibanda BL, Sternberg MJ and Thornton JM (1987). Knowledge-based prediction of protein structures and the design of novel molecules. Nature 326: 347-352.
Boeckmann B, Bairoch A, Apweiler R, Blatter MC, Estreicher A, Gasteiger E,Martin MJ, Michoud K, O'Donovan C, Phan I, Pilbout S and Schneider M (2003). The SWISS-PROT protein knowledgebase and its supplement TrEMBL in 2003. Nucleic Acids Res 31: 365-370.
Brooks BR, Bruccoleri RE, Olafson BD, States DJ, Swaminathan S, and Karplus M (1983). CHARMM: A program for macromolecular energy, minimization, and dynamics calculations. J Comp Chem 4:187-217.
Brusic V, Rudy G, Honeyman G, Hammer J and Harrison L (1998b). Prediction of MHC class II-binding peptides using an evolutionary algorithm and artificial neural network. Bioinf 14: 121-130.
Burge C and Karlin S (1997). Prediction of complete gene structures in human genomic DNA. J Mol Biol 268:78-94.
Bystroff C and Shao Y (2002). Fully automated *ab initio* protein structure prediction using I-SITES, HMMSTR and ROSETTA. Bioinf 18:S54-S61.
Castelo AT, Martins W and Gao GR (2002). TROL-Tandem Repeat Occurrence Locator. Bioinf 18:634-636.
Chen J, Le S and Maize J (2000). Prediction of common secondary structures of RNAs: A genetic algorithm approach. Nucleic Acids Res 28:991–999.
Chou KC (2001). Prediction of protein cellular attributes using pseudo-amino acid composition. Proteins 43: 246-255.
Combet C, Jambon M, Deléage G and Geourjon C (2002). Geno3D an automated protein modeling web server. Bioinf 18: 213-214.
Cuff JA, Clamp ME, Siddiqui AS, Finlay M and Barton GJ (1998). JPred: a consensus secondary structure prediction server. Bioinf 14:892-893.
Diehn M, Sherlock G, Binkley G, Jin H, Matese JC, Hernandez-Boussard T, Rees CA, Cherry JM, Botstein D, Brown PO and Alizadeh AA (2003). SOURCE: a unified genomic resource of functional annotations, ontologies, and gene expression data. Nucleic Acids Res 31:219-223.
Dieterich, C., Wang, H., Rateitschak, K., Luz, H. and Vingron, M. (2003). CORG: a database for Comparative Regulatory Genomics. Nucleic Acids Res 31: 55-57.
Dunham A, Matthews LH, Burton J, Ashurst JL, Howe KL,Ashcroft KJ, Beare DM, Burford DC, Hunt SE, Griffiths-Jones S et al. (2004). The DNA sequence and analysis of human chromosome 13. Nature 428: 522-528.
Eisenhaber F and Bork P (1998). Wanted: subcellular localization of proteins based on sequence. Trends Cell Biol 8: 69–70.
Felsenstein J PHYLIP: Phylogeny Inference Package (unpublished).
Fichant GA and Burks C (1991). Identifying potential tRNA genes in genomic DNA sequences. J Mol Biol 220: 659-671.
Fleischmann RD, Adams MD, White O, Clayton RA, Kirkness EF, Kerlavage AR,Bult CJ, Tomb JF, Dougherty BA, Merrick JM, et al. (1995). Whole-genome random sequencing and assembly of *Haemophilus influenzae* Rd. Science 269:496-512.
FlyBase Consortium (2003). The FlyBase database of the *Drosophila* genome projects and community literature. Nucleic Acids Res 31: 172-175.

Frishman D, Mokrejs M, Kosykh D, Karstenmuller G, Kolesov G, Zubrzycki I, Gruber C, Geier B, Kaps A, Volz A, Wagner C, Fellenberg M, Heumann K and Mewes HW (2003). The Pedant genome database. Nucleic Acids Res 31: 207-211.

Fuchs R (1990). Free molecular biological software available from the EMBL file server. Comput Appl Biosci 6:120-121.

Gasteiger E, Gattiker A, Hoogland C, Ivanyi I, Appel RD and Bairoch A (2003). ExPASy: the proteomics server for in-depth protein knowledge and analysis Nucleic Acids Res 31:3784-3788.

Gilbert D (2000). Free software in molecular biology for Macintosh and MS Windows computers. Methods Mol Biol 132:149-184.

Ginalski K, Elofsson A, Fischer D, Rychlewski L (2003). 3D-Jury: a simple approach to improve protein structure predictions. Bioinf 19:1015-1018.

Goldsby RA, Kindt TJ and Osborne BA (2000). Kuby Immunology, WH Freeman and Company, 4th edition.

Grillo G, Licciulli F, Liuni S, Sbisa E and Pesole G (2003). PatSearch: a program for the detection of patterns and structural motifs in nucleotide sequences. Nucleic Acids Res 31: 3608-3612

Guex N and Peitsch MC (1997). SWISS-MODEL and the Swiss-PdbViewer: An environment for comparative protein modeling. Electrophoresis 18:2714-2723.

Hall T BioEdit: Biological Sequence alignment editor for windows 95/98/NT (unpublished).

Hansen JE, Lund O, Tolstrup N, Gooley AA, Williams KL and Brunak S (1998). NetOglyc: prediction of mucin type O-glycosylation sites based on sequence context and surface accessibility. Glycoconj J 15:115-130.

Havel TF and Snow ME (1991). A new method for building protein conformations from sequence alignments with homologs of known structure. J Mol Biol 217: 1-7.

Hein J (1990). Unified approach to alignment and phylogenies. Methods Enzymol 183: 626-645.

Hopp TP and Woods KR (1981). Prediction of protein antigenic determinants from amino acid sequences. Proc Natl Acad Sci USA 78: 3824-3828.

Horn F, Vriend G and Cohen FE (2001). Collecting and harvesting biological data: the GPCRDB and nuclear information systems. Nucleic Acids Res 29:346-349.

Hulo N, Sigrist CJ, Le Saux V, Langendijk-Genevaux PS, Bordoli L, Gattiker A, De Castro E, Bucher P and Bairoch A (2004). Recent improvements to the PROSITE database. Nucleic Acids Res 32:D134-D137.

Issac B and Raghava GPS (2002). GWFASTA: A server for FASTA search in eukaryotic and microbial genomes. Biotechniques 33: 548-556.

Issac B, Singh H, Kaur H and Raghava GPS (2002). Locating probable genes using Fourier transform. Bioinf 18: 196-7.

Issac B and Raghava GP (2004). EGPred: prediction of eukaryotic genes using ab initio methods after combining with sequence similarity approaches. Genome Res 14:1756-66.

Jaakkota T, Diekhans M and Haussler D (2000). A discriminative framework for detecting remote protein homologies. J Comput Biol 7: 95-114.

Jonassen I, Collins JF and Higgins DG (1995). Finding flexible patterns in unaligned protein sequences. Protein Sci 4: 1587-1595.

Juncker AS, Willenbrock H, Von Heijne G, Brunak S, Nielsen H and Krogh A (2003). Prediction of lipoprotein signal peptides in Gram-negative bacteria. Protein Sci 12: 1652-1662.

Karchin R, Karplus K and Haussler D (2002). Classifying G-protein coupled receptors with support vector machines. Bioinf 18:147-159.

Kaur H and Raghava GPS (2003a). Prediction of beta-turns in proteins from multiple alignment using neural network. Protein Sci 12:627-634.

Kaur H and Raghava GPS (2003b). A neural network based method for prediction of gamma-turns in proteins from multiple sequence alignment Protein Sci 12:923-929.

Kaur H and Raghava GPS (2004). Prediction of alpha-turns in proteins using PSI-BLAST profiles and secondary structure information. Proteins: Structure, Function, and Bioinformatics 55:83-90.

Kogelnik AM, Lott MT, Brown MD, Navathe SB and Wallace DC (1998). MITOMAP: a human mitochondrial genome database -- 1998 update. Nucleic Acids Res 26: 112-115.

Kolaskar AS and Tongaonkar PC (1990). A semi-empirical method for prediction of antigenic determinants on protein antigens. FEBS Lett 276:172-174.

Krogh A (1997). Two methods for improving performance of an HMM and their application for gene finding. In Proc Fifth Int Conf on Intelligent Systems for Molecular Biology, ed. (Gaasterland, T. et al., Menlo Park, CA: AAAI Press), pp. 179-186.

Kuiken CL, Foley B, Hahn B, Korber B, Marx PA, McCutchan F, Mellors JW and Wolinksy S (2001). HIV sequence compendium eds. (Theoretical Biology and Biophysics Group, Los Alamos National Laboratory, Los Alamos, NM, LA-UR 02-2877).

Kurtz S, Choudhuri JV, Ohlebusch E, Schleiermacher C, Stoye J and Giegerich R (2001). REPuter: the manifold applications of repeat analysis on a genomic scale. Nucleic Acids Res 29: 4633-4642.

Lander ES, Linton LM, Birren B, et al. (2001). Initial sequencing and analysis of the human genome. Nature 409: 860-921.

Lefebvre A, Lecroq T, Dauchel H and Alexandre J (2003). FORRepeats: detects repeats on entire chromosomes and between genomes. Bioinf 19:319-326.

Lu X and Olson WK (2003). 3DNA: a software package for the analysis, rebuilding and visualization of three-dimensional nucleic acid structures. Nucleic Acids Res 31: 5108-5121.

Majoros WH, Pertea M, Antonescu C and Salzberg SL (2003). GlimmerM, Exonomy and Unveil: three *ab initio* eukaryotic gene finders. Nucleic Acids Res 31: 3601-3604.

Masignani V, Rappuoli R, and Pizza M (2002). Reverse vaccinology: a genome-based approach for vaccine development. Expert Opin Biol The. 2:895-905.

Matzura O and Wennborg A (1996). RNAdraw: an integrated program for RNA secondary structure calculation and analysis under 32-bit Microsoft Windows. CABIOS 12: 247-249.

Maxam AM and Gilbert W (1977). A new method for sequencing DNA. Proc Natl Acad Sci USA 74:560-564.

McGuffin LJ, Bryson K, and Jones DT (2000). The PSIPRED protein structure prediction server. Bioinf 16: 404-405.

Miyazaki S, Sugawara H, Ikeo K, Gojobori T and Tateno Y (2004). DDBJ in the stream of various biological data. Nucleic Acids Res 32: D31-D34.

Mulder NJ, Apweiler R, Attwood TK, Bairoch A, Barrell D, Bateman A, Binns D, Biswas M, Bradley P, Bork P, Bucher P, Copley RR, Courcelle E, Das U, Durbin R, Falquet L, Fleischmann W, Griffiths-Jones S, Haft D, Harte N, Hulo N, Kahn D, Kanapin A, Krestyaninova M, Lopez R, Letunic I, Lonsdale D, Silventoinen V, Orchard SE, Pagni M, Peyruc D, Ponting CP, Selengut JD, Servant F, Sigrist CJA, Vaughan R and Zdobnov EM (2003). The InterPro Database, 2003 brings increased coverage and new features. Nucleic Acids Res 31: 315-318.

Nakai K and Kanehisa M (1991). Expert system for predicting protein localization sites in gram-negative bacteria. Proteins 11: 95-110.

Nielsen H, Engelbrecht J, Brunak S and von Heijne G (1997). Identification of prokaryotic and eukaryotic signal peptides and prediction of their cleavage sites. Protein Eng 10:1-6.

Odorico M and Pellequer JL (2003). BEPITOPE: predicting the location of continuous epitopes and patterns in proteins. J Mol Recognit 16:20-22.

Orengo CA, Brown NP and Taylor WR (1992). Fast structure alignment for protein databank searching. Proteins 14: 139-167.

Pascarella S and Argos P (1992). A data bank merging related protein structures and sequences. Protein Eng 5:121-137.

Pearlman DA, Case DA, Caldwell JW, Ross WR, Cheatham TE III, DeBolt S, Ferguson D, Seibel G and Kollman P (1995). AMBER, a computer program for applying molecular mechanics, normal mode analysis, molecular dynamics and free energy calculations to elucidate the structures and energies of molecules. Comp Phys Commun 91: 1-41.

Pearson WR and DJ Lipman (1988). Improved tools for biological sequence comparison. Proc Natl Acad Sci USA 85:2444-2448.

Peitsch M, Schwede T, Guex N and Peitsch MC (1995). Protein modeling by e-mail. Bio/Technol 13: 658-660.

Perumal K, Gu J, Chen Y and Reddy R (1999). Small RNA database compiled by the Department of Pharmacology, Baylor College of Medicine. Ed . (C Burks, Molecular Biology Database List, Nucleic Acids Res) pp.1-9.

Puntervoll P, Linding R, Gemünd C. Chabanis-Davidson S, Mattingsdal M, Cameron S, Martin DMA, Ausiello G, Brannetti B, Costantini A, Ferrè F, Maselli V, Via A, Cesareni G, Diella F, Superti-Furga G, Wyrwicz L, Ramu C, McGuigan C, Gudavalli R, Letunic I, Bork P, Rychlewski L, Küster B, Helmer-Citterich M, Hunter WN, Aasland R and Gibson TJ (2003). ELM server: a new resource for investigating short functional sites in modular eukaryotic proteins. Nucleic Acids Res 31:3625-3630.

Quackenbush, J, Cho J, Lee D, Liang F, Holt I, Karamycheva S, Parvizi B, Pertea G, Sultana R and White J (2001). The TIGR Gene Indices: analysis of gene transcript sequences in highly sampled eukaryotic species. Nucleic Acids Res 29:159-164.

Raghava GPS (1995). DNAOPT: A computer program to aid optimization of gel conditions of DNA gel electrophoresis and SDS-PAGE. Biotechniques 18: 274-281.

Raghava GPS (1994). Improved estimation of DNA fragment lengths from gel electrophoresis. Biotechniques 17: 100-104.

Raghava GPS (2001a). PDSB: public domain software in biology. Biotech Software & Internet Report 2:154-156.

Raghava GPS (2001b). PDWSB: public domain web servers in biology. Biotech Software & Internet Report 2:152-153.

Raghava GPS and Sahni G (1994). GMAP: a multipurpose computer program to aid synthetic gene design, cassette mutagenesis and introduction of potential restriction sites into DNA sequences. Biotechniques 16: 1116-1123.

Rammensee HG, Friede T and Stevanovic S (1995b). MHC ligands and peptide motifs: first listing. Immunogenetics 41: 178-228.

Rappuoli R (2000). Reverse vaccinology. Curr Opin Microbiol 3: 445-450.

Rice P, Longden I and Bleasby A (2000). EMBOSS: The European molecular biology open software suite. Trends Genet 16: 276-277.

Robinson-Rechavi M and Laude V (2003). Bioinformatics of nuclear receptors. Methods Enzymol 364: 95-118.

Rodriguez-Tome P (1998). The BioCatalog. Bioinf 14: 469-470.

Rosenblad MA, Gorodkin J, Knudsen B, Zwieb C and Samuelsson T (2003). SRPDB: signal recognition particle database. Nucleic Acids Res 31: 363-364.

Rost B (1996). PHD: predicting one-dimensional protein structure by profile based neural networks. Methods Enzymol 266:525-539.

Rozen S and Skaletsky HJ (2000). Primer3 on the WWW for general users and for biologist programmers. In: Krawetz S and Misener S ed. (*Bioinformatics Methods and Protocols: Methods in Molecular Biology.* Humana Press, Totowa, NJ) pp 365-386.

Rutherford K, Parkhill J, Crook J, Horsnell T, Rice P, Rajandream M-A and Barrell B (2000). ARTEMIS: sequence visualization and annotation. Bioinf 16: 944-945.

Sadowski MI and Parish JH (2003). Automated generation and refinement of protein signatures: case study with G-protein coupled receptors. Bioinf 19: 727-734.

Sali A and Blundell TL (1993). Comparative protein modeling by satisfaction of spatial restraints. J Mol Biol 234:779-815.

Sali A, Overington JP, Johnson MS, Blundell TL(1990). From comparisons of protein sequences and structures to protein modeling and design. Trends Biochem Sci 15: 235-240.

Sanger F, Nicklen S and Coulson AR (1977). DNA sequencing with chain-terminating inhibitors. Proc Natl Acad Sci USA. 74:5463-5467.

Schneider TD, Stormo GD, Haemer JS and Gold L (1982). A design for computer nucleic-acid sequence storage, retrieval and manipulation. Nucleic Acids Res 10:3013-3024.

Schueler-Furman O, Altuvia Y, Sette A and Margalit H (2000). Structure-based prediction of binding peptides to MHC class I molecules: application to a broad range of MHC alleles. Protein Sci 9:1838-1846.

Schultz J, Milpetz F, Bork P and Ponting CP (1998). SMART, a simple modular architecture research tool: identification of signaling domains. Proc Natl Acad Sci USA. 95: 5857-5864.
Sharma D, Issac B, Raghava GP, Ramaswamy R (2004). Spectral Repeat Finder (SRF): identification of repetitive sequences using Fourier transformation. Bioinf 20:1405-1412.
Shimko N, Liu L, Lang BF and Burger G (2001). GOBASE: the organelle genome database. Nucleic Acids Res 29: 128-132.
Singh H and Raghava GPS (2001) ProPred: prediction of HLA-DR binding sites. *Bioinformatics* 17, 1236-7 .
Smit,A.F. (1996). The origin of interspersed repeats in the human genome. *Curr. Opin.*
Snel, B., Lehmann, G., Bork, P. and Huynen, M.A. (2000). STRING: a web-server to retrieve and display the repeatedly occurring neighborhood of a gene. Nucleic Acids Res. 28, 3442-4.
Somorjai RL. (1990). Theories and simulation of protein folding. Biotechnology. 14, 1-19.
Sonnhammer ELL and Durbin R (1995). A dot-matrix program with dynamic threshold control suited for genomic DNA and protein sequence analysis. Gene 167:, GC1-GC10.
Sprinzl M, Horn C, Brown M, Ioudovitch A and Steinberg S (1998). Compilation of tRNA sequences and sequences of tRNA genes. Nucl Acids Res 26:148-153.
Stern LJ, Brown JH, Jardefzky TS, Gogra JC, Urban RG, Strominger JL and Wiley DC (1994). Crystal structure of the human class II MHC protein HLA-DR1 complexed with an influenza virus peptide. Nature 368: 215-221.
Stoehr PJ and Omond RA (1989). The EMBL network file server. Nucleic Acids Res 17: 6763-6764.
Strachan T and Read AP (1999). *Human Molecular Genetics* 2nd Edition. John Wiley
Sutcliffe MJ, Hayes FR and Blundell TL (1987). Knowledge based modeling of homologous proteins, Part II: Rules for the conformations of substituted sidechains. Protein Eng 5: 385-392.
Tatusov RL, Fedorova ND, Jackson JD, Jacobs AR, Kiryutin B, Koonin EV,Krylov DM, Mazumder R, Mekhedov SL, Nikolskaya AN, Rao BS, Smirnov S, Sverdlov AV, Vasudevan S, Wolf YI, Yin JJ, Natale DA, (2003). The COG database: an updated version includes eukaryotes. BMC Bioinf 4: 41.
Tatusov RL, Koonin EV and Lipman DJ (1997). A genomic perspective on protein families. Science 278:631-637.
Thompson JD, Higgins DG and Gibson TJ (1994). CLUSTAL W: improving the sensitivity of progressive multiple sequence alignment through sequence weighting, positions-specific gap penalties and weight matrix choice. Nucleic Acids Res 22: 4673-4680.
Van de Peer Y (1994). A new version (3.0) of TREECON .Department of Biochemistry, University of Antwerp (UIA) Universiteitsplein 1B-2610 .
Van Gunsteren WF and Berendsen HJC (1990). Computer simulation of molecular dynamics: methodology, applications and perspectives in chemistry angew. Chem Int Ed Engl 29: 992-1023.
Watts C and Powis S (1999). Pathways of antigen processing and presentation. Rev Immunogenet 1: 60-74.
Waugh A, Gendron P, Altman R, Brown JW, Case D, Gautheret D, Harvey SC,Leontis N, Westbrook J, Westhof E, Zuker M and Major F (2000). RNAML: a standard syntax for exchanging RNA information. RNA 8: 707-717.
Wuyts J, Perriere G and Van de Peer Y (2004). The European ribosomal RNA database. Nucleic Acids Res 32: D101-D103.

Applied Mycology
and Biotechnology
An International Series
Volume 6. Bioinformatics

Creating Fungal Pathway/Genome Databases Using Pathway Tools

Suzanne M. Paley, Michelle Green, Markus Krummenacker, Peter D. Karp*
Bioinformatics Research Group, SRI International
333 Ravenswood Ave, EK207, Menlo Park, CA 94025
{paley,green,kr,pkarp}@ai.sri.com

*To whom correspondence should be addressed.

May 10, 2006

Abstract

The Pathway Tools software allows a group of scientists to create, update, and publish on the Web an evolving knowledge resource describing the genome and biochemical networks of the organism. Such a knowledge resource will minimize duplication of experimental effort, ensure that all relevant knowledge will be brought to bear on interpreting new experimental results, and permit system-level computational analyses. Creation of a new Pathway/Genome Database (PGDB) by Pathway Tools includes inference of fungal metabolic pathways and pathway hole fillers (genes that code for enzymes missing from predicted pathways). Pathway Tools also infers the transport reactions present in an organism. A collection of interactive editing tools allows refinement of a PGDB by adding or modifying gene functions or pathways to capture knowledge from the biomedical literature. Pathway Tools provides a variety of query and visualization capabilities including a genome browser, displays of biochemical pathways, and a visualization of the cellular biochemical network. The Omics Viewer paints multiple types of functional genomics data onto that cellular network diagram. Comparative genomics capabilities allow comparison with other fungal Pathway/Genome Databases.

1 Introduction

The completion of a genome sequencing project can mark the start of a systematic effort to understand the function of every gene and biochemical pathway within an organism. When multiple functional genomics technologies are brought to bear on studying an organism, it becomes increasingly critical to integrate and synthesize the results of those studies to create an evolving knowledge resource describing the genome and biochemical networks of the organism. Such a knowledge resource will minimize duplication of experimental effort, ensure that all relevant knowledge will be brought to bear on interpreting new experimental results, and permit system-level computational analyses, such as flux-balance analyses [3].

The Pathway Tools software developed in the Bioinformatics Research Group at SRI International provides a powerful and multifaceted software environment for creating fungal model-organism databases. Despite its name, Pathway Tools can manipulate *many* biological datatypes that range from genome to pathway information. A collection of computational inference modules within Pathway Tools allows a group to quickly (within days) create a new PGDB that contains inferred metabolic pathways, pathway hole fillers, and transport reactions. A collection of interactive editing tools lets a group refine a PGDB by adding or modifying gene functions or pathways to capture knowledge from the biomedical literature. Pathway Tools also provides Web publishing capabilities that allow a group to mount a PGDB on the World Wide Web for querying by the scientific community. Comparative genomics capabilities allow comparison with other fungal PGDBs, and the Pathway Tools Omics Viewer provides pathway-based visualization and interpretation of large-scale functional genomics data, such as gene expression or metabolomics data.

We advocate a model in which a group of biologists who are experts in different facets of a fungus work together to curate a PGDB by incorporating information from the biomedical literature. By curate, we mean to update and refine the PGDB to contain new information from computational and experimental analyses. Updates can be made to gene functions, pathway definitions, and regulatory networks, and should include authoring of mini-review comments and inclusion of relevant literature citations.

A fungal PGDB project could begin with creation of a new PGDB with Pathway Tools, or with adoption of an existing PGDB from the BioCyc collection of more than 200 PGDBs. Currently, two fungal PGDBs exist within the BioCyc collection: PGDBs for *Schizosaccharomyces pombe* and for *Neurospora crassa* that were created by the Computational Genomics Group at the European Bioinformatics Institute — see URL `http://biocyc.org/server.html`. Note that because these two fungal PGDBs were created from proteome rather than genome information [6], genome-related functions such as the Pathway Tools genome browser are not available for them. Another existing fungal PGDB provides a metabolic pathway component to the Saccharomyces Genome Database (SGD), and is curated on an ongoing basis by the SGD group. It is available through the Web at URL `http://pathway.yeastgenome.org/biocyc/`.

Once a PGDB has been developed, it can not only be published on the Web in a manner analogous to the BioCyc.org Web site, but it can also added to SRI's online PGDB registry (see `http://biocyc.org/registry.html`). This site allows downloading of registered PGDBs by other Pathway Tools users in a manner analogous to peer-to-peer sharing of music files on the Internet. Furthermore, a PGDB can be exported to many file formats including Genbank, BioPAX, and SBML.

The remainder of this chapter summarizes the steps in creating a new PGDB, and describes the computational inference modules within Pathway Tools. The chapter also describes some of the query and visualization tools provided by Pathway Tools, including its genome browser, comparative genomics tools, and Omics Viewer. The chapter closes with information on how to obtain Pathway Tools, and on how to learn more about the software.

2 Pathway Tools Computational Inferences

The PathoLogic component of Pathway Tools is responsible for creating a new PGDB, and for performing computational inferences within the PGDB. The input to PathoLogic is an annotated fungal genome, which can be provided as a series of Genbank files (one per replicon), or as a series of files in a format called PathoLogic format. Both formats describe all known genes of the organism, and for each gene provide information such as the name of the gene, the gene product, the nucleotide position of each gene and of its introns and exons, and an optional EC (Enzyme Commission) number for genes with enzymatic products. Pathway Tools does not attempt to re-annotate the genome, that is, to identify coding regions or predict gene functions. Rather, it takes the gene functions provided in the input file as a starting point for further analysis.

The first step performed by PathoLogic is to transform the description of the genome provided in the input file(s) into its internal object database format. It creates database objects for each replicon and gene described in the input file(s), and it creates DB objects for each protein or RNA gene product described in the input file(s).

2.1 Inference of Metabolic Pathways

Just as sequence analysis identifies gene functions by inferring the functions of newly sequenced genes by their similarity to known genes, the PathoLogic component of Pathway Tools infers the presence of known metabolic pathways by recognizing in a genome annotation the enzymes in known metabolic pathways. The PathoLogic pathway prediction algorithm is described in more detail in [11].

The reference DB of metabolic pathways employed by PathoLogic is the MetaCyc DB [2]. MetaCyc version 9.5 contains 620 experimentally elucidated metabolic pathways, which were curated from more than 7300 publications. These pathways have been experimentally demonstrated to be present in more than 500 different organisms.

PathoLogic recognizes the presence of a MetaCyc pathway in a new organism through a two-step process. In the first step, enzyme matching, it matches the protein functions listed in the annotated genome sequence for that organism to the biochemical reactions (as defined in MetaCyc) that those enzymes catalyze. In the second step, it matches the reactions thus inferred to be catalyzed by the organism against MetaCyc pathways.

That enzyme matching process is not based on sequence analysis, because we believe that any automated sequence analysis that PathoLogic could perform would most likely be less accurate than sequence analyses performed by genome center annotators who manually oversee the assignment of gene functions. Instead, we leverage the existing genome annotation by matching already-assigned enzyme functions, specifically, by matching enzymes to reactions based on EC number and enzyme name (gene product names). MetaCyc contains an extensive dictionary of enzyme names and synonyms, and our matcher employs various text processing techniques to decrease the likelihood that matches will be missed because of irregularities in how gene products are named (e.g., trimming from gene-product name suffixes such as "alpha subunit").

An example of how enzyme matching occurs is in Figure 1, which shows the two pathways that make up the pentose shunt — the oxidative and nonoxidative branches of the pentose phosphate pathway — that were inferred by PathoLogic. PathoLogic detected that for the *S. pombe* protein whose genome unique identifier is SPOM-XXX-01-004113, both the function name assigned to the protein ("probable transketolase") and the EC number assigned to the protein ("2.2.1.1") match a reaction in a step of this pathway. In some cases, matches are found to either the EC number or product name, but not both; if both are recognized, PathoLogic warns the user if they disagree (that is, if they refer to different enzyme activities).

PathoLogic pathway scoring considers every MetaCyc pathway, and computes how many reactions

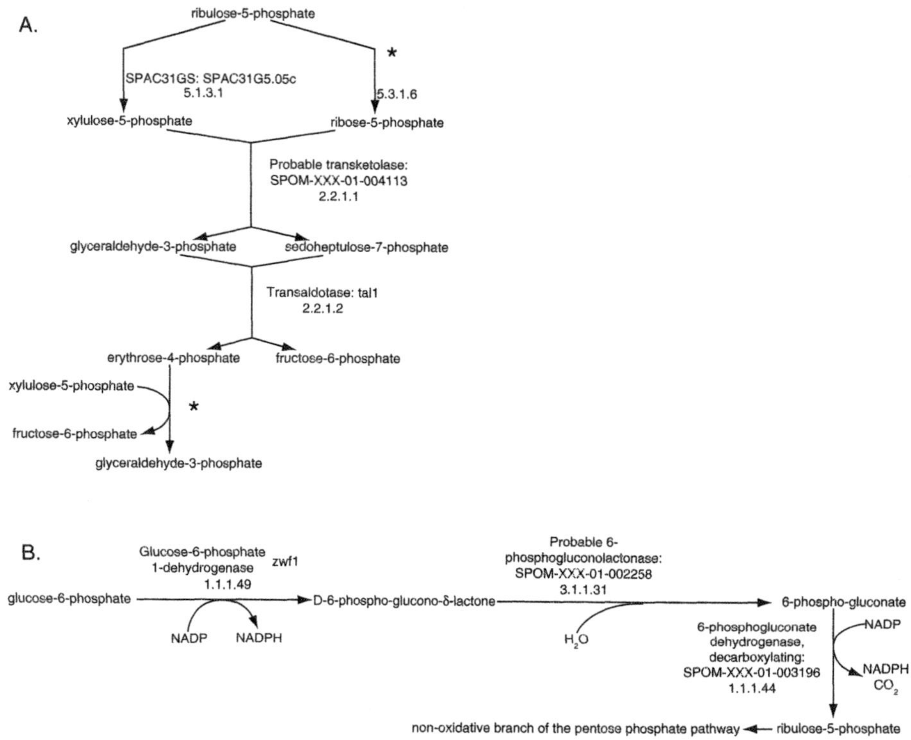

Figure 1: (A) Oxidative branch of the pentose phosphate pathway. (B) Nonoxidative branch of the pentose phosphate pathway. The two pathway holes are indicated with an asterisk (*).

within a pathway are assigned to enzymes within the annotated genome — the more reactions are assigned, the higher is the probability that the pathway is present. MetaCyc sometimes contains multiple similar variants of a given pathway, which often share enzymes in common. For example, MetaCyc contains twelve pathways for the degradation of arginine. The pathway scoring procedure seeks to differentiate among these variants by assigning higher weight in the pathway scoring to reactions that are unique to a given pathway, and therefore serve as special signatures for the presence of that pathway over its competing variants. In addition, the scoring algorithm is designed to err on the side of predicting more false positive pathways than on missing the presence of a pathway, under the hypothesis that it is better to bring possible pathways to the attention of a scientist for review.

Note that the genome sequence supplied to PathoLogic need not be complete, and can be in multiple contigs, but the more complete the sequence, the more complete will be the pathway analysis.

In evaluations of PathoLogic pathway predictions for both the *Helicobacter pylori* [11] and human genomes [12], predictions were found to agree extremely well with pathways known for these organisms, and in *H. pylori*, the algorithm discovered the presence of pathways that had been overlooked in manual analyses. PathoLogic is much faster than manual pathway analyses — a PathoLogic prediction can be completed in a few hours, and reviewed and refined in a few days, whereas a manual pathway analysis can take weeks. Furthermore, a PathoLogic analysis is likely to be more sensitive than a manual analysis because of the wide repertoire of pathways that it considers from MetaCyc.

2.2 Pathway Hole Filling

When PathoLogic infers the pathways present in an organism based on the genome annotation of the organism, many pathways are incomplete in the sense that some pathway reactions contain no assigned enzymes. Figure 1 shows the enzymes that PathoLogic has matched to the pentose phosphate pathway from the *S. pombe* genome annotation. The pathways include eight reactions altogether, but enzymes have been identified in *S. pombe* for only six of these reactions. Each reaction without an enzyme assigned is called a "missing reaction" or "pathway hole." We used the Pathway Hole Filler (PHFiller) to search the *S. pombe* genome for enzymes that might fill these holes.

PHFiller combines homology-based and pathway-based evidence to identify candidates for filling pathway holes in Pathway/Genome databases. The program not only identifies potential candidate sequences for pathway holes, but combines data from multiple, heterogeneous sources to assess the likelihood that each of those candidates has the required function. Our algorithm emulates the manual sequence annotation process, considering not only evidence from homology searches, but also evidence from genomic context (for prokaryotic organisms only) and functional context (e.g., does the candidate gene perform a second related function in the organism, such as catalyzing another reaction in the pathway?) to determine the probability that a candidate has the required function. Once all candidates for a particular pathway hole have been assigned a probability, the candidates can be filtered to eliminate those below a chosen threshold probability. The filtered results can either be entered into the database automatically, or undergo further manual review of the evidence supporting each candidate before the predictions are entered into the database.

The *S. pombe* PGDB includes 134 pathways with holes and 37 complete pathways (i.e., pathways where each reaction has an enzyme assigned by PathoLogic). Among these 134 pathways, there are 383 individual pathway holes. We used PHFiller to identify and evaluate candidates to fill these pathway holes. PHFiller identified candidate enzymes for 69 of the 383 pathway holes at its default threshold of $P > 0.9$. Thirty-two of these pathway holes were filled with proteins of unknown function — PHFiller has elucidated functions for these 32 proteins. As a result of filling these 69 pathway holes in the *S. pombe* genome, PHFiller has completed an additional 18 pathways, including the nonoxidative branch of the pentose phosphate pathway shown in Figure 1A.

The nonoxidative branch of the pentose phosphate pathway is the second half of the phosphogluconate pathway, an alternative pathway for oxidizing glucose (Figure 1A). The first half of the phosphogluconate pathway (Figure 1B) is complete; there are no pathway holes. The nonoxidative branch of the pentose phosphate pathway, however, includes two pathway holes. The enzymes for EC# 5.3.1.6, ribose-5-phosphate isomerase and the last reaction in the pathway, transketolase, were not identified by PathoLogic in the genome annotation for *S. pombe*. Pathway holes can be easily identified in Pathway Tools pathway displays because neither enzyme names nor genes are listed for these reactions.

PHFiller identified a candidate enzyme for each of these reactions. The enzyme catalyzing the conversion of erythrose-4-phosphate and xylulose-5-phosphate to glyceraldehyde-3-phosphate and fructose-6-phosphate, TKT_SCHPO, is not actually a missing reaction. In this case, PHFiller has helped identify an instance where the enzyme was present, but PathoLogic was unable to match it to the appropriate reaction because PathoLogic did not recognize its annotation. The candidate is a probable transketolase and is already assigned to another reaction in the same pathway. The candidate identified by PHFiller for EC# 5.3.1.6, Q9UTL3_SCHPO, on the other hand, had no functional annotation in the *S. pombe* genome. This pathway is just one example of how investigating protein functions in the context of an organism's predicted metabolic network can identify functions that may have been overlooked during the original annotation process.

2.3 Transport Identification Parser

The Transport Identification Parser (TIP) component of PathoLogic analyzes the gene-product names of transport proteins within a PGDB. It attempts to identify the transported substrate for each transporter, the direction of transport (influx/efflux), the names of any cotransported substrates (e.g., H^+ or Na^+), and the energy coupling mechanism used by the transporter (e.g., is it an ATP-driven transporter or a passive channel?). When TIP is able to extract all this information from the transporter name with high confidence, it creates a transport reaction object describing the transport event catalyzed by the transporter.

An example transport reaction describing the ATP-driven transport of arginine from the periplasm to the cytoplasm is as follows.

```
L-arginine [periplasm] + ATP + H2O  ==>  L-arginine + ADP + phosphate
```

Transport reactions label substrates with their cellular compartment, which defaults to the cytoplasm.

Creation of transport reactions within a PGDB is advantageous because it enables computational manipulation of transporters. For example, Pathway Tools adds all transporters for which transport reactions are defined to the Cellular Overview.

3 Pathway/Genome Editors

The Pathway/Genome Editors are a suite of forms-based tools for interactively creating and updating objects within a PGDB. For example, the pathway editor allows the user to interactively create a new pathway, or to add reactions to, or remove reactions from, an existing pathway. The reaction editor allows the user to create a new metabolic or transport reaction, or to change the substrates of an existing reaction. The compound editor allows the user to modify the name or synonyms of a chemical compound within a PGDB, or to create or alter its chemical structure using either the Marvin or JME chemical editor.

The gene editor allows updating of the genome map position of a gene, and the Gene Ontology terms assigned to it. The transcription unit editor allows one to define transcription factor binding sites for a gene, and to define interaction events between those sites and a transcription factor. The protein editor allows the user to define activators, inhibitors, and cofactors for an enzyme. For any protein, it allows the user to annotate sites or regions within the protein, such as the active site of an enzyme, phosphorylation or other chemically modified site within the protein, repeat regions, or transmembrane domains. These sites and regions are displayed graphically within the protein pages produced by the Navigator, such as shown in Figure 2.

All editors provide common edits such as updating names and synonyms of an object, entering a comment and literature citations, adding links to external databases, and entering evidence codes.

4 Analysis and Visualization

This section describes the analysis and visualization capabilities present in Pathway Tools.

4.1 Genome Browser

The Pathway Tools genome browser can be used to examine the linear arrangement of genes within a region of a chromosome. It provides several levels of semantic zooming, meaning that as the user increases the magnification level, additional details appear, such as promoters. The genome browser can be invoked from the Genome Browser section of the Pathway Tools Web query page, and from a gene display page, by clicking on the base-pair coordinates mentioned on the map position line.

An example genome browser visualization for *Escherichia coli* is shown in Figure 3. At the top of the display, the full length of the chromosome is shown at low resolution, to provide orientational context. A region of the chromosome can be selected for display at much higher magnification in the lower part of the screen. The selected region will be drawn using as many lines as will fit on the screen. The full chromosome view at the very top indicates the magnified region by means of a red, rectangular cursor.

Users can move around within the chromosome using several methods. Clicking on a position within the full chromosome line at the top shows the immediate neighborhood of that position. The tick marks within the magnified region can also be clicked on, to quickly recenter the region around the selected tick mark. Start and end base pair position numbers can be used for specifying the region to display. The region around a given gene can be shown by searching for the gene's name. The selected gene is then visually highlighted. For easy interactive navigation, the panel of navigation

E. coli K-12 Protein: ArcB

Synonyms: B3210, ArcB, unphosphorylated sensor kinase-phosphotransferase ArcB, unmodified sensor kinase-phosphotransferase ArcB, aerobic respiration control sensor protein ArcB, aerobic respiration control sensor protein, sensor protein ArcB, sensor kinase-phosphotransferase ArcB

Comment:

The ArcB protein is the sensor kinase component of the Arc two-component system which regulates the expression of many genes in response to respiratory growth conditions **[Liu04]**. D-lactate amplifies the kinase activity of ArcB *in vivo* and *in vitro* **[Rodriguez04]**. Quinone electron carriers inhibit ArcB kinase activity under aerobic conditions by oxidizing two redox-active cysteine residues that are involved in intermolecular disulfide bond formation **[Malpica04]**.

The solution NMR structure of ArcB has been determined **[Ikegami01]**, and crystal structures of parts of ArcB have been solved **[Kato97]**.

Gene:

Locations: inner membrane

Subunit composition of **ArcB** = $[ArcB]_2$
ArcB sensory histidine kinase = ArcB

Sequence Length: 778 AAs

Molecular Weight of Polypeptide (from nucleotide sequence): 87.983 kD

pI: 5.19

Unification Links: **PDB:1A0B, PDB:1BDJ, PDB:1FR0, PDB:2A0B, PIR:b3210, Swiss-Prot:P22763**

In Pathway Reactions as a Reactant:

ATP + ArcB sensory histidine kinase = ADP + ArcB-P^{his292}

In Pathway Reactions as a Product:

ArcB-P^{his292} + ArcA = ArcA-P^{asp} + ArcB sensory histidine kinase ,
ArcA + ArcB-P^{his717} = ArcA-P^{asp} + ArcB sensory histidine kinase ,
ArcB-P^{his717} = ArcB sensory histidine kinase + phosphate

Sequence Features

1 [illegible] 100
101 [illegible] 200
201 [illegible] 300
His292 Phosphorylation Site
301 [illegible] 400
401 [illegible] 500
501 [illegible] 600
Asp Phosphorylation Site
601 [illegible] 700
701 [illegible] 778
His717 Phosphorylation Site

His292 Phosphorylation Site:
Feature Class: **Phosphorylation-Modifications**

Feature Residue Number(s): 292

Asp Phosphorylation Site:
Feature Class: **Phosphorylation-Modifications**

Feature Residue Number(s): 574

His717 Phosphorylation Site:
Feature Class: **Phosphorylation-Modifications**

Feature Residue Number(s): 715

References

Ikegami01: Ikegami T, Okada T, Ohki I, Hirayama J, Mizuno T, Shirakawa M (2001). "Solution structure and dynamic character of the histidine-containing phosphotransfer domain of anaerobic sensor kinase ArcB from Escherichia coli." Biochemistry 40(2):375-86. PMID: 11148031

Kato97: Kato M, Mizuno T, Shimizu T, Hakoshima T (1997). "Insights into multistep phosphorelay from the crystal structure of the C-terminal HPt domain of ArcB." Cell 88(5):717-23. PMID: 9054511

Liu04: Liu X, De Wulf P (2004). "Probing the ArcA-P Modulon of Escherichia coli by Whole Genome Transcriptional Analysis and Sequence Recognition Profiling." J Biol Chem 279(13):12588-97. PMID: 14711822

Malpica04: Malpica R, Franco B, Rodriguez C, Kwon O, Georgellis D (2004). "Identification of a quinone-sensitive redox switch in the ArcB sensor kinase." Proc Natl Acad Sci U S A 101(36):13318-23. PMID: 15326287

Rodriguez04: Rodriguez C, Kwon O, Georgellis D (2004). "Effect of D-lactate on the physiological activity of the ArcB sensor kinase in Escherichia coli." J Bacteriol 186(7):2085-90. PMID: 15028693

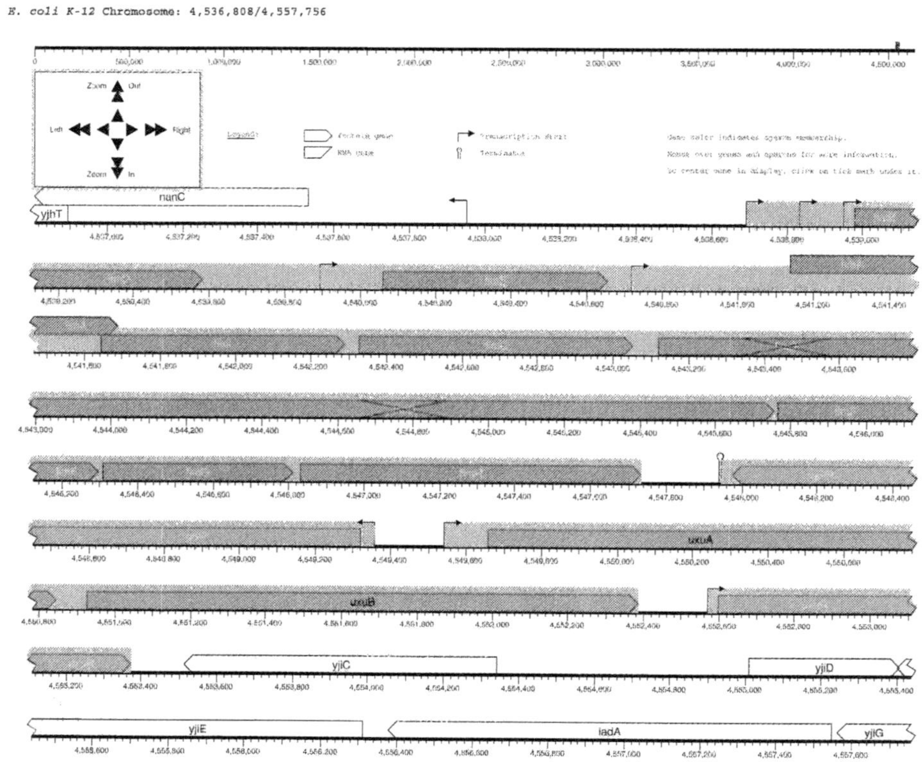

Figure 3: Pathway Tools genome browser.

arrows in the upper left region of the browser can be used to move to a nearby region. That panel allows lateral translation to the left or right, and zooming in or out.

The magnified section of the genome indicates the transcription direction of genes by rectangular blocks with an arrow at one end, pointing from the 5' to the 3' end of the gene. ORFs for actual or inferred proteins have symmetrical arrowheads (with the arrow apex in the center), whereas RNA genes have an asymmetrical arrowhead (with the apex at the top edge). Pseudo-genes are crossed out with a large X. When a gene wraps across more than one line, a zigzag at the end of the line indicates that the gene continues on the next line. If the overlap between adjacent genes is more than a small amount, the shorter gene is drawn on a second level above the longer gene, to avoid visual clashes.

We chose to show a genome browser example from EcoCyc, a very thoroughly curated bacterial PGDB, so that an additional feature can be demonstrated that is rare in eukaryotic genomes, namely poly-cistronic transcription units, also called operons. Genes shown with solid colors were assigned to a transcription unit. All the genes that are part of a given operon are assigned the same color, whereas genes that have not been assigned to any transcription unit are not colored. In addition, transcription units are indicated by a gray background area behind the genes, spanning the entire region of the operon. When zoomed in to a high level of detail, transcription start sites are indicated with small arrows, and terminators are shown as hairpin loops.

Clicking on a gene shows the corresponding gene description page. Clicking on the gray area of a transcription unit or transcription start site shows the corresponding transcription unit description page. Moving the mouse pointer over the genes reveals their product name and the length in base pairs of the intergenic regions between the chosen gene and its neighboring genes to the left and right. Moving the pointer over a gray transcription unit area reveals the transcription factors that are known to control its expression.

The comparative genome browser can display several chromosomes from multiple selected organisms side by side, aligned at the positions of a set of orthologous genes. Figure 4 shows the regions around the miaA genes of several bacteria (a bacterial example was chosen because more than two fungal PGDBs do not yet exist).

4.2 Cellular Overview Diagram and Omics Viewer

The Cellular Overview Diagram is a low-resolution view of all metabolic pathways and other metabolic and transport reactions that are inferred or curated within the PGDB for an organism. An example is shown in Figure 5. The Cellular Overview Diagram can be used as a quick visual summary of an organism, as a tool for query, analysis and navigation, and for visualizing the results of high-throughput experiments using the Omics Viewer.

Each node within the Overview (such as the small circles and squares) represents a metabolite, with each connecting line representing a reaction. Thick blue lines represent reactions for which an enzyme has been identified, whereas thin gray lines are pathway holes. Pathways are grouped in the diagram by class — each gray shaded background region contains a set of functionally related pathways, such as those for cofactor biosynthesis or amino acid biosynthesis. The overall organization of the diagram positions biosynthetic pathways on the left side of the diagram, energy metabolism pathways in the middle, and catabolic pathways on the right.

Metabolic reactions that have not been assigned to any pathway are tabulated at the far right. A border representing the cell's plasma membrane surrounds the diagram, and any transport reactions

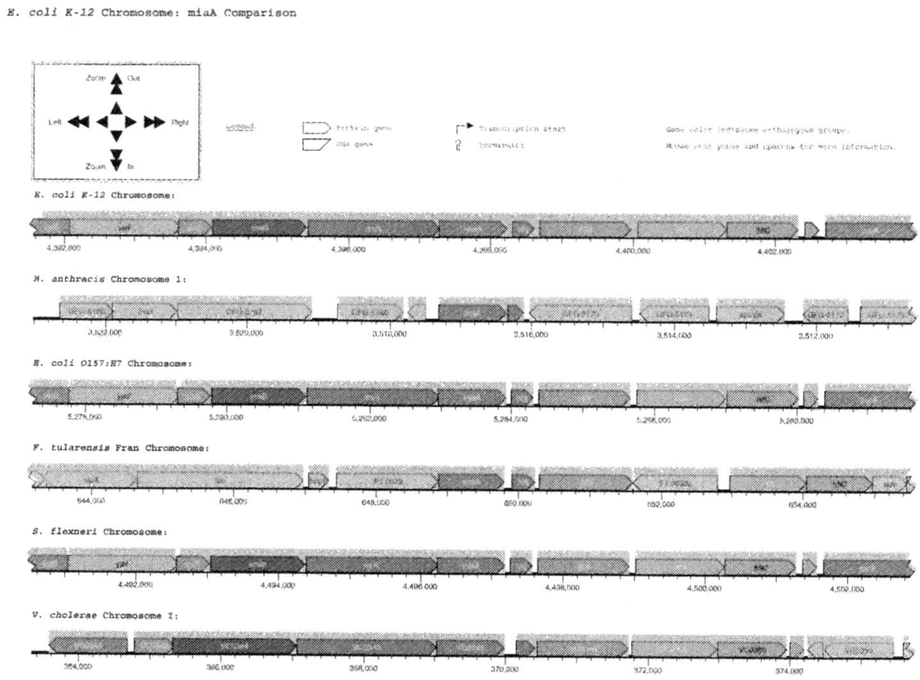

Figure 4: Comparative genome browser aligned at the miaA genes of several bacteria.

that cross this membrane, as well as non-transport proteins located n this membrane, are superimposed on the border (protein cellular locations are not inferred automatically by PathoLogic, but may be entered manually by curators).

The Cellular Overview Diagram is generated entirely automatically upon creation of a PGDB, and can be regenerated at any time to reflect changes, such as the addition of new pathways.

Users can mouse over any display element to display the identity of a particular metabolite, reaction, pathway, or pathway class. Clicking on an object navigates to the detailed display page for that object. Users can query a particular metabolite, gene, protein, and so on, to highlight all the places in the Overview where that object appears. Assuming that the relevant data is present in the PGDB, users can search and highlight, for example, all proteins with a particular cellular location, all reactions whose enzymes are activated or inhibited by a particular compound, all reactions with multiple isozymes, or any one of a number of other predefined queries. The Cellular Overview Diagram can be used to compare PGDBs for related organisms by highlighting all reactions shared or not shared among the selected organisms. Note that some of the query options described here are available only to users who have installed the Pathway Tools software on their own machines,

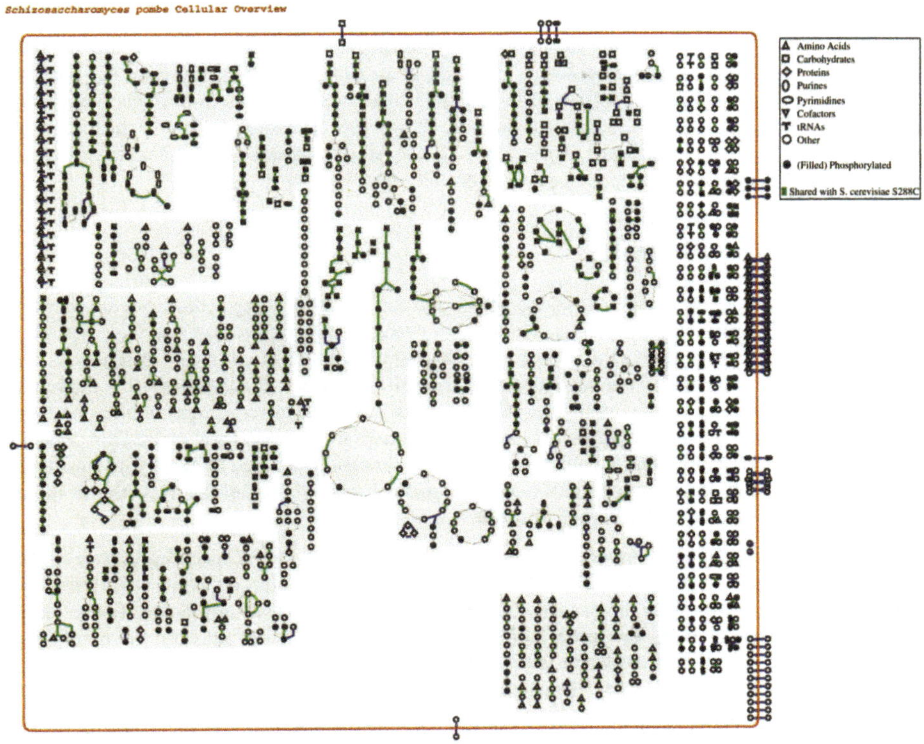

Figure 5: Cellular Overview Diagram for *Schizosaccharomyces pombe*, highlighted to show a comparison with *Saccharomyces cerevisiae*. Reactions shared between the two organisms are green, whereas reactions unique to *S. pombe* are blue. Thin gray lines are pathway holes.

and are not yet available through the Web.

The Omics Viewer, a popular tool, allows Pathway Tools users to upload their own files of high-throughput data, such as gene expression, proteomics, or metabolomics data, onto the Cellular Overview Diagram. Each metabolite icon or reaction line is colored according to the data value in the supplied file and the selected color scheme. Users can also zoom in to see their data superimposed on individual pathway diagrams. This tool allows users to view their experimental high-throughput data in a metabolic pathway context.

4.3 Comparative Genomics

The Pathway Tools software offers a wide range of tools for comparing the genomes and biochemical machinery of multiple organisms. The comparative genome ortholog browser has been described above, as has the use of the Cellular Overview Diagram to obtain an overview of reactions and pathways shared or not shared among organisms. In addition, a newly developed set of comparative tools generates tables that compare a user-selected list of organisms across a variety of dimensions. The list of selected organism PGDBs can be any combination of user-generated PGDBs, PGDBs that were generated at SRI and included in the software distribution, and PGDBs that have been imported from other sites via our PGDB registry. Users can select from sets of tables that compare Reactions, Pathways, Compounds, and Proteins.

The tables in the Reactions comparison provide statistics about the set of reactions in each PGDB by type of substrate, by top-level EC category, by number of isozymes, and so on. The Pathways comparison provides statistics on the set of reactions by pathway class, and describes the distribution of pathway holes. At the initial lowest level of detail, the tables list only the numbers of reactions or pathways in each category. However, users can click on any table, row, or cell to see it at the next level of detail, in which the complete list of relevant objects is shown.

Finally, users can request a visual comparison of a reaction or a pathway across a selected set of organisms. The detailed comparison of a reaction includes the list of enzymes and genes for that reaction, and the list of pathways in which the reaction participates. The detailed comparison of a pathway includes a schematic of the pathway indicating which steps have enzymes assigned, and lists the enzymes and genes for every step. Figure 6 shows an example progression from a top-level table to the next level of detail and finally to a detailed comparison of a single pathway. The detailed comparisons of a single reaction or pathway across organisms can also be reached from the regular display page for that reaction or pathway in any organism by clicking the "Species Comparison" button.

The Compounds comparison provides statistics on different cellular roles of the small molecules within a PGDB, for example, as substrates, enzyme activators or inhibitors, and cofactors. The Proteins comparison lists proteins shared by the selected organisms or unique to a single organism, and breaks down protein complexes by number of unique subunits and multifunctional proteins by number of reactions.

5 Pathway Tools Ontologies

Pathway Tools includes support for several ontologies. Curators can assign genes within a PGDB to Gene Ontology classes [1]. Metabolic pathways can be assigned to classes within the Pathway Tools Pathway Ontology. Reactions are queryable through the Enzyme Commission system.

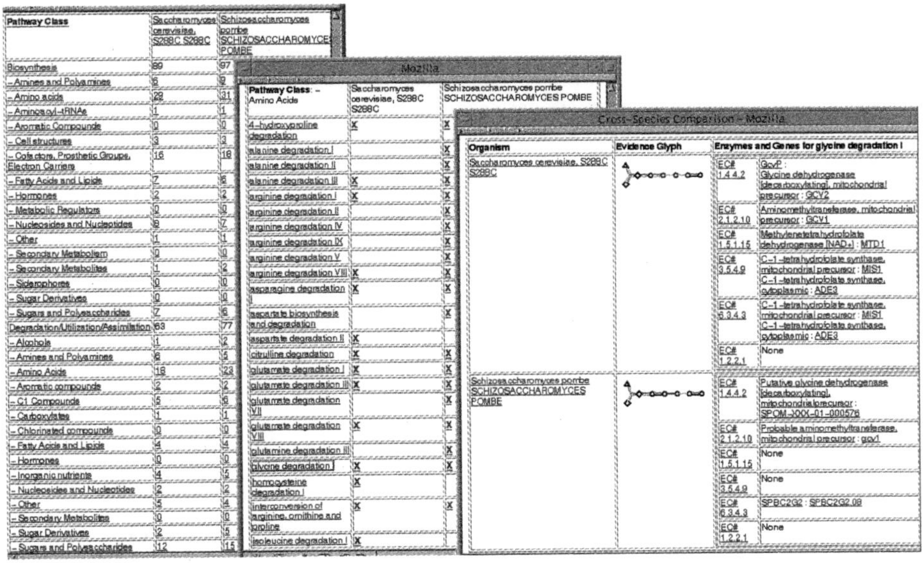

Figure 6: Pathway comparison between *Saccharomyces cerevisiae* and *Schizosaccharomyces pombe*. The leftmost browser window contains the top-level comparison, showing the number of pathways in each pathway class for the two organisms. Clicking on the Amino Acids class (under Degradation/Utilization/Assimilation) displays the list of amino acid degradation pathways and shows which are present in each organism (middle browser window). From there, clicking on the glycine degradation I pathway displays a detailed comparison of the pathway across the two organisms. The *S. cerevisiae* schematic shows evidence for five of the reactions (green), with one pathway hole (black), whereas the *S. pombe* schematic shows evidence for three of the reactions, with three pathway holes. Reactions that are unique to a pathway have an additional orange line.

An evidence ontology can be used to record the type(s) of evidence supporting different assertions within a PGDB [7]. For example, PathoLogic assigns a computational evidence code to computationally predicted metabolic pathways. A curator could later add an experimental evidence code to such a pathway if its presence was confirmed experimentally. Pathway Tools contains a cell component ontology to record protein locations and transport events. It also contains an ontology of protein sites and regions.

6 Obtaining Pathway Tools and System Requirements

Pathway Tools runs on the following platforms: PC/Windows, PC/Linux, and Sun/Solaris. Furthermore, the software can run as both a desktop application, and in a Web server mode. Note that the capabilities of these two modes are largely the same, but each mode has some capabilities that the other lacks. For example, the desktop mode contains many more operations for the Cellular Overview, whereas the Web mode has more comparative operations.

PGDBs can be stored within disk files, or within the Oracle or MySQL database management system (DBMS). We recommend use of a DBMS when multiple curators will update a PGDB concurrently, however, it is easier to start a PGDB project without a DBMS.

The software is freely available to academic groups, and can be obtained through an online click-through license agreement from `http://biocyc.org/download.shtml`. Source code is available. Commercial users should contact ptools-support@ai.sri.com.

7 How to Learn More

A number of publications exist for Pathway Tools [8] and explain details of its pathway prediction algorithm [11], pathway hole filling algorithm [4], and how to query PGDBs through Perl and Java APIs [10]; additional information is available at the Pathway Tools home page at `http://bioinformatics.ai.sri.com/ptools/`.

MetaCyc is described in several publications [2, 9, 5], as well as through an online MetaCyc User's Guide at `http://MetaCyc.org/MetaCycUsersGuide.shtml`. The preceding publications compare Pathway Tools and MetaCyc to related software and DBs.

8 Summary

Pathway Tools provides three types of computational inference tools for inferring new information from a genome: it infers metabolic pathways, pathway hole fillers, and transport reactions. It provides a suite of interactive editing tools that allow curators to update and refine the description of an organism's genome and metabolic network within a PGDB. It provides many query and visualization tools that can be used both from the software's desktop mode and from its Web mode. A new suite of comparative tools allow scientists to compare multiple PGDBs. The Pathway Tools Omics viewer enables visualization and analysis of multiple types of functional genomics data in the context of the cellular biochemical network. Together, these tools ensure that the maximal information has been extracted from a genome, and from the biomedical literature, to create a central knowledge resource to accelerate further discoveries about the organism.

9 Acknowledgments

Tom Lee and Ian Paulsen contributed to the Transport Identification Parser. This work was supported by grants GM70065, GM75742, and RR07861 from the National Institutes of Health, and by the Department of Energy under grant DE-FG03-01ER63219. The contents of this publication are solely the responsibility of the authors and do not necessarily represent the official views of the National Institutes of Health or the Department of Energy.

References

[1] M. Ashburner, C.A. Ball, J.A. Blake, D. Botstein, H. Butler, J.M. Cherry, A.P. Davis, K. Dolinski, S.S. Dwight, J.T. Eppig, M.A. Harris, D.P. Hill, L. Issel-Tarver, A. Kasarskis, S. Lewis, J.C. Matese, J.E. Richardson, M. Ringwald, G.M. Rubin, and G. Sherlock. Gene Ontology: Tool for the unification of biology. *Nature Genetics*, 25:25–29, 2000.

[2] R. Caspi, H. Foerster, C.A. Fulcher, R. Hopkinson, J. Ingraham, P. Kaipa, M. Krummenacker, S. Paley, J. Pick, S. Y. Rhee, C. Tissier, P. Zhang, and P. D. Karp. MetaCyc: A multiorganism database of metabolic pathways and enzymes. *Nuc Acids Res*, 34:D511–6, 2006.

[3] J.S. Edwards, R. Ramakrishna, C.H. Schilling, and B.O. Palsson. Metabolic flux balance analysis. In *Metabolic Engineering*, pages 13–57. Marcel Dekker, 1999.

[4] M.L. Green and P.D. Karp. A Bayesian method for identifying missing enzymes in predicted metabolic pathway databases. *BMC Bioinformatics*, 5(1):76, 2004. http://www.biomedcentral.com/1471-2105/5/76.

[5] P.D. Karp. The MetaCyc metabolic pathway database. In *Metabolic Engineering*. Horizon Scientific Press, 2003.

[6] P.D. Karp, C.A. Ouzounis, C. Moore-Kochlacs, L. Goldovsky, P. Kaipa, D. Ahren, S. Tsoka, N. Darzentas, V. Kunin, and N. Lopez-Bigas. Expansion of the BioCyc collection of pathway/genome databases to 160 genomes. *Nuc Acids Res*, 33(19):6083–89, 2005.

[7] P.D. Karp, S. Paley, C.J. Krieger, and P. Zhang. An evidence ontology for use in pathway/genome databases. In R. Altman and T. Klein, editors, *Proc Pacific Symposium on Biocomputing*, pages 190–201, Singapore, 2004. World Scientific.

[8] P.D. Karp, S. Paley, and P. Romero. The Pathway Tools Software. *Bioinformatics*, 18:S225–S232, 2002.

[9] C.J. Krieger, P. Zhang, L. A. Mueller, A. Wang, S. Paley, M. Arnaud, J. Pick, S. Y. Rhee, and P. D. Karp. MetaCyc: A multiorganism database of metabolic pathways and enzymes. *Nuc Acids Res*, 32:D438–42, 2004.

[10] M. Krummenacker, S. Paley, L. Mueller, T. Yan, and P.D. Karp. Querying and Computing with BioCyc Databases. *Bioinformatics*, 21:3454–5, 2005.

[11] S. Paley and P.D. Karp. Evaluation of computational metabolic-pathway predictions for *H. pylori*. *Bioinformatics*, 18(5):715–24, 2002.

[12] P. Romero, J. Wagg, M.L. Green, D. Kaiser, M. Krummenacker, and P.D. Karp. Computational prediction of human metabolic pathways from the complete human genome. *Genome Biology*, 6(1):1–17, 2004.

Applied Mycology
and Biotechnology
An International Series
Volume 6. Bioinformatics

Comparative Genomic Analysis of Glycoylation Pathways in Yeast, Plants and Higher eukaryotes

Shoba Ranganathan[1,2], Sangdao Wongsai[1] and K.M. Helena Nevalainen[1]
[1]Department of Chemistry and Biomolecular Sciences; [2]Biotechnology Research Institute, Macquarie University, Sydney, NSW 2109, Australia (shoba@els.mq.edu.au)

N-linked glycosylation is an essential modification of secretory and membrane proteins in all eukaryotic cells. Here, we review the current metabolic pathways of N-linked oligosaccharide biosynthesis in the endoplasmic reticulum and in the Golgi apparatus for yeasts: *Saccharomyces cerevisiae* and *Schizosaccharomyces pombe*, and higher eukaryotes: plants and human. The evolutionarily conserved proteins, processed in the cytosolic and the luminal side of the ER membrane, and the unique genes and their specific functions, occurring in the Golgi complex, for each selected organism, will be collated and discussed. This precise knowledge of the glycosylation pathway contributes to better understanding of the N-linked glycoprotein biosynthesis among different species, resulting in the recently successfully engineered strains for heterologous gene expression systems for industrial and therapeutic protein production.

1. INTRODUCTION

Glycosylation is a major post-translational modification process in eukaryotes that adds carbohydrates to nascent proteins to form glycoproteins (Herscovics and Orlean 1993; Ziegler et al. 1994; Lerouge et al. 1998; Bobrowicz et al. 2004; Wildt and Gerngross 2005). The structure and function of secretory and membrane proteins in eukaryotes have been regulated and controlled throughout the process of co-translational modification in the lumen of the endoplasmic reticulum (ER) (Helenius and Aebi 2004) and post-translational modification in the Golgi apparatus (Chen et al. 2005). There are two types of protein glycosylation, resulting in N- and O-linked oligosaccharides. O-linked glycosylation is the attachment of monosaccharides to the hydroxyl groups of serine or threonine residues of nascent proteins. N-linked glycosylation is the attachment of carbohydrates directly onto the asparagine residues of the polypeptides. About 70% of all secreted proteins are glycosylated and these modifications are essential for determining the final destinations of the proteins, whether transported outside the cell or integrated into cellular organelles

Corresponding Author: Shoba Ranganathan

and the plasma membrane. Protein localization and protein folding are important processes for the transportation of glycoproteins to their final destinations. The glycoprotein folding in the ER begins when the nascent polypeptides are transferred across the ER membrane to link with the proper N-glycans in the ER lumen. Proteins that do not attain their native conformation cannot pass the quality control in the ER due to the blocking of N-glycosylation and will remain misfolded and may get aggregated (Parodi 2000). Appropriate N-linked oligosaccharides are thus essential for protein folding, transport out of the ER and efficient protein secretion.

Secreted proteins will be released to the external medium via the secretory pathway consisting of the ER, *cis*-Golgi, *medial*-Golgi, *trans*-Golgi, and the medium itself (Conesa et al. 2001). They are not merely controlled by protein folding but also regulated by protein localization. Protein localization plays a critical role in controlling the quality and quantity of the secretion products, of both homologous and heterologous proteins. Yeasts, filamentous fungi, and plants have been widely exploited as hosts for expression of heterologous proteins, producing a great variety of glycoproteins used in industrial applications and therapeutic productions (Giddings et al. 2000; Joosten et al. 2003). Although their functionality has been unraveled over the past decades, comprehensive knowledge of the secretory pathway, particularly in the identification of proper and suitable localization, is still very preliminary and localization of a gene product remains very difficult to predict.

To date, recombinant DNA techniques, coupled with the high-throughput technologies of genome sequencing, transcriptomics and proteomics, have opened the gateway for scientists who are interested in glycoprotein biotechnology to gain an insight into the mechanism of protein localization and folding, for producing large amounts of glycoproteins in eukaryotic cells. In this way, new metabolic pathways consisting of several enzymatic steps can be introduced into organisms, to generate novel non-native products. Pathway mapping can be accomplished by using comparative genomic approaches. Armed with these techniques, we are now able to mine the enormous resource of genome sequences for both genes conserved during evolution and novel or unique genes in specific organisms and elucidate their genetic roles and biological functions. Understanding of enzymatic activity involved in this metabolic pathway has contributed to the identification of the bottle neck for protein secretion, suggesting possible engineering of strains at the molecular and metabolic levels. The methanotrophic yeast, *Pichia pastoris*, for example, has been engineered and "humanized" as an artificial glycosylation pathway by blocking a critical reaction step in the core oligosaccharide assembly, resulting in the production of purified glycoproteins for human use (Bobrowicz et al. 2004). By contrast, it has been recently reported that baker's yeast, Saccharomyces cerevisiae, an attractive fermentation organism and a host for recombinant protein expression, cannot be used to generate heterologous proteins due to non-human N-glycosylation reactions that are associated with yeast protein expression (Wildt and Gerngross 2005). So, what are the key factors that distinguish secreted proteins among species? What are specific roles for glycoproteins in each organism? Which genes in the glycosylation pathway are conserved and which ones are species-specific? A combination of genomic and transcriptomic data into metabolic pathway will provide a deeper understanding of the mechanism of glycoprotein biogenesis in the ER and Golgi complex.

In this article, we review the current understanding of the metabolic pathway of N-glycan biosynthesis in three compartments: the cytosol, the ER and the Golgi apparatus for different organisms that are lower eukaryotes, *S. cerevisiae* and fission yeast, *Schizosaccharomyces pombe*; and the higher eukaryotes, plants and human. We also discuss the common enzymes shared in each compartment and highlight the distinct enzymes unique to each selected organism. This review aims to provide an informatics perspective of engineering the N-glycosylation pathway, which has enormous potential in biotechnology for heterologous protein expression.

2. METABOLIC PATHWAY OF N-GLYCAN BIOSYNTHESIS

The biosynthesis of N-linked oligosaccharides has been highly conserved during evolution (Helenius and Aebi 2004). The oligosaccharide residues are assembled on the cytoplasmic face of the ER membrane and then flipped into the ER lumen to create the final structure of core oligosaccharide which is then transferred to a

Table 1. Glycosylation genes completely conserved in *Saccharomyces cerevisia, Schizosaccharomyces pombe*, plants and human, in the cytosol and the Endoplasmic Reticulum compartments.

Gene	EC number	Sugar Donor	Product	Function
Cytosol				
Sec59	2.7.1.106	-	Dol-P	Dolichol kinase
Alg7	2.7.8.15	UDP-GlcNAc	Dol-PP-GlcNAc	Dol-PP-GlcNAc -1-P transferase
Alg13/ Alg14	2.4.1.141	UDP-GlcNAc	Dol-PP-GlcNAc$_2$	Dol-PP-GlcNAc transferase
Alg1	2.4.1.142	GDP-Mannose	Dol-PP-GlcNAc$_2$Man	β-1,4-mannosyltransferase
Alg2	2.4.1.32	GDP-Mannose	Dol-PP-GlcNAc$_2$Man$_2$	β-1,4-mannosyltransferase
Alg2	2.4.1.a*	GDP-Mannose	Dol-PP-GlcNAc$_2$Man$_3$	α-1,6-mannosyltransferase
Alg11	2.4.1.b*	GDP-Mannose	Dol-PP-GlcNAc$_2$Man$_4$	α-1,2-mannosyltransferase?
Alg11	2.4.1.c*	GDP-Mannose	Dol-PP-GlcNAc$_2$Man$_5$	α-1,2-mannosyltransferase
Endoplasmic Reticulum				
Alg3	2.4.1.d*	Dol-P-Mannose	Dol-PP-GlcNAc$_2$Man$_6$	α-1,3-mannosyltransferase
Alg9	2.4.1.e*	Dol-P-Mannose	Dol-PP-GlcNAc$_2$Man$_7$	α-1,2-mannosyltransferase
Alg12	2.4.1.f*	Dol-P-Mannose	Dol-PP-GlcNAc$_2$Man$_8$	α-1,6-mannosyltransferase
Alg9	2.4.1.g*	Dol-P-Mannose	Dol-PP-GlcNAc$_2$Man$_9$	α-1,2-mannosyltransferase
Alg6	2.4.1.h*	Dol-P-Glucose	Dol-PP-GlcNAc$_2$Man$_9$Glc	α-1,3-glucosyltransferase
Alg8	2.4.1.i*	Dol-P-Glucose	Dol-PP-GlcNAc$_2$Man$_9$Glc$_2$	α-1,2-glucosyltransferase
Alg10	2.4.1.j	Dol-P-Glucose	Dol-PP-GlcNAc$_2$Man$_9$Glc$_3$	α-1,2-glucosyltransferase
Ost4	2.4.1.119	-	Asn-GlcNAc$_2$Man$_9$Glc$_3$	oligosaccharyltransferase
Gsd1	3.2.1.106	-	Asn-GlcNAc$_2$Man$_9$Glc$_2$	glucosidase I
Gsd2	3.2.1.84	-	Asn-GlcNAc$_2$Man$_9$	glucosidase II
Mns1	3.2.1.113	-	Asn-GlcNAc$_2$Man$_8$	mannosidase I
Dpm1	2.4.1.83	GDP-Mannose	Dol-PP-Mannose	dolichol phosphate mannose synthase
Alg5	2.4.1.117	GDP-Glucose	Dol-PP-Glucose	UDP-glucose:dolichyl phosphate glucosylatransferase

*EC. 2.4.1.x, where x = a, b, ...j, represents our nomenclature for enzymes not uniquely identified by KEGG (Hasimoto et al. 2005). This nomenclature has also been used in Figures 1 and 2.

specific site on the nascent protein. The entire process is shown in Figure 1. We notice that a number of enzymes in the glycan biosynthesis pathway have not been specific site on the nascent protein. The entire process is shown in Figure 1. We notice that a number of enzymes in the glycan biosynthesis pathway have not been uniquely defined by KEGG (Hashimoto et al. 2005) and have therefore sequentially numbered these EC.2.4.1-a to EC.2.4.1-j in Table 1 and Figure 1.

N-glycan biosynthesis begins with the stepwise addition of the first seven monosaccharides using nucleotide-sugar donors, uridine-diphosphate-N-acetylglucosamine (UDP-GlcNAc) and guanosine-diphosphate-mannose (GDP-mannose), to a lipid carrier dolichol phosphate (dol-P) on the cytosolic face of the ER membrane. Next the oligosaccharides are translocated across the membrane into the

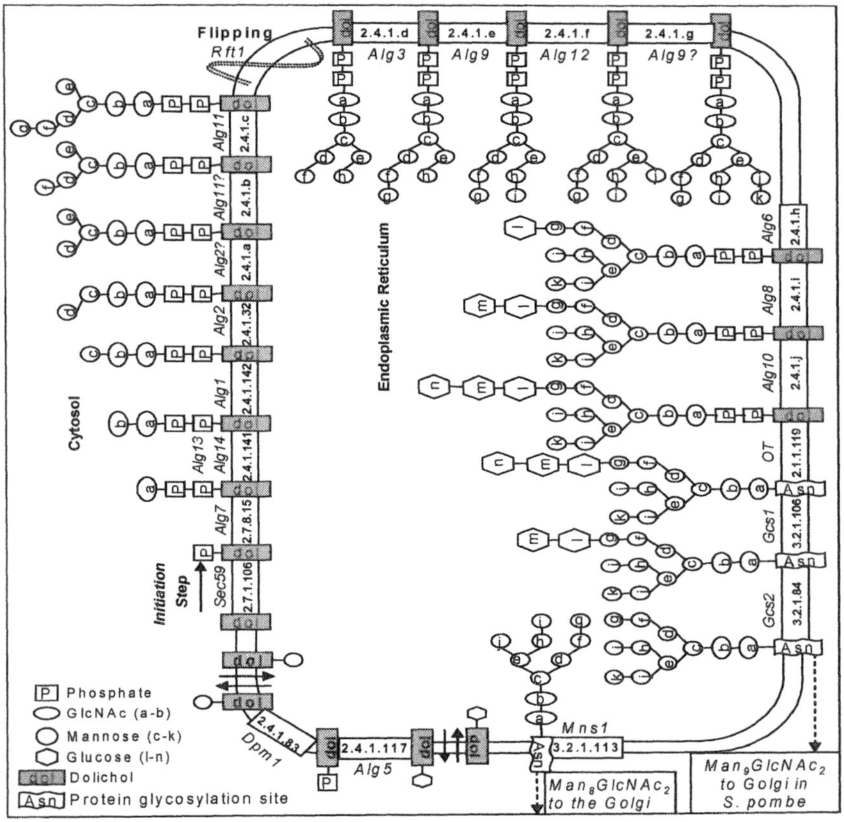

Fig. 1. Metabolic Pathway of core oligosaccharide synthesis in the ER. EC 2.4.1.x, where x = a-j represents our nomenclature in Tables 1 and 2, for enzymes not uniquely identified by KEGG (Hashimoto et al. 2005) Abbreviations: *Alg1*, β-1,4-mannosyltransferase; *Alg2*, α-1,3-mannosylransferase; *Alg3*, α-1,3-mannosylransferase; *Alg5*, dolichol phosphate β-glucosyltransferase; *Alg6*, α-1,3-glycotransferase; *Alg8*, α-1,3-glycotransferase; *Alg9*, α-1,2-mannosyltransferase; *Alg10*, α-1,2-glycotransferase; *Alg11*, α-1,6-mannosyltransferase; *Alg12*, α-1,6-mannosyltransferase; *Alg13*, β-1,4-acetylglucosaminyltransfer-ase; *Alg14*, β-1,4-acetylglucosaminyltransferase; dol, dolichol; *Dpm1*, N-acetylglucosaminephospho-transferase; *Gcs1*, glycosidase 1; *Gcs2*, glycosidase 2; *Mns1*, mannosidase1; *OT*, dolichyl-diphosphooligosaccharide protein glycosyltransferase; *Rft1*, flippase; *Sec59*, dolichol kinase.

ER lumen. In the ER lumen, four mannose and three glucose residues are added onto the lipid-bound oligosaccharide by mannosylation and glucosylation processing, respectively. At this stage, the lipid sugar donors, dolichyl-phosphate-glucose (dol-P-glucose) and dolichyl-phosphate-mannose (dol-P-mannose), take the role of nucleotide-sugar donors in the ER lumen. The final glycan structure contains two GlcNAc units, nine mannose units and three glucose units, resulting in $Glc_3Man_9GlcNAc_2$, prior to being transferred to the protein. Following this, the enzyme oligosaccharyltransferase catalyzes the transfer of the core oligosaccharide from the dolichol phosphate anchor to the identified asparagine residues on the nascent protein. At this point, the N-linked glycoprotein is trimmed by the removal of three terminal glucose residues by glycosidases, followed by the clipping of mannose moiety by mannosidase. The resulting asparagine-containing oligosaccharide $Man_8GlcNAc_2$ is then transported to the Golgi apparatus. Although the N-glycan biosynthesis of the core oligosaccharide and the initial stage of the glycoprotein in the ER are similar in lower and higher eukaryotes, the processing in the Golgi apparatus is subtly different in the generation of the final glycoprotein.

The following sections explain the specific steps involved in growing the core oligosaccharide chain in the cytosol and the lumen, followed by its transfer to the nascent protein at the N-glycosylation site and subsequent trimming, prior to export to the Golgi. At each step, the specific enzyme involved and its role is discussed.

2.1 Processes in the Cytosol

2.1.1 Initial steps of N-glycan biosynthesis

At the outset, a phosphate is added to dolichol molecule (dol) catalyzed by dolichol kinase (EC. 2.7.1.106), which is encoded by *Sec59* gene (Heller et al. 1992). The production of dol-P serves as an anchor in the ER membrane for the initiation and elongation of N-linked oligosaccharide biosynthesis. The first seven monosaccharides are transferred sequentially and specifically to dol-P by specific glycosyltransferase enzymes (details available from Table 1) using the nucleotide sugar donors, UDP-GlcNAc and GDP-mannose. The process enables the build-up of the proximal domain, $Man_5GlcNAc_2$, anchored to the cytosolic face of the ER. The first gene *Alg7* encoding UDP-N-acetylglucosamine-dolichyl-phosphate N-acetylglucosamine-phosphotransferase (EC.2.7.8.15) transfers a GlcNAc residue (marked a, in Fig. 1) from UDP-GlcNAc to dol-P to form dolichyldiphospho-N-acetylglucosamine (dol-PP-GlcNAc) (Kukuruzinska and Robbins, 1987). The *S. pombe Gpt* gene is essential to the cell to initiate the dolichol pathway of N-glycosylation and encodes a protein with 50% sequence identity to *S. cerevisiae* protein encoded by *Alg7* gene (Zou et al. 1995). The second GlcNAc substrate (Fig. 1, b) is transferred to dol-PP-GlcNAc to generate dol-PP-$GlcNAc_2$ by the activity of UDP-GlcNAc transferase (EC. 2.4.1.141). The mechanism of this enzyme was previously unclear until in early 2005 when Bickel and colleagues (2005) demonstrated that the *S. cerevisiae* gene for this step is a complex enzyme of *Alg13* and *Alg14*. Both genes are localized in the membrane although an inactive form of *Alg13* is also found in the cytosol. The first mannose (Fig. 1, c) is then transferred from GDP-mannose to dol-PP-$GlcNAc_2$ to synthesize dol-PP-$GlcNAc_2$Man, catalyzed by *Alg1*, a β-1,4-mannosyltransferase (EC. 2.4.1.142) (Albright and Robbins 1990), whose function is conserved between yeast and human (reviewed by Takahashi et al. 2000). Subsequently, the second and the third mannose units (Fig. 1, d and e) are added to dol-PP-$GlcNAc_2Man$. Presumably the gene *Alg2* catalyzes these two steps, forming dol-PP-$GlcNAc_2Man_2$ and dol-PP-$GlcNAc_2Man_3$, respectively. Firstly, an α-1,3-mannose (Fig. 1, d) linkage is created on dol-PP-$GlcNAc_2Man$ (catalyzed by EC. 2.4.1.32) (Huffaker and Robbins 1983; Samuelson et al. 2005). Then, α-1,6-mannose (Fig. 1, e) is added to dol-PP-$GlcNAc_2Man_2$, (Jackson et al. 1993; Gao et al. 2004) catalyzed by an enzyme not classified in KEGG, which we depict as EC.2.4.1.a. Although its function has been studied over the past decades, the specific role of *Alg2* is still unclear. In contrast to cell lethality of *Alg2* mutants in *S. cerevisiae*, *Alg2* mutants in zygomycete fungus *Rhizomucor pusillus* are viable, with a lower growth rate than the wild-type strain (Takeuchi et al. 1999). The elongation of dol-PP-$GlcNAc_2Man_3$ to $GlcNAc_2Man_5$ is achieved by the activity of α-1,2-mannosyltransferases (EC. 2.4.1.b and EC. 2.4.1.c) in the presence of *Alg11*. Again, *Alg11* may be required for the addition of either the fourth or the fifth mannose residue (Fig. 1, f and g) in the normal glycosylation process. The formation of dol-PP-$GlcNAc_2Man_5$ is the minimal intermediate structure required for the successful N-linked glycosylation (Cipollo et al. 2001). An *Alg11* mutant makes it possible to translocate both lipid-linked $GlcNAc_2Man_4$ and $GlcNAc_2Man_3$ to the ER lumen, probably causing the truncated oligosaccharides to be transferred to the protein. *Alg11* is markedly conserved throughout the evolution of eukaryotes: fission yeast, worms, flies and plants. It also shows sequence similarity to *Alg2* in *S. cerevisiae* and *C. elegans*. However, cells carrying the *Alg11* mutant remain deficient in cell viability, even while overexpressing *Alg2*, suggesting that these two genes may not perform overlapping functions. On the other hand, the *S. pombe gmd3* gene is a functional homologue of the *S. cerevisiae Alg11* gene, suggesting that *gmd3* is involved in the early stages of N-linked oligosaccharide biosynthesis (Umeda et al. 2000).

2.1.2 Flippase activity

The dolichol-bound $Man_5GlcNAc_2$ is flipped from the cytosolic face to the luminal face of the ER membrane with the help of a membrane protein *Rft1*. This protein is evolutionarily conserved in all eukaryotes. Depletion of *Rft1* leads to the inability of the proximal domain to be translocated across the ER membrane, resulting in a significant accumulation of the $Man_5GlcNAc_2$ precursor in the cytosol (Helenius et al. 2002). Overexpression of *Rft1* in the *Alg11* mutant alone results in a decrease in $Man_3GlcNAc_2$ in the cytosol and an increase in $Man_7GlcNAc_2$ in the ER lumen, whereas the *Alg3Alg11* double mutant results in the largest amount of lipid-bound $Man_3GlcNAc_2$ in the cytoplasm. However, these incompletely assembled lipid-linked glycan forms such as $Man_3GlcNAc_2$ and $Man_4GlcNAc_2$ could still be translocated into the ER lumen. These are then transferred to protein, indicating that the flippase activity is not restricted to the dolichol-linked $Man_5GlcNAc_2$ but also extends to other substrates (Cipollo et al. 2001).

2.2 Processes in the Endoplasmic Reticulum

In the ER lumen, the lipid-bound oligosaccharide, dol-PP-$Man_5GlcNAc_2$, is extended by the addition of four mannose (Fig. 1, h-k) and three glucose (Fig. 1, l-n) residues with the respective enzymes, mannosyltransferase and glucosyltransferase. Here, the core oligosaccharide, $Glc_3Man_9GlcNAc_2$, is completely synthesized.

Unlike sugar donors used in the initial stage of N-glycan biosynthesis in the cytosol and in the elaboration of complete glycoprotein in the Golgi, lipid sugar donors are selected to be substrates for the addition of monosaccharides onto the lipid-bound glycan in the ER lumen. Dol-P-mannose is synthesized from dol-P and GDP-mannose by the transferase enzyme, *Dpm1* (dolichol phosphate mannose synthase, EC. 2.4.1.83), while dol-P-glucose is produced from dol-P and UDP-glucose by the transferase enzyme *Alg5* (UDP-glucose:dolichyl-phosphate glucosyltransferase, EC. 2.4.1.117). Orlean (1992) demonstrated that at its non-permissive temperature, the *Dpm1* mutant exhibits blocked N-linked oligosaccharide biosynthesis. Also, this mutant bears a similar phenotype to the *Sec59* mutant, suggesting a defect in the dolichol kinase domain. These hypoglycosylated proteins might accumulate in the ER and get degraded back to the cytoplasm or are secreted, provided they have an N-linked saccharide of a minimum size.

2.2.1 Mannosylation

Four specific mannose substrates (Fig. 1, h-k) are added to dolichol-linked $Man_5GlcNAc_2$ to form dolichol-linked $Man_9GlcNAc_2$ by the action of *Alg* family members: *Alg3, Alg9, Alg12*. Aebi and coworkers (1996) presented evidence that mutations in *Alg3* reduce the activity of the transferase adding the sixth α-1,3-mannose (h) linkage to dolichol-linked $Man_5GlcNAc_2$ (EC. 2.4.1-d). *Alg3* deletion further leads to the accumulation of lipid-bound glycan $Man_5GlcNAc_2$ in the cytoplasmic face of the ER membrane and underglycosylation of secretory proteins but no growth defect. Frank and Aebi (2005) showed that *Alg9* is not only required for the addition of α-1,2 mannose (i) linkage to lipid-linked $Man_6GlcNAc_2$ (EC. 2.4.1-e) but is also needed to extend the terminal α-1,2 mannose (k) linkage to lipid-linked $Man_8GlcNAc_2$ (EC. 2.4.1-g) in the lumen of the ER. This work demonstrates the high specificity of glycosyltransferases in this pathway and their ability to augment the lipid-linked glycan by one monosaccharide unit at a time. *Alg9* deletion is analogous to the deletion of *Alg3* in producing significant accumulation of the lipid-bound oligosaccharide $Man_6GlcNAc_2$ and hypoglycosylation of secretion proteins (Burda et al. 1996). Also, the addition of α-1,3-mannose by *Alg3* is a prerequisite for the *Alg9*-dependent addition of the next mannose residue. $Man_7GlcNAc_2$ takes on the eighth mannose residue (j) due to the activity of *Alg12*, an α-1,3-mannosyltransferase (EC. 2.4.1-f), forming the lipid-linked $Man_9GlcNAc_2$. *Alg12* deletion resulted in the incomplete lipid-linked core oligosaccharide $Man_8GlcNAc_2$ with a large accumulation of dolichol-linked $Man_7GlcNAc_2$ (Burda et al. 1999).

2.2.2 Glucosylation

After the ordered transfer of nine mannoses, three glucose residues (l-n) are transferred from dol-P-glucose to lipid-linked $Man_9GlcNAc_2$, forming the complete lipid-linked core oligosaccharide, $Glc_3Man_9GlcNAc_2$. The addition of the first two glucose units (l and m) are catalyzed by the *Alg6* and *Alg8*, α-1,3-glucosyltransferases (EC. 2.4.1-h and EC. 2.4.1-j, respectively). *Alg6* was isolated with sequence similarity to *Alg8* and its mutant results in the buildup of lipid-bound

$Man_9GlcNAc_2$ in the ER lumen (Reiss et al. 1996). *Alg8* mutation blocks the addition of the second α-1,3-glucose unit. This mutant causes the transfer of improper lipid-linked $GlcMan_9GlcNAc_2$ to the protein, instead of the common oligosaccharide $Glc_3Man_9GlcNAc_2$ (Runge and Robbins 1986). The terminal glucose residue (n) is activated by the α-1,2-glucosyltransferase, *Alg10* (EC. 2.4.1-j), and serves as an important signal for cotranslational glycosylation as it is specifically recognized by the protein across the membrane (Burda and Aebi, 1998). The normal core oligosaccharide contains two GlcNAc units, nine mannose units, and three glucose units, $Glc_3Man_9GlcNAc_2$. This structure is highly conserved in lower and higher eukaryotic cells, with the exception in *P. falciparum* reporting a shortage of N-glycosylation (Davidson and Gowda 2001). In contrast to yeast and mammalian cells, trypanosomatids in plants and insects catalyze the transfer of monosaccharides containing fewer mannose residues in the final structure and lacking the terminal glucose residue, essential for efficient recognition by the oligosaccharyltransferase (Dempski and Imperiali 2002).

2.2.3 Oligosaccharyltransferase activity

Following glucosylation, the core oligosaccharide is transferred from the dol-P anchor to the conserved sequence motif of Asn-X-Ser/Thr in the nascent protein to initiate N-glycoprotein formation, in a cotranslational modification process. This step needs the terminal glucose residue (n) as a signal that can be recognized by oligosaccharyltransferase (OT), the specific enzyme localized in the luminal ER that transfers the finished oligosaccharide from the lipid-bound precursor to the polypeptide. OTs are membrane-associated enzyme complexes, many of which are remarkably homologous from protozoan to mammalian species. Silberstein and Gilmore (1996) showed that incomplete core oligosaccharides lacking the terminal glucose unit cannot participate in protein targeting due to lack of recognition by OT, leading to misfolding and hypoglycosylation. The OT complex in budding yeast *S. cerevisiae* contains nine different protein subunits; Nlt1p/Ost1p, Ost2p, Stt3p, Swp1p, Wbp1p, Ost3p, Ost4p, Ost5p, and Ost6p (Zubkov et al. 2004). The first five proteins are essential for cell growth, while deletions of the remaining non-essential genes lead to viable cells with inefficient glycosylation. The essential genes have significant homologies with the protozoan and higher eukaryotic counterparts. Ost4p has recently been matched with high sequence similarity to orthologues in worm, mouse, and humans (Dempski and Imperiali 2002).

2.2.4 Glycosidase and mannosidase activity

Following its transfer to the polypeptide, the oligosaccharide structure is trimmed to $Man_8GlcNAc_2$, an initial precursor for the elongation of glycoprotein in the Golgi complex. The specific enzymes used in this process are glucosidase I, glucosidase II and mannosidase I. Glucosidase I (EC. 3.2.1.106), a membrane-bound protein, removes the outmost α-1,2-glucose residue (n) to produce protein-containing $Glc_2Man_9GlcNAc_2$. Glucosidase II (EC. 3.2.1.84), an ER-residing membrane protein, removes the other two α-1,3-glucose moieties (m and l) to form polypeptide-linked $Man_9GlcNAc_2$. Glucosidase inhibition blocks the natural course of N-linked glycoprotein modification such that cells were able to produce a variety of the glycan structures including high mannose, hybrid, and complex glycans (Elbein 1991).

In yeast *S. cerevisiae*, a single mannose residue (k) is removed from $Man_9GlcNAc_2$ to $Man_8GlcNAc_2$ by mannosidase I, *Mns1* (3.2.1.113) in the ER, before being conveyed to the Golgi via transport vescicles to be converted to high mannose glycans (Jelinek-Kelly and Herscovics 1988). Overexpression of *Mns1* resulted in an increase of specific α-1,2-mannosidase activity, while disruption resulted abolition of enzyme activity with no effect on cell growth (Camirand et al. 1991). The properties and specificity of human α-1,2-mannosidase are identical to the *S. cerevisiae* homologue resident in the ER (Tremblay and Herscovics 1999). However, these ER enzymes differ from the previously cloned α-1,2-mannosidases found in the Golgi that remove four mannose moieties from $Man_9GlcNAc_2$ from the N-linked glycoproteins. In *S. pombe*, the absence of mannosidase I causes $Man_9GlcNAc_2$ to transfer directly to the Golgi complex, for elongation of the final glycoprotein by the addition of mannose and galactose residues to form "galactomannan" (Ziegler et al. 1994). Protein-linked oligosaccharide $Man_8GlcNAc_2$ (or $Man_9GlcNAc_2$ in *S. pombe*) is transported to the Golgi complex by *Sec18* in yeast, human and plants. The arrival of N-linked glycoproteins sets in motion the post-translational modification processes by Golgi glycosyltransferases to form a diversity of N-linked oligosaccharides with high mannose, hybrid and complex glycans, depending on the activity of glycosyltransferase enzymes in that particular organism.

2.3 Processes in the Golgi

While N-glycan biosynthesis of the core oligosaccharide and the initial stages of the glycoprotein formation in the ER are similar in yeast and most other eukaryotes, the processing in the Golgi apparatus is significantly diverse leading to different final glycoproteins. Figure 2 shows the processing in the Golgi complex for selected organisms. In budding yeast (*S. cerevisiae*, Figure 2a), the final glycoprotein formed is the highest mannose-containing glycoprotein, "hypermannan", using α-1,2-, α-1,3-, and α-1,6-mannosyltransferases as well as mannosylphosphate transferases (Herscovics and Orlean 1993). In fission yeast (*S. pombe*, Figure 2b), the N-linked oligosaccharides contain large amounts of galactose in addition to mannose known as "galactomannan" (Ziegler et al. 1994). In human (Figure 2c), the protein glycosylation is post-translationally modified to three types of glycan-linked proteins; high-mannose, hybrid, and complex glycans (Wildt and Gerngross 2005). In plants (Figure 2c), N-linked glycosylation is classified into four types; high mannose, hybrid, paucimannosidic and complex glycans, with markedly different final structures (Lerouge et al 1998; Chen et al. 2005).

2.3.1 In budding yeast *S. cerevisiae*

N-linked glycoprotein processing in budding yeast *S. cerevisiae* finally undergoes post-translational modification in the Golgi apparatus (Figure 2a). Unlike in human and plants, the final oligosaccharide structure in yeast is glycosylated to maturation and is present as the highest mannose glycan containing up to 100 mannose residues called "hypermannan". The protein-bound $Man_8GlcNAc_2$ is transported from the ER lumen to the *cis*-Golgi by *Sec18*. In the Golgi compartment, α-1,6-mannose is linked

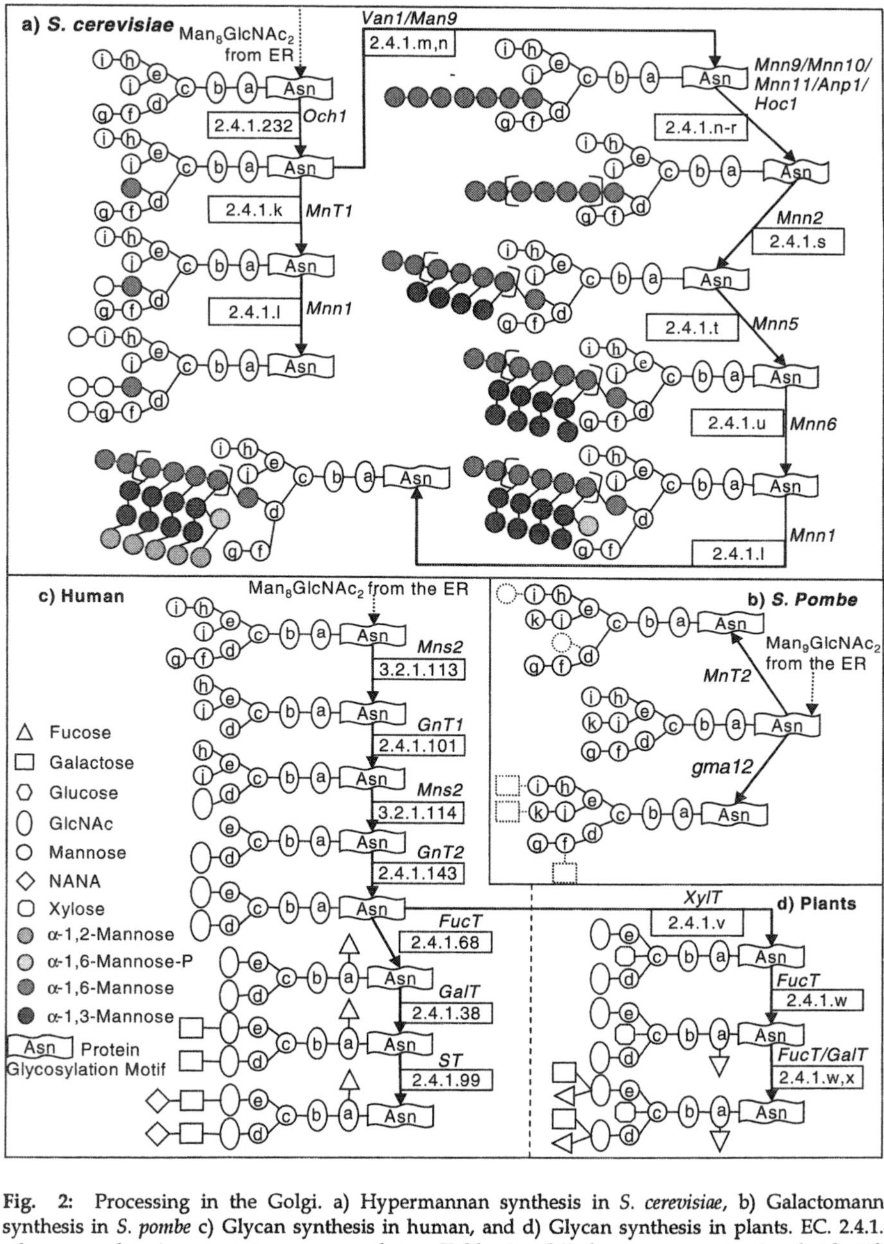

Fig. 2: Processing in the Golgi. a) Hypermannan synthesis in *S. cerevisiae*, b) Galactomannan synthesis in *S. pombe* c) Glycan synthesis in human, and d) Glycan synthesis in plants. EC. 2.4.1.-x, where x = a, b, ...j, represents our nomenclature (Tables 1 and 2), for enzymes not uniquely identified by KEGG. Abbreviations: Fuc, fucose; *FucT*, fucosyltransferase; *GalT*, galactosyltransferase; GlcNAc, N-acetylglucosamine; GnT1, N-acetylglucosaminyl transferase 1, *GnT2*, N-acetylglucosaminyl transferase 2; Man, mannose; *MnT1*, mannosyltransferase 1; *MnT2*, mannosyl transferase 2; *Mnn5*, mannosyltransferase; *Mnn6*, mannosyltransferase; *Mnn9*, mannosyltransferase; *Mnn10*, mannosyltransferase; *Mnn11*, mannosyltransferase; *Mns2*, mannosidase 2; NANA, N-acetylneuraminic acid; ST, sialyltransferase; Xyl, xylose; *XylT*, xylosyltransferase.

Table 2. Glycosylation Genes in the Golgi compartments.

Gene	EC number*	Sugar Donor	Product#	Function	Organism SC	SP	P	H
Och1	2.4.1.232	GDP-Mannose	$RMan_9$	α1,6-mannosyltransferase	+	-	-	-
MnT2	2.4.1.-k*	GDP-Mannose	$RMan_{10}$	α1,2-mannosyltransferase	+	+	-	-
Mnn1	2.4.1.-l*	GDP-Mannose	$RMan_{13}$	α1,3-mannosyltransferase	+	-	-	-
Van1	2.4.1.-m*	GDP-Mannose	$RMan_{15}$	Complex mannosyltransferase	+	-	-	-
Mnn9	2.4.1.-n*	GDP-Mannose	$RMan_{15\sim n}$	Complex mannosyltransferase	+	-	-	-
Mnn10	2.4.1.-o*	GDP-Mannose	$RMan_{15\sim n}$	Complex mannosyltransferase	+	-	-	-
Mnn11	2.4.1.-p*	GDP-Mannose	$RMan_{15\sim n}$	Complex mannosyltransferase	+	-	-	-
Anp1	2.4.1.-q*	GDP-Mannose	$RMan_{15\sim n}$	Complex mannosyltransferase	+	-	-	-
Hoc1	2.4.1.-r*	GDP-Mannose	$RMan_{15\sim n}$	Complex mannosyltransferase	+	-	-	-
Mnn2	2.4.1.-s*	GDP-Mannose	$RMan_{15\sim n}$	α1,2-mannosyltransferase	+	-	-	-
Mnn5	2.4.1.-t*	GDP-Mannose	$RMan_{15\sim n}$	α1,2-mannosyltransferase	+	-	-	-
Mnn6	2.4.1.-u*	GDP-Mannose	$RMan_{15\sim 100}$	Mannosylphosphate transferase	+	-	-	-
GnT1	2.4.1.101	UDP-GlcNAc	$RMan_5GlcNAc$	N-acetyl-glucosaminyl-transferase I	-	-	-	+
Mns2	3.2.1.114	-	$RMan_3GlcNAc$	Mannosidase II	-	-	-	+
GnT2	2.4.1.143	UDP-GlcNAc	$RMan_5GlcNAc_2$	N-acetyl-glucosaminyl-transferase II	-	-	-	+
FucT	2.4.1.68	GDP-Fucose	$RMan_5GlcNAc_2Fuc_2$	Fucoslytransferase	-	-	+	+
GalT	2.4.1.38	UDP-Galactose	$RMan_5GlcNAc_2 Fuc_2 Gal_2$	Galactosyltransferase	-	+	+	+
ST	2.4.1.99	CMP-NANA	$RMan_5GlcNAc_2 Fuc_2 Gal_2$	Sialyltransferase	-	-	-	+
XylT	2.4.1.-v*	GDP-Xylose	$RMan_5GlcNAc_2 Xyl$	Xylosyltransferase	-	-	+	+

* EC. 2.4.1.-x, where x = a, b, ...j, represents our nomenclature for enzymes not uniquely identified by KEGG. This nomenclature has also been used in Figure 1 and 2.

\# R represents the protein-linked oligosaccharide $AsnGlcNAc_2$ moiety.

to α-1,3-mannose (e) in the tri-core glycan (c, d, e) by a mannosyltransferase (EC. 2.4.1.232) encoded by *Och1*. This transfer results in the formation of $Man_9GlcNAc_2$. It is currently unclear why the cellular machinery removes a mannose moiety prior to transport to the Golgi and then adds it back to the glycan chain. Several research efforts indicate that α-1,6-mannose may be a key precursor to initiate the polymerase outer chain elongation of N-linked glycans, but this specificity has not been demonstrated for *Och1*. The maturation of glycans, reviewed by Herscovics and Orlean (1993), involves two possible routes based on its final glycan structure.

Each route requires the *Mnn* family with or without the activity of *Mnt1/Kre2* (EC. 2.4.1.k). The maturation of $Man_{8-13}GlcNAc_2$ has been identified in the route containing *Mnt1/Kre2* protein. *Mnt1/Kre2* protein catalyzes the addition reaction of α-1,2-mannose to the outer chain α-1,6-mannose. Then, *Mnn1* (EC. 2.4.1.l) adds α-1,3-mannose to α-1,2-mannose in the core glycan structure. The final product being $Man_{8-13}GlcNAc_2$. $Man_{8-100}GlcNAc_2$ is confined to the route involving only the *Mnn* family, *Mnn1* to *Mnn11*. A series of α-1,6-mannoses are added to the outer chain by the complex proteins *Van1* (EC. 2.4.1.m) *Mnn9* (EC. 2.4.1.n), *Mnn10* (EC. 2.4.1.o), *Mnn11* (EC. 2.4.1.p), *Anp1* (EC. 2.4.1.q), and *Hoc1* (EC. 2.4.1.r). After that α-1,2-mannose units are added to the outer branch as α-1,6-mannose-α-1,2-mannose and α-1,2-mannose-α-1,2-mannose by *Mnn2* (EC. 2.4.1.s) and *Mnn5* (EC. 2.4.1.t), respectively. Then, *Mnn6* mannosylphosphotransferase (EC. 2.4.1.u) adds α-1,6-mannosephosphate onto the α-1,2-mannose as a prerequisite for the elongation of α-1,3-mannoses by the action of *Mnn1*. The production of "hypermannan" is a unique to the *S. cerevisiae* N-glycosylation pathway.

2.3.2 In fission yeast *S. pombe*

In fission yeast (*S. pombe,* Figure 2b), the steps in core oligosaccharide modification to the final glycan structures are quite different from those occurring in budding yeast *S. cerevisiae,* although the ordered assembly of N-linked oligosaccharides is remarkably similar. *S. pombe* glycosylation processing has been reported in the absence of mannosidase in the ER, an enzyme that trims the $Man_9GlcNAc_2$ to $Man_8GlcNAc_2$, making it possible to elongate the lipid-linked oligosaccharide $Man_9GlcNAc_2$ with both mannose and galactose residues to uniquely generate "galactomannan" (Ziegler et al. 1994). However, the structures of processing intermediates need to be identified and the activity of sugar transferases remain uncharacterized. While mannosylation is carried out by *Mnt2,* galactosylation in *S. pombe* is effected by *gmd12* (Ziegler et al. 1999). Huang and Snider (1995) showed that four mutations result in defective terminal oligosaccharide modification in *S. pombe. Gps1* mutation affected Golgi mannosyltransferase, with decreased amounts of mannose in the modified core glycan, but no change in the galactose content. On the other hand, *gps2* mutation causes a reduced amount of galactose due to the decreased activity of UDP-glucose-4-epimerase synthesizing the nucleotide-sugar donor, UDP-galactose. In addition, *gps1* and *gps2* mutants have no adverse effect on cell viability; albeit with a reduced growth rate. Ziegler and coworkers (1999) isolated novel *S. pombe* N-linked $GalMan_9GlcNAc$ isomers by gel filtration from the endo H-released N-glycans. This work showed the range of the monosaccharide composition and the structural linkages of mannose and galactose to the *S. pombe* core oligosaccharide. Four major isomers were discussed. Firstly, we have a $Man_{10}GlcNAc$ with a new α-1,2 linked mannose to the upper arm of $Man_9GlcNAc$. Secondly, there is a $GalMan_9GlcNAc$ with a terminal galactose added either to the upper or the middle arm of $Man_9GlcNAc$. Thirdly, $Man_{10}GlcNAc$ has a new α-1,6 linked mannose onto the lower arm of α-1,3 linked core residue, found previously in *S. cerevisiae* and *P. pastoris.* The last is a $Man_9GlcNAc$ with the low molar stoichiometry of both galactose

(GalMan$_9$GlcNAc) and glucose (Glc$_{1\text{-}2}$Man$_9$GlcNAc). This study paved the way for systematic understanding of the modifications in the processing of N-linked galactomannan in *S. pombe.*

2.3.3 In human

In contrast to yeast N-glycan biosynthesis, which is limited to the addition of mannose and mannosylphosphate sugars in the early stage of N-glycan processing, N-linked glycoprotein in human and other mammals occurring in the Golgi is a post-translational modification of three major types: high mannose, hybrid, and complex glycans (Figure 2c) (Wildt and Gerngross 2005). Early stages of N-glycosylation taking place in the *cis*-Golgi, involve the trimming of protein-containing oligosaccharides Man$_8$GlcNAc$_2$ to Man$_5$GlcNAc$_2$, catalyzed by α-1,2-mannosidase (EC. 3.2.1.113). This enzyme belongs to the mannosidase I enzyme family of membrane proteins localized both in the ER (late stage) and in the Golgi (early stage) and affects the removal of three mannose residues (i, g, f) outward. The high mannose oligosaccharide is a substrate for the linkage of a GlcNAc moiety onto the terminal α-1,3-mannose (e) of the tri-mannose core (c, d, e) by the action of N-acetylglucosaminyltransferase (GnT I; EC. 2.4.1.101). The structure of the tri-mannose core with two GlcNAc residues (a-e) is highly conserved among species as the minimal structure for post-translational modification in N-glycan processing of secretory and membrane proteins. The resulting glycan has the structure, GlcNAcMan$_5$GlcNAc$_2$. Two remaining terminal mannose residues, which are α-1,3-(h) and α-1,6-mannose (j) moieties linked to the tri-mannose core (c, d, and e), are then removed by mannosidase II (EC. 3.2.1.114) to the initial hybrid glycan GlcNAcMan$_3$GlcNAc$_2$. Following this trimming, N-acetylglucosaminyltransferase II (GnT II; EC. 2.4.1.143) adds a GlcNAc sugar to the terminal α-1,6-mannose arm of the tri-mannose core, consequential in the formation of GlcNAc$_2$Man$_3$GlcNAc$_2$, the first complex glycan structure. Further post-translational modification plays a fundamental role in producing typically complex glycan structures by the attachment of the additional sugar moieties; fucose, galactose, and N-acetylneuraminic acid (NANA). Fucosyltransferase (EC. 2.4.1.68) catalyzes the addition of α-1,6-fucose to proximal GlcNAc residue (a) which lays near the protein backbone. Two galactose units are added to generate Gal$_2$FucGlcNAc$_2$Man$_3$GlcNAc$_2$ by using β-1,4-galactosyl-transferases (EC. 2.4.1.38). In the final step, the Golgi type II transmembrane protein, sialyltransferase (EC. 2.4.1.99), terminates the reaction by adding α-2,6-NANA residues to produce NANA$_2$Gal$_2$FucGlcNAc$_2$Man$_3$GlcNAc$_2$. This structure is known as the Lewis antigen usually found on mammalian cell surface glycoconjugates. The final structures of glycoproteins vary from high mannose to complex glycans and terminally sialyated glycans. The sialyated glycans are unique structures in the human N-glycosylation biosynthesis. Several efforts to engineer yeast strains to produce a high amount of these glycoproteins using heterologous gene expression systems have met with limited success, due to the cascade of enzymes required for generating these varied structures. The relationship between particular glycosylation structures and their biological activity has been investigated over the past decade and much remains to be discovered in this area of glycoprotein research.

2.3.4 In plants

Plants show very similar steps in glycan synthesis compared to human and other mammalian species upto fucosylation. At this branch point, plants uniquely undergo xylosylation, followed by fucose and galactose addition. Plants also lack sialyltransferase, which catalyzes the terminal step in N-glycan synthesis in humans.

The difference in the final steps of glycan trimming and its modification is detected in the Golgi complex, even though the N-linked oligosaccharide processing is highly evolutionarily conserved from yeast to mammalian and plant (Lerouge et al. 1998; Lerouge et al. 2000; Maia and Leite 2001; Ma et al. 2003; Chen et al. 2005). In the early stages of Golgi processing, high mannose and hybrid glycans are formed in the *cis-* and *middle*-Golgi apparatus, respectively. The structures of these glycoproteins are homologous to those processed in other higher organisms, such as human. The core structure, which is a substrate for the further modification in the *trans*-Golgi, is $GlcNAcMan_3GlcNAc_2$ and the first complex glycan is the transfer of a GlcNAc residue to the α-1,6-mannose branch (d) in the core structure to form $GlcNAc_2Man_3GlcNAc_2$ (Figure 2d). In addition another specific structure of N-linked glycoprotein was proposed as paucimannosidic type N-linked glycan, which results from the elimination of terminal residues from the complex glycans and directs the protein to the vacuole compartment as its final destination. Despite the conservation of the intial steps in N-glycan biosynthesis, the complex N-linked oligosaccharides occurring in the *trans*-Golgi of plants are distinctive in their final patterns compared to those processed in the *late*-Golgi and *trans*-Golgi network found in the mammalian and other eukaryotic cells.

In plants, the complex N-glycan biosynthesis and mechanism is posttranslationally modified by the activity of xylosyltransferase (EC. 2.4.1.v) adding β-1,2-xylose to the β-1,4-mannose unit (c) in the tri-core mannose (c, d, and e), followed by the action of fucosyltransferase (EC. 2.4.1.w) adding α-1,3-fucose to the proximal GlcNAc unit (a) located next to the polypeptide backbone, resulting in the formation of $FucXylGlcNAc_2Man_3GlcNAc_2$. These two steps occur in the *medial* and *trans*-Golgi, respectively. The former is absent in mammals although present in invertebrates (Ma et al. 2003). The latter has a distinctive α-1,3-fucose linkage, instead of α-1,6-fucose in mammalian and animal cells (Lerouge et al. 1998). The xylosylation and fucosylation of the common core $Man_3GlcNAc_2$, with GlcNAc as a prerequisite for substrate specificity, may contribute to yield plant specific N-linked glycans. At this stage, the final complex glycan in plants is due to the addition of β-1,3-galactose and α-1,4-fucose molecules at the terminal GlcNAc residue by galactosyltransferases (EC. 2.4.1.x) and fucosyltransferases (EC. 2.4.1.w) to generate $Gal_2Fuc_2XylGlcNAc_2Man_3GlcNAc_2$. The final structure of complex plant N-glycan is characterized by the absence of terminal sialic acids, which mark human and other mammalian complex glycans.

However, most of enzymes involved in the N-glycan processing are still unknown and the regulation of its mechanism at each step is also unclear. To gain a better understanding of the N-glycosylated protein biosynthesis and mechanism in plants, a genomic approach along with bioinformatics tools may contribute to build the glycosyltransferase sequence database. The first attempt to gain insight into N-linked oligosaccharide biosynthesis in a plant species has successfully been done by Maia and Leite (2001), by mining the Sugarcane Expressed Sequence Tag (SUCEST) database for sugarcane gene products involved in the N-glycosylation pathway. In this landmark study, 90 sugarcane expressed sequence tag (EST) entries in the SUCEST database, which share significant homology with the enzymes involved in the N-glycosylation process, have been identified and assigned their functions. The hitherto unknown novel genes introduced from this analysis are described. For example, although ER mannosidase had not been detected earlier (Lerouge et al. 1998), it was successfully identified in this study and the alignment of amino acid sequences of ER-resident mannosidase provided the homologue between sugarcane and mammals. Also, detailed analysis further revealed significant sequence similarity to β-1,4-N-acetylglucosaminyltransferase and α-1,4-fucosyltransferase, whose homologues were not found by routine BLAST searches. However, there are many genes in the N-glycan biosynthesis pathway that are still to be identified and characterized and thus remain a hot area of current research.

3. DEFECTIVE MUTANTS

Several studies have identified and characterized mutations of the *Alg* family in the glycosylation pathway. Yeast mutations that interfere with the synthesis of sugar donors or the assembly of oligosaccharide in the early steps occurring in the cytosolic face of the ER membrane are essential for cell viability whereas the later steps are not. Essential genes include *Alg1* (Albright and Robbins 1990; Takahashi et al. 2000), *Alg2* (Huffaker and Robbins 1983; Jackson et al. 1993; Takeuchi et al. 1999; Gao et al. 2004; Samuelson et al. 2005), *Alg7* (Zou et al. 1995), *Alg11* (Umeda et al. 2000; Cipollo et al. 2001) and *Sec59* (Hashimoto et al. 2005). In contrast, the mutations in the later step occurring in the luminal side of the ER membrane accumulate lipid-linked core oligosaccharides in which five or more mannose residues have been attached, causing hypoglycosylation of secretory proteins, but no growth effect. The non-essential genes involve *Alg3* (Aebii et al. 1996), *Alg6* (Reiss et al. 1996), *Alg8* (Runge and Robbins 1986), *Alg9* (Burda et al. 1996; Frank and Aebi 2005), *Alg10* (Burda and Aebi 1998), *Alg12* (Burda et al. 1999) as well as *Dmp1* (Helenius et al. 2002). Furthermore, several proteins remain unglycosylated due to a decrease in the efficiency of the oligosaccharyltransferase towards incompletely assembled lipid-bound oligosaccharides. However, they are still inefficiently transferred to protein. The resulting hypoglycosylation is the difference in the final structure of glycoproteins and its structure in normal conditions and this factor may be employed to distinguish glycosylation processing in particular organisms, under different environmental conditions. The mutations leading to improper core oligosaccharide may affect the folding process of proteins and thus lead to varied localization of secretion products. In addition the functions of the *Alg* genes appear to be conserved among all eukaryotes; therefore, it is likely that these genes play a similar role in the mammalian cell.

In humans, mutations to *Alg* genes cause diseases termed the congenital disorder of glycosylation (CDG) with defects in the assembly of dolichol-linked glycan or its transfer to proteins. There are two types of CDG. Type I CDG is associated with the mutations that affect the assembly of the core oligosaccharide, containing 14 sugars, on the lipid carrier dolichol phosphate. Type II CDG is the mutations in proteins that affect N-linked glycosylation. For examples, the mutations of *Alg7* gene blocking the enzyme activity in the early step of glycan synthesis cause a novel CDG type Ij (Xu et al. 2003) whereas the mutations of *Alg1* gene with a deficiency of the first step of adding a mannose residue to the lipid-linked glycan cause a novel CDG type Ik (Schwarz et al. 2004; Grubenmann et al. 2004). The mutations of *Alg6* gene blocking the addition of the first glucose residue to the glycan structure cause CDG type Ia and Ic (Westphal et al. 2002).

4. THE DIVERSITY OF N-GLYCAN STRUCTURES

There are several particular and specific reasons that cause the final complex glycans in each eukaryotic cell to be totally different from each other. Three questions mentioned in the early section of this review are significant in the explanation of the diversity of complex glycan structures. What are the key factors that distinguish secreted proteins among species? What are specific roles for glycoproteins among species? Which genes in the glycosylation pathway are conserved and which one are species-specific? Two of the major roles for glycoproteins are protein folding in the ER lumen and protein localization in the secretory pathway. The protein conformation is typically regulated by the quality control mechanism in the ER. Proteins that do not attain their native conformations are not transfereed to the Golgi. These misfolded proteins are retained in the ER by the quality control mechanism known as the unfolded protein response (UPR), detecting protein misfolding and inducing the transcription of chaperones and other genes involved in the ER to achieve the native fold or get degraded to the cytosol via the endoplasmic reticulum associated protein degradation process, ERAD (reviewed in Parodi 2000; Helenius and Aebi 2004). Thus glycosylation affects the native conformation and stability of the protein, and further determines its ultimate destination *via* protein trafficking signals. The specific localization of each protein is also important to the secretion in the secretory pathway: ER, cis-, middle-, and trans-Golgi. Improper and incomplete glycans will be transported to the different organelles such as lysosome, vacuole, cytosol, plasma membrane, and external medium via the secretory pathway depending on the particular glycan structures recognized by specific signal in each compartment. Therefore, the controlling process of proper protein folding and the prediction of precise protein localization are primarily important approaches in the synthesis of the diversity of complex glycan in the N-linked glycoprotein either in the same organism or in the different organism.

The structures of the core oligosaccharide conserved in all eukaryotes and the extensions of glycoproteins unique to particular organisms, are shown in Figure 3. These core oligosaccharides are generated by the stepwise reactions elaborated previously and represented in Figures 1 and 2. Two conserved patterns are distinguished. The first pattern is the evolutionarily conserved core oligosaccharide,

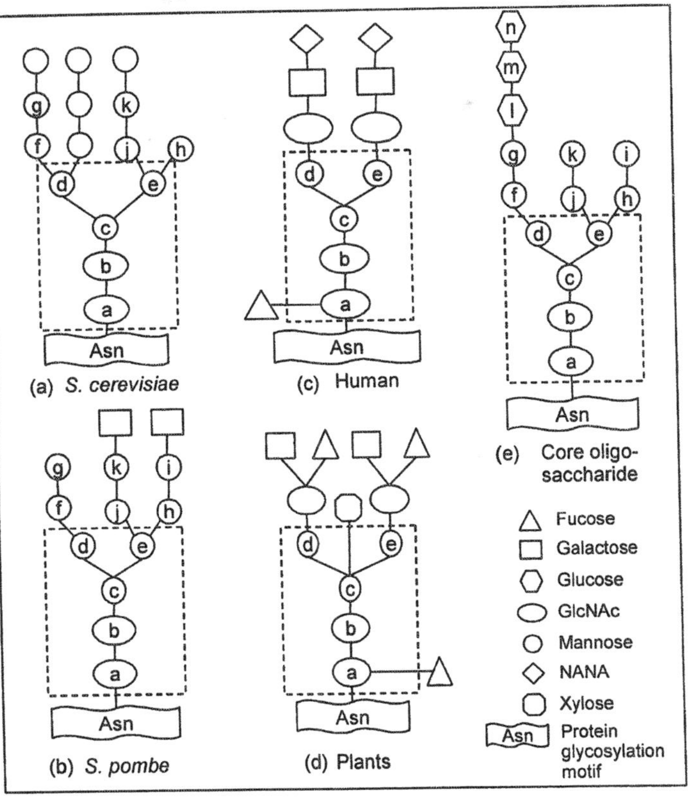

Fig. 3: Glycoprotein diversity in different organisms: (a) hypermannan in *S. cerevisiae*, (b) halactomannan in *S. pombe*, (c) complex glycan in human, (d) complex glycan in plants and (e) the core oligosaccharide. The common five-sugar structure for outer chain elongation in all eukaryotes, containing the mannose tri-core and two GlcNAc residues, is shown in dotted lines.

$Glc_3Man_9GlcNAc_2$ (Figure 3 e), in the co-translational modification processing in the ER. The core oligosaccharide contains fourteen monosaccharide residues comprising two GlcNAc units, nine mannose units, and three glucose units. The synthesis of the core oligosaccharide is by the orderly assembly in the N-glycan biosynthetic process and substrate specificity is required for the individual enzyme involved in each step of the pathway. The second discernable pattern is the evolutionarily conserved tri-core mannose with two GlcNAc, which are located immediately adjacent to the protein, $Man_3GlcNAc_2$, in the post-translational modification processing in the Golgi complex. These five sugar residues are the minimal structure for the different elongation of the glycan outer chain of N-linked glycoprotein, relying on the specific enzymes involved in particular organism. The conserved structure has been preserved in all eukaryotes shown here (dotted lines in Figure 3 a-d). The species-specific enzymes indicate the specific roles for glycoproteins in each organism, causing the diversity of complex glycan structures. In yeast *S. cerevisiae*, the glycoprotein formed is known as "hypermannan" (Figure 3 a), whereas "galactomannan" (Figure 3 b) has been produced in the yeast *S. pombe*. Even though the final structures are different in the two yeasts, they still share commonalities compared with the final glycans in the higher eukaryotes which are strikingly different. In human and plants, the final complex structures are subtly different, both in the linkage position and in the sugar moieties (Figure 3 c-d). Xylose is a specific substrate that is transferred to the glycan of N-glycoprotein by a specific enzyme xylosyltransferase, making the sugar branch of outer chain in plant glycosylation different from the human glycan. Sialic acid is a unique substrate, catalyzed by sialyltransferase, specific to human (and other mammalian) glycoproteins. This substrate is a key for the humanization of N-linked glycoproteins using heterologous gene expression system.

5. ENGINEERING THE GLYCOSYLATION PATHWAY

Engineering the N-glycosylation pathway needs an in-depth understanding of enzymatic activity involved in each step of the pathway. From our current knowledge of metabolic pathways coupled with recombination DNA technology, we will be able to engineer yeast strains for heterologous gene expression systems, resulting in correctly glycosylated protein products of therapeutic and commercial value. Many organisms have been studied as alternative hosts for the expression of foreign proteins. In the humanization of N-glycosylation pathway, the synthesis of $Man_5GlcNAc_2$ is a main key for elongation of specific monosaccharides into the human N-glycoprotein, especially the generation of sialyated glycans. Therefore, many researchers have been attempted to engineer selectively specific strains to introduce the formation of $Man_5GlcNAc_2$. Two examples of successful efforts at the humanization of the N-glycan pathway are presented here.

The humanization of N-glycoprotein pathway in yeast *S. cerevisiae* has been studied with limited success due to its inability to modify proteins in human-like N-glycosylation reactions (Wildt and Gerngross 2005). Jigami and coworkers (Nakanishi-Shindo et al. 1993) reported the first successful attempt to eliminate yeast-specific N-glycan processing and to provide a human intermediate of N-glycan biosynthesis. Their approach to humanize glycoproteins in *S. cerevisiae* focussed on the elimination of yeast-specific hypermannosylation by abolishing the activity of α-1,6-mannosyltransferase *Och1*, which initiates the synthesis of outer chain in yeast glycan and of 1,3-mannosyltransferase *Mnn1*. The *Och1 Mnn1* double deletion mutant was able to produce predominantly a single ER-form oligosaccharide species ($Man_8GlcNAc_2$). This achievement suggested a potential use of this strain as a host cell to produce glycans containing mammalian high mannose type oligosaccharides. Furthermore, the triple deletion *Och1 Mnn1 Alg3* mutant accumulated $Man_5GlcNAc_2$ and $Man_8GlcNAc_2$ in total cell mannoprotein, confirming the lack of outer chain addition to the incomplete core-like oligosaccharide and the leaky phenotype of the *Alg3* mutation.

The optimization of these gene deletion efforts was achieved by Wildt and coworkers (Bobrowicz et al. 2004) in the yeast *Pichia pastoris*. A double mutant of *P. pastoris, alg3 and och1,* resulted in the complete blockage of dolichol-linked oligosachharide synthesis, resulting in unnatural $Man_5GlcNAc_2$ structures that only contain 1,2-mannoses attached to the tri-mannose core. The key point to note here is that rather than aim for complete humanization, these researchers arrived at an optimum intermediate structure, for further enhancement by introduced genes. The introduction of active α-1,2-mannosidase, 1,2-N-acetylglucosaminyltransferase I, 1,2-N-acetylglucosaminyltransferase II and a UDP-dependent N-acetylglucos-amine transporter resulted in the production of uniform complex glycoproteins with a terminal N-acetylglucosamine group.

Taken together, these examples demonstrate that the artificial *in vivo* glycoengineering of yeast represents a major advance in the production of glycoproteins and will emerge as a practical tool to systematically elucidate the structure-function relationship of N-glycans. This research is the starting point for glycoengineering for humanization of glycosylation pathway in yeast and paves the way to engineered strains as heterologous expression systems for therapeutic and biotechnological applications.

6. CONCLUSION

The N-glycosylation pathway has been conserved evolutionarily across species, including yeast, protozoa, plants, animals and human. The exceptional absence of N-linked glycoprotein is found only in the malarial parasite, *P. falciparum*. N-linked protein glycosylation in most eukaryotic cells is initiated with the transfer of the oligosaccharide from the lipid carrier dolichol phosphate to selected asparagine residues, occurring in the cytosolic and luminal face of the ER membrane. The assembly of core oligosaccharide is synthesized in a stepwise reaction scheme, suggesting that the biosynthesis and mechanism of common oligosaccharides in the ER stipulate substrate specificity for each individual catalytic enzyme. The genes in the early steps of N-linked glycan biosynthesis pathway have been reported critical to cell viability, whereas, those in the later stages of the synthesis have no effect on cellular growth, but cause cells to secrete hypoglycosylated forms of proteins. *Alg* mutations, for instance, affect the assembly of the lipid-linked oligosaccharide at the membrane of the ER, resulting in the accumulation of lipid-linked intermediates and hypoglycosylation of proteins. Although the ordered assembly of core oligosaccharide in the ER compartment is remarkably conserved in lower and higher eukaryotes, the processing of post-translational modification of its glycan structure to generate the final complex glycoproteins in the Golgi apparatus is subtly different. In this review we have compared the N-glyoprotein biosynthesis in two species of yeast, plants and humans. The highest mannose structure is usually found in budding yeast *S. cerevisiae*, while the galactomannans containing both mannose and galactose residues has been specifically discovered in fission yeast *S. pombe*. Unlike N-linked glycoproteins found in yeasts, a striking variety of structures of the complex glycans have been detected in plants and human, albeit distinctive in their compositions and linked structures. To date, genomic approaches and recombinant DNA technologies provide us a very good opportunity to identify and predict the gene functions. These are rapid fundamental approaches for providing the novel directions to biologists who are working on gene and protein production. A deep understanding of the metabolic pathway of N-linked oligosaccharide biosynthesis coupled with those techniques would facilitate the development of novel microbial strains as cell factories for the production of therapeutic and biopharmaceutical glycoproteins of human and mammalian origin. Until now, the humanization of the glycosylation pathway is limited by the combinatorial task of gene deletion experiments, although many enzymatic reactions and metabolic steps have been identified and characterized. With the availability of genome sequences for several fungi, the identification of secretion blocks for heterologous protein production by comparative genome analysis.

REFERENCES

Aebi M, Gassenhuber J, Domdey H, and te Heesen S (1996). Cloning and characterization of the *ALG3* gene of *Saccharomyces cerevisiae*. Glycobiology 6:439-444.

Albright CF, and Robbins PW (1990).The sequence and transcript heterogeneity of the yeast gene *ALG1*, an essential mannosyltransferase involved in N-glycosylation. J Biol Chem 265:7042-7049.

Bickel T, Lehle L, Schwarz M, Aebi M, and Jakob CA (2005). Biosynthesis of lipid-linked oligosaccharides in Saccharomyces cerevisiae: *Alg13p* and *Alg14p* form a complex required for the formation of $GlcNAc_2$-PP-dolichol. J Biol Chem 280:34500-34506.

Bobrowicz P, Davidson RC, Li H, Potgieter TI, Nett JH, Hamilton SR, Stadheim TA, Miele RG, Bobrowicz B, Mitchell T, Rausch S, Renfer E, and Wildt S (2004). Engineering of an artificial glycosylation pathway blocked in core oligosaccharide assembly in the yeast *Pichia pastoris*: production of complex humanized glycoproteins with terminal galactose. Glycobiology 14:757-766.

Burda P, te Heesen S, Brachat A, Wach A, Dusterhoft A, and Aebi M (1996). Stepwise assembly of the lipid-linked oligosaccharide in the endoplasmic reticulum of *Saccharomyces cerevisiae:* identification of the *ALG9* gene encoding a putative mannosyl transferase. Proc Natl Acad Sci USA 93:7160-7165.

Burda P, and Aebi M (1998). The *ALG10* locus of *Saccharomyces cerevisiae* encodes the 1,2 glucosyltransferase of the endoplasmic reticulum: the terminal glucose of the lipid-linked oligosaccharide is required for efficient N-linked glycosylation. Glycobiology 8:455-462.

Burda P, Jakob CA, Beinhauer J, Hegemann JH, and Aebi M (1999). Ordered assembly of the asymmetrically branched lipid-linked oligosaccharide in the endoplasmic reticulum is ensured by the substrate specificity of the individual glycosyltransferases. Glycobiology 9:617-625.

Camirand A, Heysen A, Grondin B, and Herscovics A (1991). Glycoprotein biosynthesis in *Saccharomyces cerevisiae*. Isolation and characterization of the gene encoding a specific processing mannosidase. J Biol Chem 266:15120-15127.

Cipollo JF, Trimble RB, Chi JH, Yan Q, and Dean N (2001). The yeast *ALG11* gene specifies addition of the terminal 1,2-Man to the $Man_5GlcNAc_2$-PP-dolichol N-glycosylation intermediate formed on the cytosolic side of the endoplasmic reticulum. J Biol Chem 276:21828-21840.

Chen M, Liu X, Wang Z, Song J, Qi Q, Wang PG (2005). Modification of plant N-glycans processing: the future of producing therapeutic protein by transgenic plants. Med Res Rev 25:343-360.

Conesa A, Punt PJ, van Luijk N, and van den Hondel CA (2001). The secretion pathway in filamentous fungi: a biotechnological view. Fungal Genet Biol 33:155-171.

Davidson EA, and Gowda DC (2001), Glycobiology of *Plasmodium falciparum*. Biochimie 83:601-604.

Dempski RE, and Imperiali B (2002). Oligosaccharyl transferase: gatekeeper to the secretory pathway. Curr Opin Chem Biol 6:844-850.

Elbein AD (1991). Glycosidase inhibitors: inhibitors of N-linked oligosaccharide processing. FASEB J. 5:3055-3063.

Frank CG, and Aebi M (2005). *ALG9* mannosyltransferase is involved in two different steps of lipid-linked oligosaccharide biosynthesis. Glycobiology 15:1156-1163.

Gao XD, Nishikawa A, and Dean N (2004). Physical interactions between the *Alg1*, *Alg2*, and *Alg11* mannosyltransferases of the endoplasmic reticulum. Glycobiology 14:559-570.

Giddings G, Allison G, Brooks D, and Carter A (2000). Transgenic plants as factories for biopharmaceuticals. Nat Biotechnol 18:1151-1155.

Grubenmann CE, Frank CG, Hulsmeier AJ, Schollen E, Matthijs G, Mayatepek E, Berger EG, Aebi M, and Hennet T (2004). Deficiency of the first mannosylation step in the N-glycosylation pathway causes congenital disorder of glycosylation type Ik. Hum Mol Genet 13:535-42.

Hashimoto K, Goto S, Kawano S, Aoki-Kinoshita KF, Ueda N, Hamajima M, Kawasaki T, and Kanehisa M (2005). KEGG as a glycome informatics resource. Glycobiology 10.1093/glycob/cwj010

Helenius J, Ng DT, Marolda CL, Walter P, Valvano MA, and Aebi M (2002). Translocation of lipid-linked oligosaccharides across the ER membrane requires *Rft1* protein. Nature 415:447-450.

Helenius A, and Aebi M (2004). Roles of N-linked glycans in the endoplasmic reticulum. Annu Rev Biochem 73:1019-1049.

Heller L, Orlean P, and Adair WL Jr (1992). *Saccharomyces cerevisiae sec59* cells are deficient in dolichol kinase activity. Proc Natl Acad Sci USA 89:7013-7016.

Herscovics A, and Orlean P (1993). Glycoprotein biosynthesis in yeast. FASEB J 7:540-550.

Huang KM, and Snider MD (1995). Isolation of protein glycosylation mutants in the fission yeast *Schizosaccharomyces pombe*. Mol Biol Cell 6:485-496.

Huffaker TC, and Robbins PW (1983). Yeast mutants deficient in protein glycosylation. Proc Natl Acad Sci USA 80:7466-7470.

Jackson BJ, Kukuruzinska MA, and Robbins P (1993). Biosynthesis of asparagine-linked oligosaccharides in *Saccharomyces cerevisiae*: the *alg2* mutation. Glycobiology 3:357-364.

Jelinek-Kelly S, and Herscovics A (1988). Glycoprotein biosynthesis in *Saccharomyces cerevisiae*. Purification of the mannosidase which removes one specific mannose residue from $Man_9GlcNAc$. J Biol Chem 263:14757-14763.

Joosten V, Lokman C, Van Den Hondel CA, and Punt PJ (2003). The production of antibody fragments and antibody fusion proteins by yeasts and filamentous fungi. Microb Cell Fact 2:1-15.

Kukuruzinska MA and Robbins PW (1987). Protein glycosylation in yeast: transcript heterogeneity of the *ALG7* gene. Proc Natl Acad Sci USA 84:2145-2149.

Lerouge P, Cabanes-Macheteau M, Rayon C, Fischette-Laine AC, Gomord V, and Faye L (1998). N-glycoprotein biosynthesis in plants: recent developments and future trends. Plant Mol Biol 38:31-48.

Lerouge P, Bardor M, Pagny S, Gomord V, and Faye L (2000). N-glycosylation of recombinant pharmaceutical glycoproteins produced in transgenic plants: towards an humanisation of plant N-glycans. Curr Pharm Biotechnol 1:347-354.

Ma JK, Drake PM, and Christou P (2003). The production of recombinant pharmaceutical proteins in plants. Nat Rev Genet 4:794-805.

Maia IG, and Leite A (2001). N-glycosylation in sugarcane. Genet.Mol Biol 24:231-234.

Nakanishi-Shindo Y, Nakayama K, Tanaka A, Toda Y, and Jigami Y (1993). Structure of the N-linked oligosaccharides that show the complete loss of a-1,6-polymannose outer chain from och1, och1 mnn1, and och1 mnn1 alg3 mutants of Saccharomyces cerevisiae. J. Biol. Chem. 268:26338-26345.

Orlean P (1992). Enzymes that recognize dolichols participate in three glycosylation pathways and are required for protein secretion. Biochem Cell Biol 70:438-447.

Parodi AJ (2000). Role of N-oligosaccharide endoplasmic reticulum processing reactions in glycoprotein folding and degradation. Biochem J 348:1-13.

Reiss G, te Heesen S, Zimmerman J, Robbins PW, and Aebi M (1996). Isolation of the ALG6 locus of *Saccharomyces cerevisiae* required for glucosylation in the N-linked glycosylation pathway. Glycobiology 6:493-498.

Runge KW, and Robbins PW (1986). A new yeast mutation in the glucosylation steps of the asparagine-linked glycosylation pathway. Formation of a novel asparagine-linked oligosaccharide containing two glucose residues. J Biol Chem 261:15582-90.

Samuelson J, Banerjee S, Magnelli P, Cui J, Kelleher DJ, Gilmore R, and Robbins PW (2005). The diversity of dolichol-linked precursors to Asn-linked glycans likely results from secondary loss of sets of glycosyltransferases. Proc Natl Acad Sci USA 102:1548-1553.

Schwarz M, Thiel C, Lubbehusen J, Dorland B, de Koning T, von Figura K, Lehle L, and Korner C (2004). Deficiency of GDP-Man:$GlcNAc_2$-PP-dolichol mannosyltransferase causes congenital disorder of glycosylation type Ik. Am J Hum Genet 74:472-481.

Silberstein S, and Gilmore R (1996). Biochemistry, molecular biology, and genetics of the oligosaccharyltransferase. FASEB J 10:849-858.

Takahashi T, Honda R, and Nishikawa (2000). Cloning of the human cDNA which can complement the defect of the yeast mannosyltransferase I-deficient mutant *alg 1*. Glycobiology 10:321-327.

Takeuchi K, Yamazaki H, Shiraishi N, Ohnishi Y, Nishikawa Y, and Horinouchi S (1999). Characterization of an *alg2* mutant of the zygomycete fungus *Rhizomucor pusillus*. Glycobiology 9:1287-1293.

Tremblay LO, and Herscovics A (1999). Cloning and expression of a specific human 1,2-mannosidase that trims $Man_9GlcNAc_2$ to $Man_8GlcNAc_2$ isomer B during N-glycan biosynthesis. Glycobiology 9:1073-1078.

Umeda K, Yoko-o T, Nakayama K, Suzuki T, and Jigami Y (2000). Schizosaccharomyces pombe *gmd3(+)/alg11(+)* is a functional homologue of *Saccharomyces cerevisiae ALG11* which is involved in N-linked oligosaccharide synthesis. Yeast 16:1261-1271.

Westphal V, Kjaergaard S, Schollen E, Martens K, Grunewald S, Schwartz M, Matthijs G, and Freeze HH (2002). A frequent mild mutation in ALG6 may exacerbate the clinical severity of patients with congenital disorder of glycosylation Ia (CDG-Ia) caused by phosphomannomutase deficiency. Hum Mol Genet 11:599-604.

Wildt S, and Gerngross TU (2005). The humanization of N-glycosylation pathways in yeast. Nat Rev Microbiol 3:119-128.

Ziegler FD, Gemmill TR, and Trimble RB (1994). Glycoprotein synthesis in yeast. Early events in N-linked oligosaccharide processing in *Schizosaccharomyces pombe.* J Biol Chem 269:12527-12535.

Ziegler FD, Cavanagh J, Lubowski C, and Trimble RB (1999). Novel Schizosaccharomyces pombe N-linked $GalMan_9GlcNAc$ isomers: role of the Golgi *GMA12* galactosyltransferase in core glycan galactosylation. Glycobiology 9:497-505.

Zou J, Scocca JR, and Krag SS (1995). Asparagine-linked glycosylation in *Schizosaccharomyces pombe*: functional conservation of the first step in oligosaccharide-lipid assembly. Arch Biochem Biophys 317:487-496.

Zubkov S, Lennarz WJ, and Mohanty S (2004). Structural basis for the function of a minimembrane protein subunit of yeast oligosaccharyltransferase. Proc Natl Acad Sci USA 101:3821-3826.

Applied Mycology
and Biotechnology
An International Series
Volume 6. Bioinformatics

Genomic Rearrangement and Disease LARaLINK 2.0: Datamining for Clinical Cytogenetics

Adrian E. Platts[1], Dawei Wang[2], Brian Fayz[3], Robert Lennie[4], Bin Yao[5] and Stephen A. Krawetz[6]
[1]Department of Obstetrics and Gynecology. Wayne State University, Detroit, MI, 48201 (plattsa@genetics.wayne.edu); [2] Bioinformatics Facility, Applied Genomics Facility, Wayne State University, Detroit, MI, 48201 (dwang@genetics.wayne.edu); [3]Bioinformatics Facility, Applied Genomics Facility, Wayne State University, Detroit, MI, 48201 (bfayz@genetics.wayne.edu); [4]Bioinformatics Facility, Applied Genomics Facility, Wayne State University, Detroit, MI, 48201 (lennie@genetics.wayne.edu); [5]Bioinformatics Facility, Applied Genomics Facility, Wayne State University, Detroit, MI, 48201 (byao@genetics.wayne.edu);[6] Department of Obstetrics and Gynecology, Center for Molecular Medicine and Genetics, Institute for Scientific Computing, Wayne State University, Detroit, MI, 48201 (steve@compbio.med.wayne.edu).

A growing pool of genomic data is being archived to online public databases. These have the potential to impact both the diagnosis of genetically linked diseases as well as to aid in defining their genesis. The current *in-silico* tools for cytogenetic analysis have generally been targeted to the academic and industrial communities for use in research into the origins of disease. In comparison, few tools have been developed to integrate the multiple online resources for the clinical setting. We have addressed this deficit through a web-delivered application, LARaLINK 2.0: Loci Analysis for Rearrangements Link version 2.0 *http://LARaLINK.bioinformatics.wayne.edu:8080/unigene* controlled hierarchical vocabulary for mining cDNA and microarray expression data. This tool now provides researchers and clinicians with the means to effectively use cytogenetic data to rapidly assess disease association. The investigator is delivered a defined set of candidate disease genes together with the supporting evidence for their expression and disrupted phenotypes.

1. INTRODUCTION

The barriers presented to the clinician seeking to effectively utilize the wealth of online data on genomic function and disease are considerable. The resources required to address a clinical question are frequently located across several databases that possess different interfaces and indexing mechanisms. Equally the query process itself is often designed to aid the academic researcher whose questions are precise and their data structured to testing a specific hypothesis. By contrast the clinician frequently encounters patients who present multiple and sometimes discordant histories. Working in a time sensitive environment the clinician requires a global diagnostic and prognostic assessment rather than the intimate detail sought by the academic researcher. These divergent needs present a challenge when creating a system useful to both the basic research scientist and the practicing clinician. To meet these needs we have developed the web-delivered application, LARaLINK 2.0: Loci Analysis for Rearrangements Link version 2.0 (Fayz et al, 2005 & Platts et al., 2005). This brings together publicly available data from sources described in Table 1, including Ensembl's MART (http://www.ensembl.org/Homo_sapiens/martview), the Stanford SOURCE (http://source.stanford.edu/cgi-bin/source/sourceSearch) system, the NCBI's UniGene (Pontius et al., 2003), dbEST (Boguski et al., 1993), GEO (Barrett et al., 2005) and OMIM (http://www.ncbi.nlm.nih.gov/omim/) databases as well as the Chromosomal Variation in Man database (Borgaonkar, 1997). The system employs the controlled hierarchical vocabulary eVOC from SANBI (Kelso et al., 2003) for mining cDNA and microarray expression data. Using a series of simple questions, our approach has been to over-engineer flexibility during the initial specification and then deliver a graphical response that immediately shows global trends among the data queried. The information is layered, permitting the user to drill down from the trend to the underlying evidence. By implementing both web services and local archives we have sought to optimize the balance between the timeliness of the data, through live links and the reliability of the system, enhanced by local caching.

The clinical cytogeneticist's efforts are directed towards identifying the causative genomic aberrations underlying the presentation of a disease. Prognostic outcomes and treatment options can be inferred from observations made amongst patients presenting similar syndromes. For example, an ongoing effort to link genetic disruption to the cancer phenotype by karyotype analysis, heterozygosity mapping or with more detailed array CGH and expression mapping, offers considerable advantages and holds much promise compared to traditional comparative cell and nuclear morphology strategies. The complexity of the task is exemplified in breast cancer. This cancer presents over 400 karyotypes with subtle but different prognoses for each genomic deletion, rearrangement and duplication (Mitelman, 1995). This necessitates an analysis that can embrace the impact of multiple deletions and chimeric rearrangements.

The development of tools to aid in the identification of chromosomal anomalies has been progressing for several decades. Stand alone catalogues of cytogenetic data have been published in print, CD-ROM, and most recently in Web formats. For example, McKusick's seminal *Mendelian Inheritance in Man: A Catalog of Human Genes and Genetic Disorders* (http://www.ncbi.nlm.nih.gov/omim/) is in its twelfth print edition and has now been integrated into the NCBI data suite. Both Mitelman's *Catalog of Chromosome Aberrations in Cancer* (Mitelman, 1995). and Schinzel's *Human Cytogenetics Database* (http://www.oup.com/isbn/0-19-268623-2) are now also available in CD-ROM format. These publications provide a useful but limited clinical resource that can become dated as the human genome project continues to rapidly evolve. In addition, the original databases were inherently limited in their capacity to link to supporting resources.

Table 1. List of web sources, databases, abbreviations and definitions.

BAY GENOMICS	San Francisco consortium supplying mouse cell lines: http://baygenomics.ucsf.edu/
CGAP	Cancer genome anatomy project: http://cgap.nci.nih.gov/
CGH	Comparative genomic hybridization: http://www.ncbi.nlm.nih.gov/sky/
RECOMBINATION	A break in either one or more usually both DNA strands that is repaired incorrectly
CONTIG	A sequenced fragment of DNA, typically joined with other overlapping sequences
CYTOBAND	A sub-region of a chromosome made visible by staining that produces consistent contrast bands.
CYTOGENETIC	Relating to the physical characterization of chromosomes
DBEST	NCBI Database of Expressed Sequence Tags:: http://www.ncbi.nlm.nih.gov/dbEST/index.html
DECILE	An ordered grouping based on ten percentiles of a population
EBI	The European Bioinformatics Institute (a division of EMBL: the European Molecular Biology Laboratory): http://www.ebi.ac.uk
ENCODE	A large consortium of groups gathering detailed data on functional elements across a small subset of the human genome.: http://genome.ucsc.edu/ENCODE/
ENSEMBL	The EBI's genome browser and web-based annotation tool: http://www.ebi.ac.uk/ensembl/
ENSEMBL MART	A tool for accessing sequence and feature data at ENSEMBL
ENTREZ QUERY	The NCBI database query: http://www.ncbi.nlm.nih.gov/gquery/gquery.fcgi
EST	A short strand of sequenced DNA used to identify expressed DNA
EUTILITIES	NCBI XML-data query tool : http://eutils.ncbi.nlm.nih.gov/entrez/query/static/eutils_help.html
FACD	The Dutch Familial Cancer Database: http://facd.med.rug.nl/
FISH/MFISH	Fluorescent in-situ Hybridization/Multicolor Fluorescent *in-situ* Hybridization: http://www.ncbi.nlm.nih.gov/sky/
GENOTYPING	The mapping of short DNA mutations, particularly those associated with disease
GEO	The NCBI's Gene Expression Omnibus: http://www.ncbi.nlm.nih.gov/geo/
HETEROZYGOSITY	A measure of the allelic variation around a genomic locus.
EPIGENOME PROJECT	A project to map the non-sequence based information of chromatin: : http://www.epigenome.org/
IGTC	The International Gene Trap Consortium:Website: http://www.igtc.org.uk/
KARYOTYPE	The layout of an individual's chromosomal complement imaged during metaphase, stained to show cytobands and arranged in a standard order. Typically used to identify large-scale abnormalities.
LARaLINK	Loci Analysis for ReArrangement Linkhttp://LARaLINK.bioinformatics.wayne.edu:8080/unigene
MAP VIEWER	The NCBI's genome exploration tool: http://www.ncbi.nlm.nih.gov/mapview/
NCBI	The US National Center for Biotechnology Information: http://www.ncbi.nlm.nih.gov/
OMIM	Online Mendelian Inheritance in Man: http://www.ncbi.nlm.nih.gov/entrez/query.fcgi?db=OMIM
PHENOTYPES	The physical manifestation resulting from a combination of genetic, environmental and disease factors
POSSUM	A multimedia database used in disease diagnosis: http://www.possum.net.au/
PROBESET	On Affymetrix microarray platforms the set of related although not collocated perfect match and slightly mismatched oligonucleotide probes used to determine gene expression
PUBMED	NCBI medical literature database: http://www.ncbi.nlm.nih.gov/entrez/query.fcgi?db=pubmed
RESEQUENCING	Precise sequencing an individual's DNA to determine single nucleotide changes
SAGE	Serial Analysis of Gene Expression. An approach to gene expression profiling by joining many short cDNA sequences together and sequencing the resulting DNA for the frequency of the transcripts
SANBI eVOC	Orthogonal hierarchical vocabularies to order gene expression data: http://www.evocontology.org/
SKY	Spectral Karyotyping: http://www.ncbi.nlm.nih.gov/sky/
SNP	A Single Nucleotide polymorphism:: http://www.ncbi.nlm.nih.gov/entrez/query.fcgi?db=snp
SOURCE	The Stanford Online Universal Resource for Clones and ESTs: http://source.stanford.edu/cgi-bin/sourceSearch
TRYPSIN-GIEMSA	A staining technique used to reveal chromosomal cytobands
UNIGENE	The NCBI database of Genbank, mRNA and EST sequences Website: http://www.ncbi.nlm.nih.gov/entrez/query.fcgi?db=UniGene
UNIGENE CLUSTER	Transcripts that likely arise from expression of a single gene and given a unique UniGene cluster ID.

The *POSSUM* database (http://www.possum.net.au/about.htm) launched in 1987 and the *London Dysmorphology Database* (Baraitser and Winter (2001), a combination of *The Winter-Baraitse Dysmorphology Data-base* and *The Baraitser-Winter Neurogenetics Database,* launched in 1990 began addressing this need. The former, a locally installed database transitioning to a web-format and the latter, a CD-ROM based distribution, both link into Pubmed and the online edition of *Mendelian Inheritance in Man,* OMIM (http://www.ncbi.nlm.nih.gov/entrez/query.fcgi?db=OMIM). These resources exploit the electronic medium to present the user with a large volume of diagnostic image data and a smaller video archive. By contrast Borgaonkar's *Chromosomal Variation in Man* database (Borgaonkar, 1997) is aimed at providing the user with the most relevant links to the supporting literature. Continuously updated since 1974 and currently containing 24,000 entries, *Chromosomal Variation in Man* highlights the first reference in the literature to an aberration as well as the most relevant contemporary references. The Dutch *Familial Cancer Database* (FaCD) system launched in 2001 continues to be actively developed (Sijmons and Burger, 2003). It was one of the first free downloadable systems available for cancer genetics. However, FaCD's current release has only a limited capability to bring web resources together, employing static rather than dynamic aggregations of data from resources such as OMIM and Current Contents.

Both the EMBL and NCBI systems provide gateways into centralized archives, acting as third tier data warehouses of large subject-wide databases. Information within these systems can be readily accessed if the exact area of interest is known. For example, the Entrez tool (http://www.ncbi.nlm.nih.gov/gquery/gquery.fcgi) can query both the UniGene (Pontius et al., 2003) and OMIM (http://www.ncbi.nlm.nih.gov/entrez/query.fcgi?db=OMIM) systems for available data associated with a gene or genomic location. Despite its flexible approach to composing and limiting queries, the capability to formulate complex clinical-type queries is less forthcoming. It is challenging to design a query that simultaneously gathers data from an organ inclusive of all its constituent tissues while at the same time restricting the results to a group of genes around a chromosomal break-point or inversion.

Recently the NCBI resource has been enhanced with the addition of the GEO (Gene Expression Omnibus) high throughput gene expression database. This database includes data from multiple platforms, such as single and dual channel microarrays, SAGE and high throughput mass spectrometry. For the well-trained specialist, this data collection provides exceptional access to over 100 GBase of sequence data ://www.nlm.nih.gov/news/press_releases/dna_rna_100_gig.html) and tens of thousands of expression studies. However, for the clinician or the biologist needing to distill a systems view, the databases can remain a hostile mix of standards and disparate nomenclatures.

LARaLINK (Fayz et al., 2005 and Platts et al., 2005) was developed for the clinician to complement and extend the functionality of these systems, addressing their perceived limitations by federating databases through unified data translation. For example, LARaLINK translates genomic locations described by cytoband or by markers into a single nucleotide coordinate system. In addition it employs a hierarchical vocabulary to associate each term with a structure at a defined level, thereby establishing a relationship between different terminologies. This is illustrated in the following. When a tissue is described in one database as amygdala and more generally in another as brain, both results and their inter-relationship are reflected in LARaLINK's analysis that tracks amygdala data up to the level of brain. Furthermore, the analysis is standardized to the number of studies undertaken in that tissue when a gene's profile of expression across tissues is displayed. This markedly lessens the undesirable bias towards well-studied tissues. Together these tools have been assembled to provide a seamless unification of databases and a *top-down* view of the data.

2. LARaLINK: THE TECHNICAL FRAMEWORK

LARaLINK is a web application based on the J2EE technology stack (http://www.sun.com). The RDBMS is Oracle 9i (http://www.oracle.com) and MySQL (http://www.mysql.com). Data queries are accessed through JDBC (http://www.sun.com). The GUI was implemented with Servlet/JSP using the Apache Struts (http://jakarta.apache.org/struts) web application framework. J2EE design patterns like Page-by-Page Iterator and Data Access Objects were used (http://www.sun.com). JavaScript libraries (http://www.bosrup.com/web/overlib) overLIB and hvmenu (http://www.dynamicdrive.com) were used to construct the user interface. The Apache Batik SVG toolkit (http://xml.apache.org/batik/) was used to create the chromosome graphics. JFreeChart (http://www.jfree.org/jfreechart/index.php) and Cewolf JSP tag library (http://crewolf.sourceforge.net) were used to render the gene expression graphs. The Apache FOP library (http://xmlgraphics.apache.org/fop/) was used for PDF (http://adobe.com) file generation. The ANTLR (http://www.antlr.org), compiler toolkit, was used to generate lexical analyzer and parser for logical expression analysis.

LARaLINK uses UniGene and OMIM data sourced from NCBI. This is identical to that in the EMBL system. UniGene release 186 has been parsed using custom Perl scripts for import into a custom Oracle instance (http://www.oracle.com). OMIM data was treated in a similar manner. The vocabulary filter was built from eVOC data version 2.7. The eVOC data files were parsed using SQL*PLUS then loaded into the custom Oracle instance in a relational table format. SNP and marker data are assembled through real-time queries using the EBI's Ensembl-Mart (http://www.ensembl.org/Homo_sapiens/martview). The NCBI's GEO database is updated daily as authors' publications go to press. This is accessed as a live feed using the XML-based Entrez Programming Utilities (eUtilities) (http://eutils.ncbi.nlm.nih.gov/entrez/query/static/eutils_help.html). Experiment description files containing tissue information are cached locally to speed access. SOURCE data is accessed via direct queries into the Stanford SOURCE (http://source.stanford.edu/cgi-bin/source/sourceSearch) database. Access to the *Chromosomal Variation in Man* database (http://www.ncbi.nlm.nih.gov/entrez/query.fcgi? db=OMIM) has been made available by Wiley at www.wiley.com/borgaonkar.

3. CALCULATIONS

The following demonstrates the process by which the specificity of a gene's expression is standardized in each tissue relative to its global expression. The procedure is essentially as described by SOURCE (Diehn et al., 2003). The standardized expression of gene Z relative to the set of all genes (M_s) measured in tissue S located in a set of N other tissues is defined in Equation 1 for S_Z. $C_{x,y}$

$$S_z = \frac{C_{S,z}}{\left(\sum_{j=1}^{m_S} C_{S,j}\right)\left(\sum_{i=1}^{n} \frac{C_{i,z}}{\sum_{k=1}^{m_i} C_{i,k}}\right)}$$

(Equation 1)

denotes the expression evidence of gene y in tissue x. The approach standardizes the expression of the gene in a specific tissue relative to all genes in that tissue, in conjunction with a correction for the evidence of global expression in that tissue relative to expression in all tissues.

In general, the number of ESTs identified in dbEST (Borguski et alk.1993) for a given UniGene cluster can be taken as a proxy for its level of expression as represented by its frequency in a library. Several other standardization protocols, e.g., normalization or subtraction, have been established to address both the limited dynamic range of EST library technology as well as to correct for the under representation of lower expressing genes. These protocols moderate the frequency of highly abundant transcripts while the frequency of low abundance transcripts is amplified. Accordingly, the number of ESTs is no longer simply related to the level of expression of that gene. This data is excluded from this calculation since it cannot be readily combined with libraries that have not been subjected to these protocols. Exclusion is informed by the descriptions of the protocols employed for cDNA library creation available from the CGAP (http://www.antlr.org & Cancer Genome Anatomy Project). Pooled libraries that include tissues from unrelated eVOC sub-hierarchies were also excluded to eliminate ambiguity as to the source of the expressed genes. Finally, libraries with fewer than 1,000 sequenced clones were excluded since they were often created and targeted for specific gene hunting projects.

4. USING LARaLINK: AN OVERVIEW OF THE SPECIFICATION PROCESS

The six stage workflow design of LARaLINK is illustrated in Figure 1. An active context-sensitive help system that includes tan colored suggestion boxes, as well as a video tutorial for more structured instruction has been integrated into the system. The program initiates when the number of patients is selected. In response to several requests this limit has been set to 99, but can be increased. It is followed by the construction of a matrix of search methods describing the information recorded for each patient.

Consider the typical clinical scenario where a phenotype is presented by a group of patients that have been subject to different karyotyping or sequence analyses. For each patient the chromosome of interest is selected and the available location data entered.

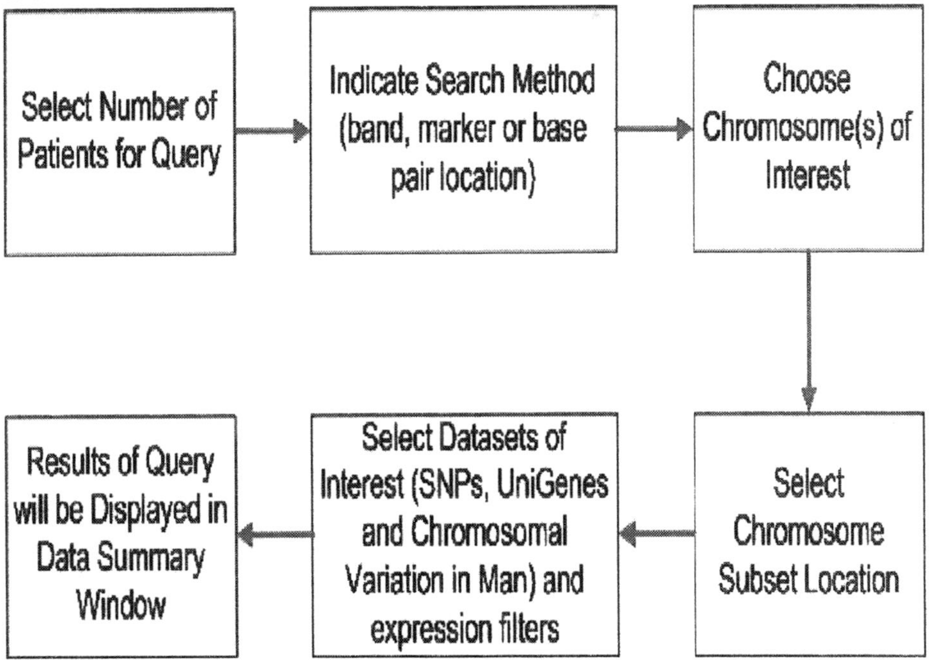

Fig. 1. An overview of the basic workflow encountered when using LARaLINK. The six stage specification process is outlined. Each stage denotes a point at which user input is required in order to proceed to the next stage.

The clinician or researcher is given the option to refine the output information displayed, e.g., to include only UniGenes or SNPs, and/or to specify tissue type or developmental stage. The user may elect to include all patients in the results or to explore the results with respect to selected subgroups of patients. This is likely to be a stage that is iteratively revisited as hypotheses are generated and discarded. In this manner, result specificity can be moderated.

4.1 Stage 1

As shown in Figure 2 up to 99 patients can be included in any one session and these can be subgrouped within the system as needed. For example, a large group of patients can be partitioned to explore correlations on the basis of gender or other categorical lifestyle variables. Where more than 99 patients are being investigated they must be partitioned in advance into subgroups on the basis of an appropriate discriminant variable.

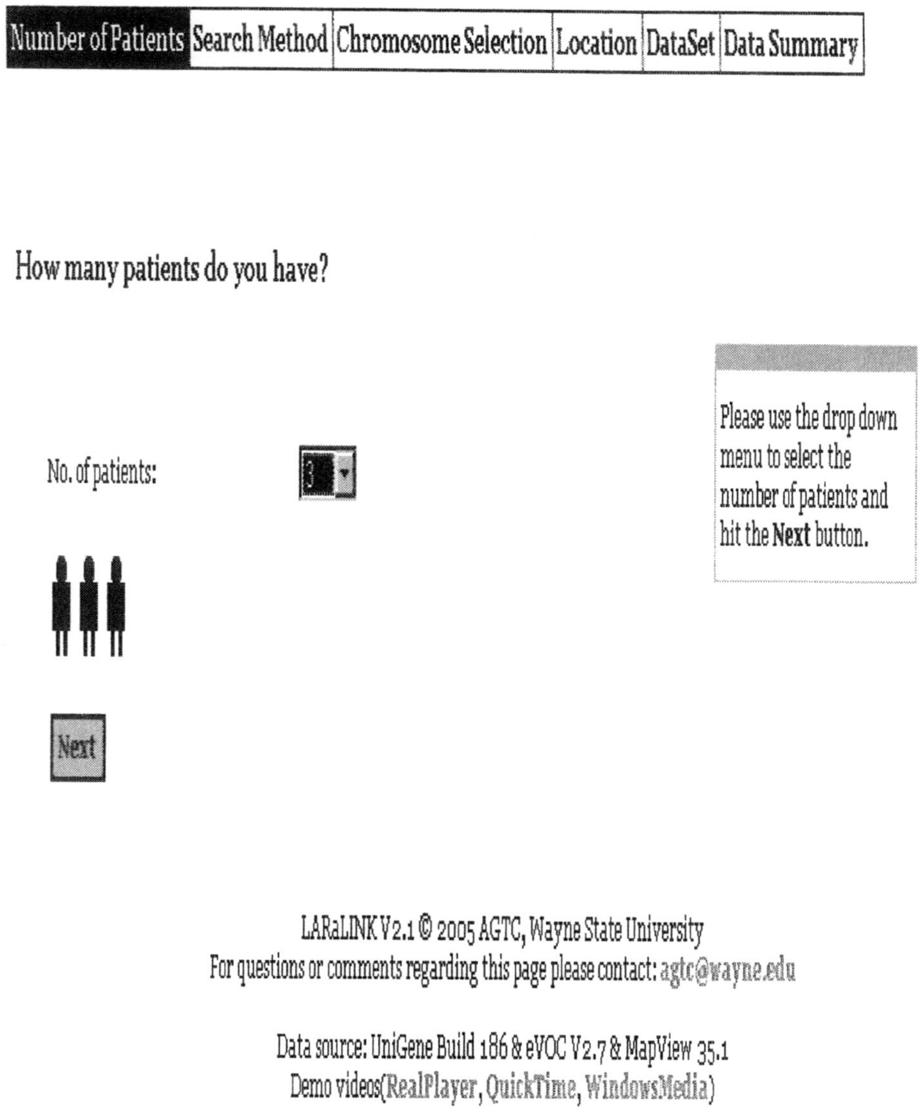

Fig. 2. The first input screen. Initially the number of patients is selected. This can range up to 99 patients in any single group.

4.2 Stage 2

There are three methods shown in Figure 3 by which a user can specify the chromosomal location for the patients, each of which is characteristic of a span of DNA. The chromosome band is inherently a multi-megabase selection. This may be most appropriate for studies of significant rearrangements that lead to broad genomic deletions. In comparison, using the genomic marker, resolution equivalent to approximately that of a locus of 50 kb or up to 5 Mb on either side of the selected marker can be achieved. Care must be taken when entering this data since the current system is not tolerant of typographical errors. The utility of the marker lies in its potential to locate sites of recombination or other genetic hotspots e.g., fragile sites. Finally where detailed genotyping or resequencing studies are available, precise base pair positions may be specified. To ensure that the correct sequence is located, care must be taken that the the genome build-version within the system and that of the query are concordant.

4.3 Stage 3

The chromosome of interest is then selected for each patient as shown in Figure 4. In cases of interchromosomal recombination, regions of interest on at least two chromosomes that define the recombination center will be studied. In those diseases that are associated with high levels of chromosomal instability, for example P53 associated cancers (van Gent et al., 2001), regions of note may exist for each patient on several chromosomes. The system therefore permits multi-chromosome selections. In those instances where more than one chromosome is selected the user will receive data that is formatted by both patient and chromosome.

4.4. Stage 4

As shown in Figure 5, the user is given a selection of input tools to describe their data. This reflects the matrix of chromosomes per patient and marker type per patient specified. When a chromosome band is selected an ideogram of the chromosome is displayed. The specific chromosome band range can be selected by either picking a region on the ideogram or by using the drop-down box to the right of the ideogram, Figure 5 - Patient 1. When a marker is selected as the search method, an entry box appears. One can then input the specific marker of interest as well as use a drop-down box to select the range, Figure 5 - Patient 2. Valid ranges are 50 kbps, 100 kbps, 200 kbps, 500 kbps, 1 Mbps, 2 Mbps or 5 Mbps. This represents the number of bases immediately flanking each side of the marker that will be queried. The system design seeks to be adaptive to the scale of disruption or the sensitivity of the diagnostic technique. For example, array CGH provides a far greater level of resolution than either MFISH or SKY. If chromosome location was selected as the search method, then two entry boxes will appear that enable a specific base pair range to be entered, as shown in Figure 5 - Patient 3. For queries arising from SNP data the locations should be identical.

4.5 Stage 5

Once the patient-chromosome data matrix has been populated the user is then required to consider which patients to include in the analysis and which analyses are relevant given the investigative context. The analyses that can be returned include any combination of SNPs in the regions selected, UniGene clusters or Chromosomal

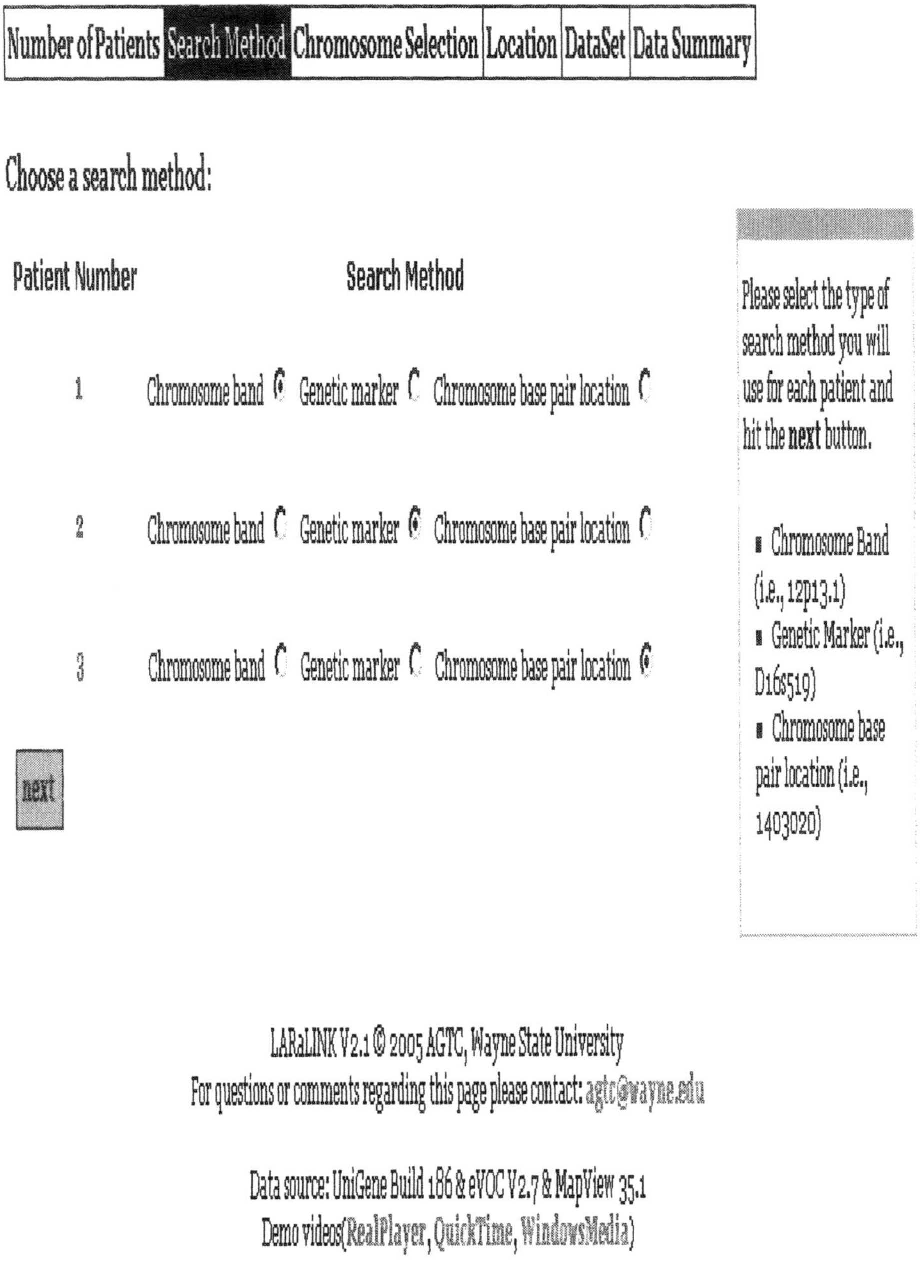

Fig. 3. A matrix of information to initially describe the chromosomal location

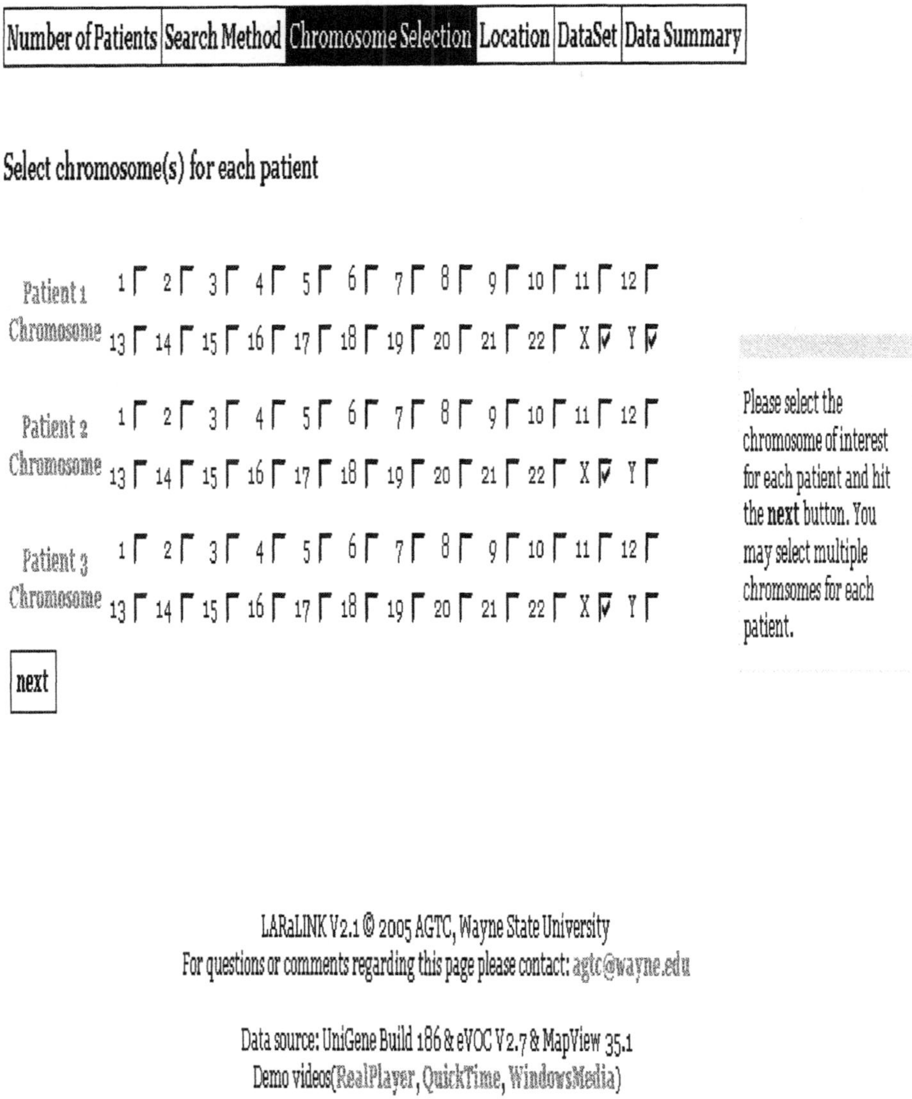

Fig. 4. Chromosome location. Multiple chromosomes for each patient can be selected.

variations in Man reported and associated OMIM data (Figure 6). The patients for inclusion can also be selected at this point. This process can be useful for investigating whether cohort is a cohesive group with respect to the variables by which it could be disaggregated. To enhance specificity, the analyses are then typically limited by organ system. This can be refined further to include developmental stage or another property of interest. Specification is achieved using a vocabulary filter derived from eVOC (Kelso et al., 2003), a standardized vocabulary of terms used to define expression mapping. The advantage of using a standardized vocabulary such as eVOC to describe the clone libraries is that an exact expression terms match is not required. Nine categories of expression ontology are included. They are anatomical site, developmental stage, pathology, cell type, experimental technique, tissue preparation, treatment, pooling and the general *'associated with'*. While the eVOC filter terminology can be entered directly, the *'choose'* link executes a *'pop-up'* box that will display the terminology in a hierarchical tree view. The search link permits one to hunt across all categories for the instance where it may not be clear where a filter term is located. This implementation of eVOC permits the use of multiple filters by using AND/OR logical operators that can be loaded from the system menu as shown in Figure 6.

4.6 Stage 6

The most important consideration when reviewing the results is the level of search specification, i.e., whether the search was too relaxed or too restrictive. If only a few genes are returned it may be best to immediately return to the search filter page and relax the filter parameters.

In cases where thousands of genes are returned, search specification may be honed in order to review the results in a timely manner. As shown in the data summary window of Figure 7, the number of UniGene clusters, SNPs or Chromosomal Variations found for each patient are summarized. A separate *'view'* link, as well as the eVOC data filter terms that are applied to dataset selection is included. Two separate links *'mapped to single chromosome'* and in those cases where ambiguity remains *'mapped to multiple chromosomes'* are provided for each UniGene cluster. A link to the *'Common Genes'* found among all patients representing the intersection of overlapping regions is displayed at the bottom of the page.

5. LARaLINK'S OUTPUT AND DATA RENDERING

The relationships between LARaLINK's output and rendering systems are summarized in Figure 8. Available directly from the summary of output page are the main findings, the complete lists of genes of interest, their collocated SNPs or the diseases to which aberrations around these genes have been linked. From these summary pages detailed graphical information can then be explored depicting the distribution of ESTs across tissues, ranked distribution by gene specificity and expression distributions extracted from the GEO database (Barrett et al., 2005). The evidence for expression, i.e., the library or reporter or experimental study is available.

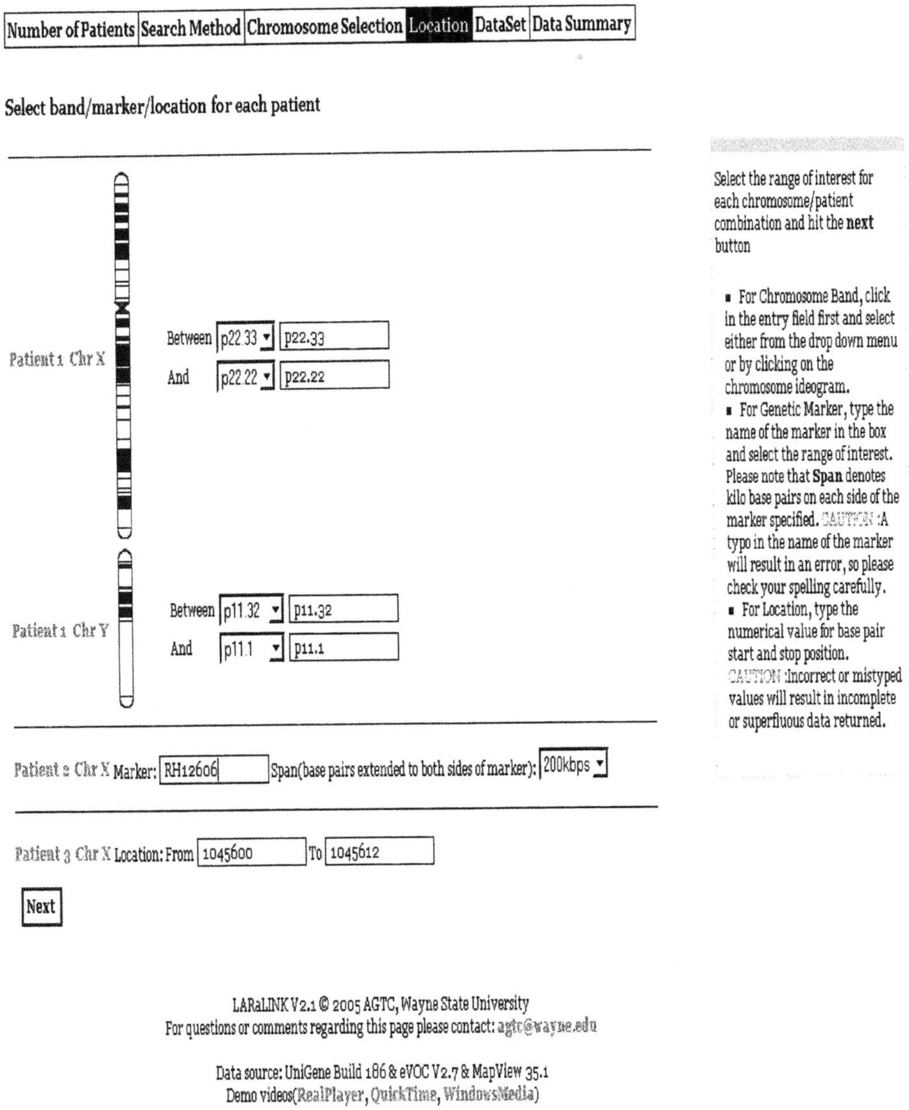

Fig. 5. Specification of chromosomal regions. Chromosomal regions of interest are specified either as cytoband, marker or base pair coordinate. Specification reflects the form and specificity of the investigation undertaken

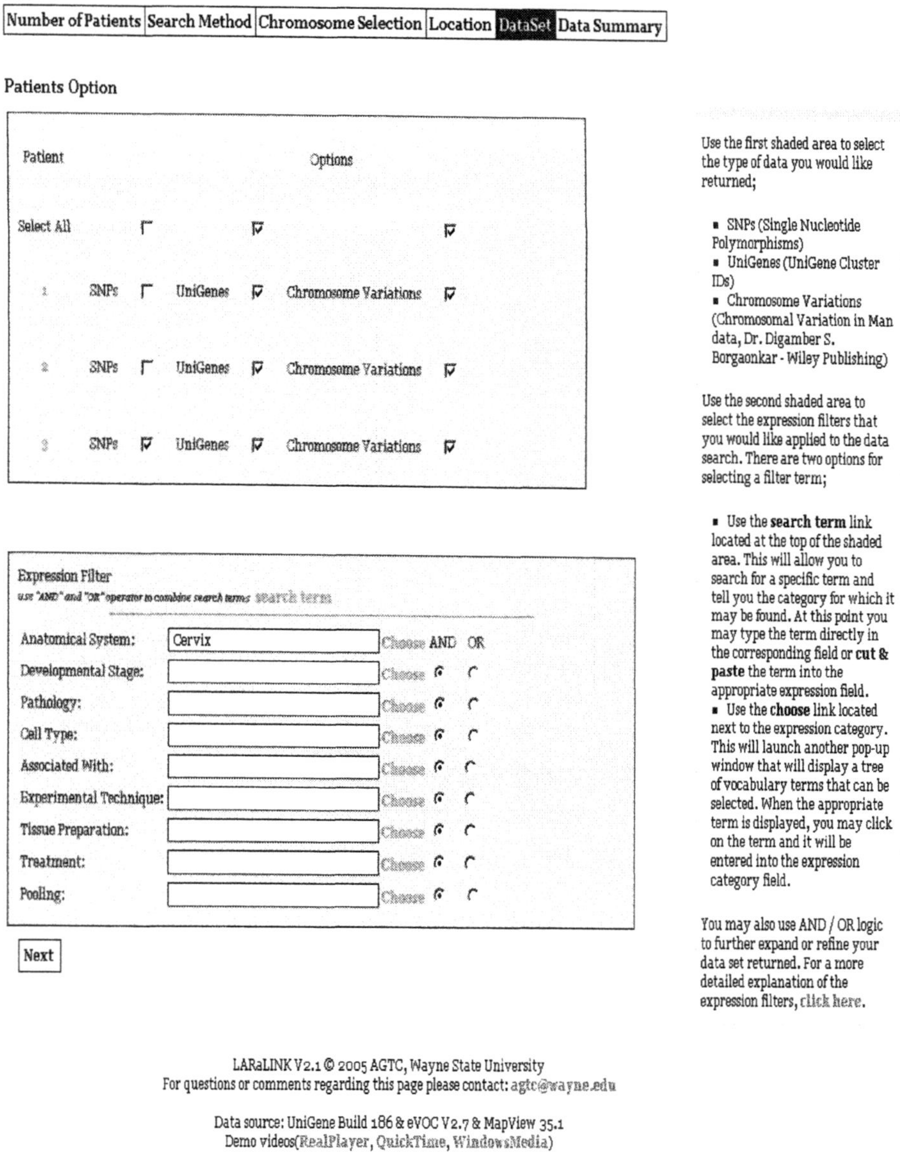

Fig. 6. Dataset selection and expression filtering. Multiple consecutive filters can be employed at the final pre-filtering stage to focus the data returned to a manageable set. These can be combined with Boolean logic operations to restrict results based upon compound criteria. The types of output can be chosen from SNPs, UniGene Cluster or Chromosomal Variations and results restricted to subgroups of patients.

5.1 UniGene Data

Perhaps the most useful information about any genomic region is the catalogue of genes it contains, the tissues in which these genes are expressed and the diseases to which they have been linked. The associated gene data is available through the '*view*' link on the data summary page as shown in Figure 9. The UniGene clusters that match the search criteria previously defined by the user are displayed. The gene data page presents the user with the gene's cytoband localization, UniGene cluster number and functional name if known. This is in addition to a summary of the expression distribution among the various libraries. Expression data from array studies can be viewed by following the '*GEO*' link to the GEO tool. The window incorporates a smart listing feature and a sort function that assists one with determining how the data is displayed. The sort function allows the user to list the top N clusters that are expressed in the filter criteria or any sub-branch of the search. For example if brain was a filter term then the genes returned could be sorted by their specificity to whole brain, hypothalamus or any other of the organ's sub-tissues. Within the gene data page are links to the ESTs contributing the evidence to each UniGene cluster, NCBI's MapView of the cluster and the OMIM entry for each gene. The results can be collated as either a PDF report or a TXT report enabling their incorporation into most text-editing programs.

5.2 WSU & SOURCE Expression

In some analytical schemas it may be sufficient to only view the summary of genes and their expression. For others the evidential basis for the distribution as well as the pattern of expression derived from evidence other than UniGene will be equally important. Two methods are utilized to display the specificity of expression. The first is SOURCE expression (http://source.stanford.edu/cgi-bin/source/sourceSearch), which retrieves the normalized expression information for the UniGene cluster directly from SOURCE and displays it in a separate window using the literal tissue. That is, brain and cerebrum will be considered two different tissues. The other method is WSU expression.

The WSU implementation calculates normalized expression in a manner similar to SOURCE but also determines specificity using the controlled vocabulary of eVOC to group libraries into tissues with a set of standard names. Thus, depending on the search criteria, brain and cerebrum can be considered as one or two different tissues. For example, if a low expressing gene is only expressed in eye, WSU expression will return a value of 100% because the gene is exclusively expressed in that tissue. In comparison, SOURCE will return the expression relative to all values determined. As shown in Figures 10 & 11, both expression systems display the normalized expression profile of a given UniGene cluster as a bar graph. Each bar represents a different tissue source with relative expression expressed as percentage displayed at the top of the bar. The expressed tissue is displayed underneath each bar. By default, each expression window allows a user to view the top 10 tissues in which that gene was expressed or view all tissues expressed in a given unigene

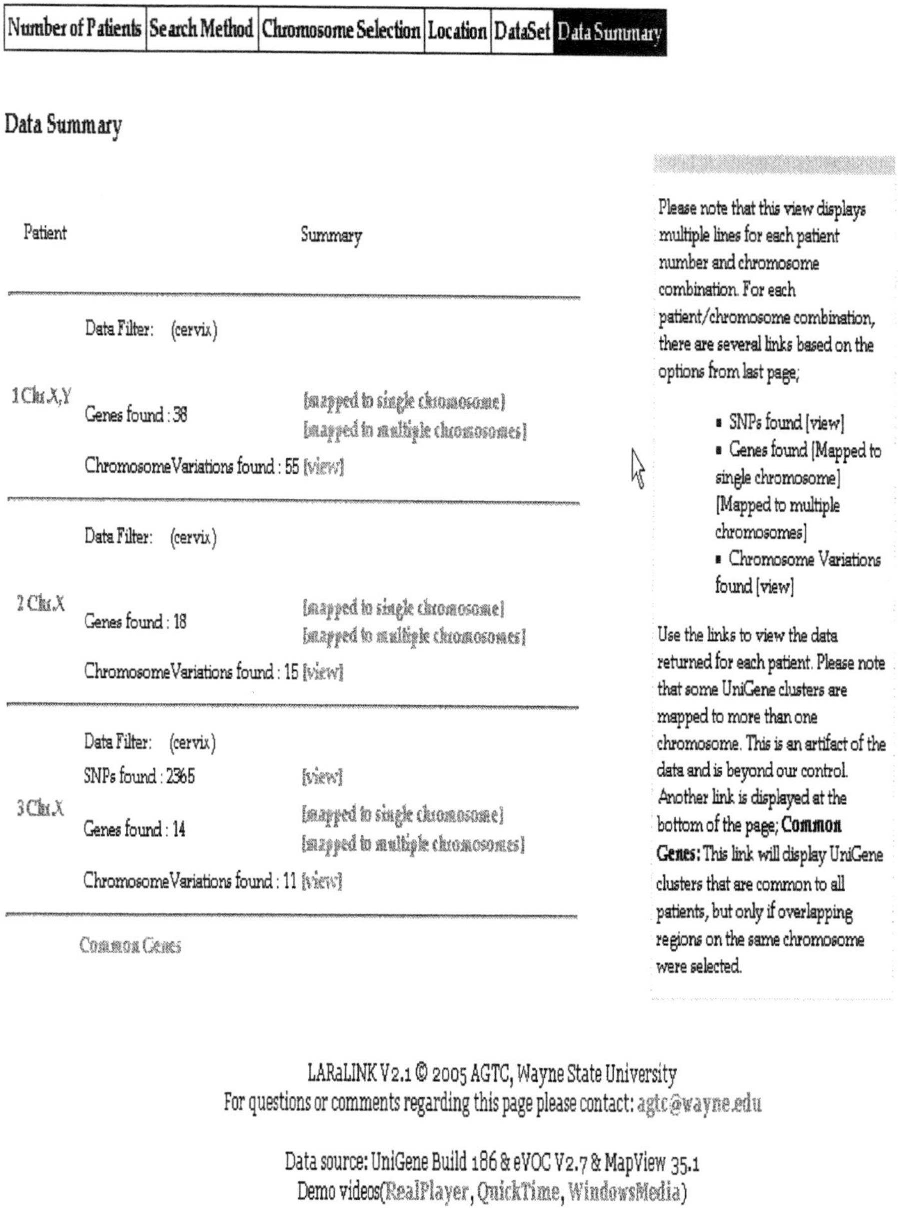

Fig. 7. The output summary page. Subsequent to mining, the results are presented and sorted by patient and chromosome. A link to genes common among all patients is provided at the bottom of the output.

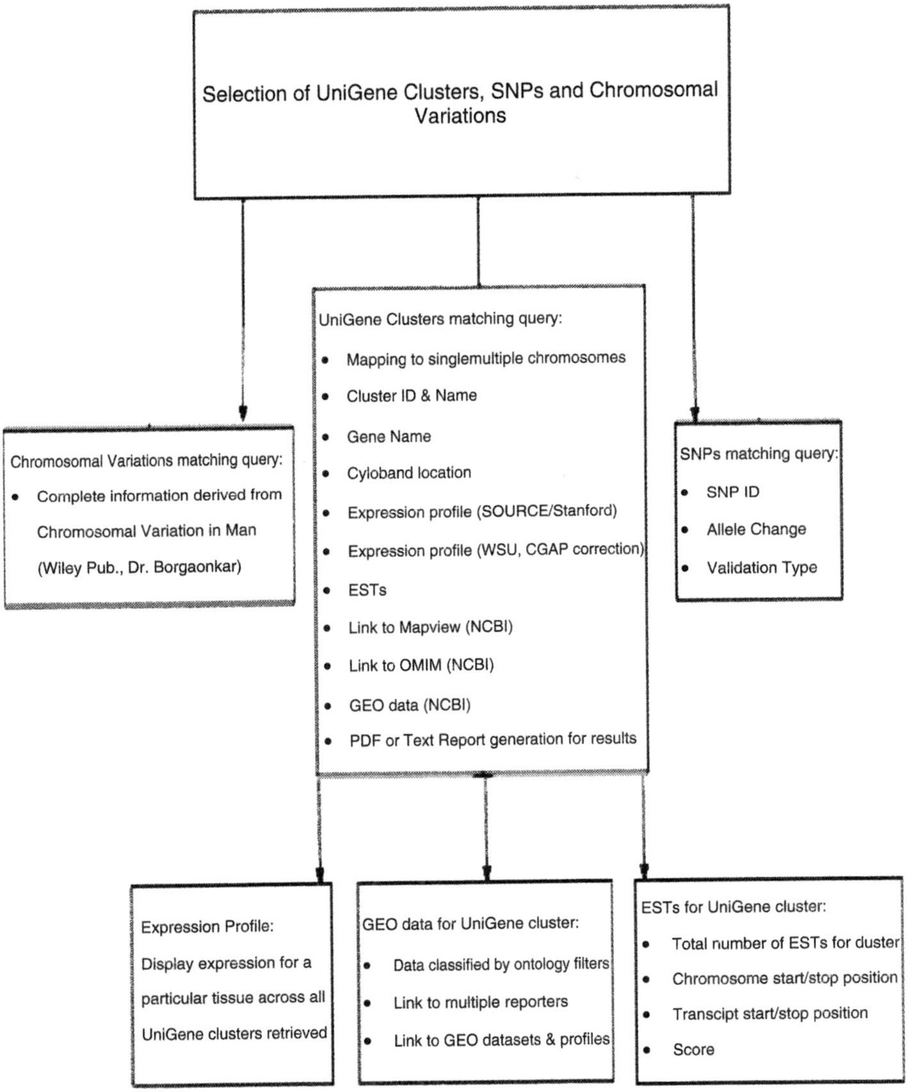

Fig. 8. Hierarchy of data gathered from the distributed datahosts. LARaLINK was designed with multiple output systems. It continues to evolve as other resources are accessed and suggestions from the user community are incorporated.

Gene Data (total 23 single-chromosome mapped)

Page: 1

1 2 PDF Report TXT Report

Top 10 in cervix ▼ sort

1 **Hs.264** **GS2 gene** **Xp22.3**

Expression:
Testis Tongue Bone Stomach Blood mixed Skin Cervix Bladder Eye Lymph_Node Mammary_Gland Liver Embryo Uterus Muscle Heart Colon Kidney Placenta Prostate Brain Lung other

[Source Expression] [WSU Expression] [ESTs] [Map View] [Omim Entry] [GEO]

2 **Hs.6483** **oral-facial-digital syndrome 1** **Xp22.2-p22.3**

Expression:
Cervix Blood Skin Heart Soft_Tissue Bladder Ovary Prostate Lymph_Node Testis Placenta Larynx Liver Mammary_Gland Pancreas Uterus Muscle Colon Lung Vascular Kidney mixed Brain other

[Source Expression] [WSU Expression] [ESTs] [Map View] [Omim Entry] [GEO]

3 **Hs.12913** **KIAA1280 protein** **Xp22.32**

Expression:
Spleen Vascular Eye Stomach Blood Cervix Testis Larynx Mammary_Gland Uterus Liver Skin Ovary Colon Prostate Heart Pancreas Brain Lymph_Node Kidney Tongue Placenta Lung mixed other

[Source Expression] [WSU Expression] [ESTs] [Map View] [GEO]

4 **Hs.19404** **ankyrin repeat and SOCS box-containing 9**

Expression:
Bone_Marrow Liver Lung Stomach Testis Tongue Ovary Brain Cervix Prostate Kidney mixed Skin other

[Source Expression] [WSU Expression] [ESTs] [Map View] [GEO]

5 **Hs.31535** **similar to carbonic anhydrase VB, mitochondrial precursor; carbonic dehydratase** **Xp22.31**

Expression:
Peripheral_Nervous_System Bone Eye Muscle Stomach Testis Larynx Uterus Liver mixed Lung Cervix Colon Mammary_Gland Kidney Blood Lymph_Node Prostate Skin Brain other

[Source Expression] [WSU Expression] [ESTs] [Map View] [GEO]

Fig. 9. Data Display. UniGene data is displayed together with map locations, gene ID and a summary of expressed tissues for each gene. Links to expression and mapping data are provided for each gene.

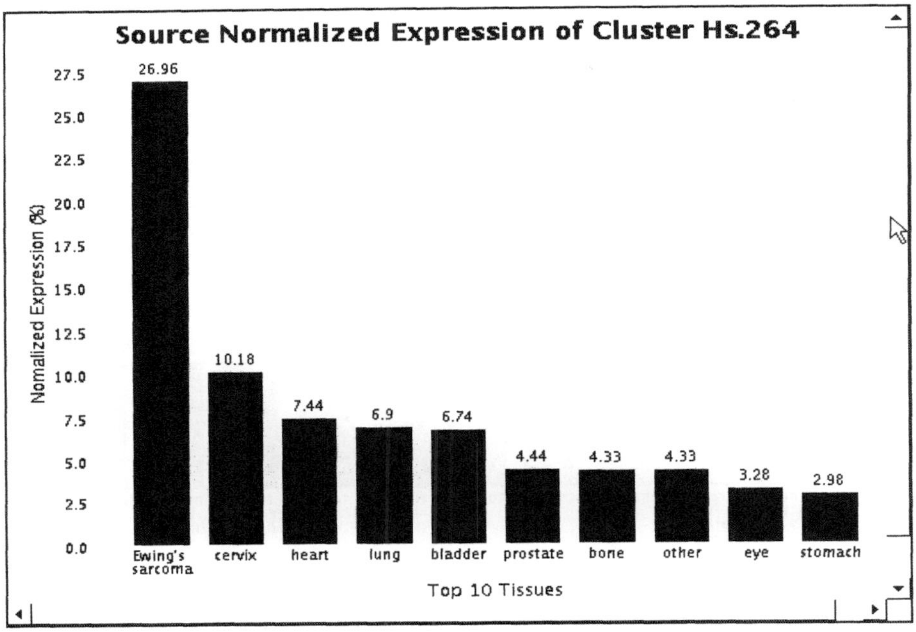

Fig. 10. SOURCE expression distribution data. The graph shows the SOURCE distribution of normalized expression for each tissue within UniGene cluster Hs.264. Each bar represents a different tissue. The SOURCE attribution of tissue origin allows mixed tissues and general terms such as *other*. Relative expression, as a percentage of total, is displayed at the top.

cluster. An additional feature within the WSU expression window is the option to select an expressed tissue in the bar graph and display the expression values for each UniGene cluster across the sub-set of data defined within the original search (Figure 12).

5.3 Assimilating GEO Expression Data

Access to the GEO gene expression data analysis suite is gained from the UniGene data page. This data management utility re-analyzes data from NCBI's high throughput Gene Expression Omnibus, within the context of eVOC's structured hierarchy. The GEO viewer represents the range and specificity of a gene's expression as determined through high throughput experimental approaches. Ambiguity in the terminology used in the eVOC vocabulary can occur when used in conjunction with the GEO descriptions of tissue. For example, while the term tract is unique within the eVOC hierarchy, it can match multiple instances of *'tract'* that investigators may use in their descriptions. It is thus amended, e.g., to *'major inter-hemispheric tract'* albeit at the risk of losing some less precisely worded contributions.

Expression data is cascaded through the eVOC hierarchy within the viewer from tissue sub-groups to higher-order structures termed parents. Thus, a microarray experiment that shows a high level of expression of a gene in the hypothalamus

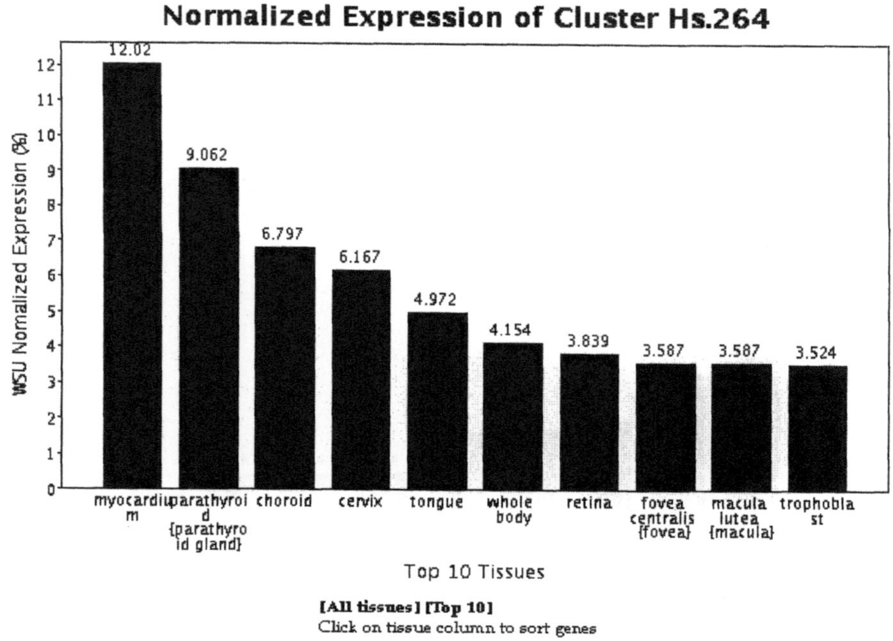

Fig.11. WSU normalized expression data. The graph presents the distribution of WSU normalized expression for each tissue within UniGene cluster Hs.264. These values were calculated after removing libraries created by normalization or subtraction protocols as well as libraries that contained small numbers of sequences. The graph presents expression using the eVOC controlled vocabulary to group libraries into tissues with a standardized nomenclature.

Would have this result combined with all other substructures of the brain and non-specific brain expression surveys to provide an expression distribution at the level of the term brain. This data would then be combined with other nervous tissue to represent expression at the level of the CNS. While the default eVOC hierarchy is the anatomical system,_ expression can also be mapped by orthogonal hierarchies such as pathology and cell type. A non-orthogonal user defined hierarchy is also an option. This might be constructed in the instance where a hierarchy of tissues infiltrated through the progression of a disease is required. Figure 13 shows a typical output from the viewer when presented with the growth hormone 1 gene GH1 on Affymetrix chips (P) Present-calls only and in human. On the vertical axis is the eVOC hierarchy of tissues, with only those tissues and their parent structures in which array expression is recorded. On the horizontal scale are a set of ten bins into which rank ordered data from the experiments are placed. Rank data allow an approximate comparison of expression to be made between otherwise non-normalized experiments. For example, in an experiment using an Affymetrix U133+2 chip with 54,000 probesets the 5,400 probesets with the highest signal (as determined by the original investigator using any of a number of algorithms) would

Total 42 unigenes are expressed in **pituitary gland {pituitary}**

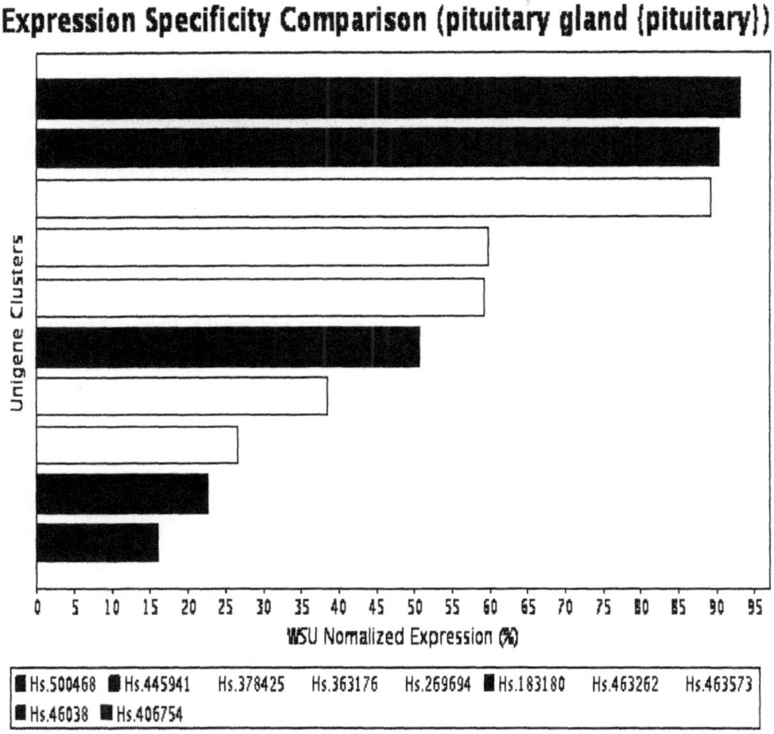

1 **Hs.500468 growth hormone 1** **17q24.2**

Expression:
Placenta Adult Brain other

[Source Expression] [WSU Expression] [ESTs] [Map View] [Omim Entry] [GEO]

2 **Hs.445941 Transcribed sequences**

Fig.12. WSU expression ranked by gene specificity to the pituitary gland and targeted to those genes between 17p22 and 17p24. GH1 is the most pituitary specific RNA, although it is also found in non-specific placental tissue. Each color bar represents a distinct UniGene cluster returned by the previous search.

be clustered into the 90th-100th percentile bin, the next 5,400 in signal ordered into the 80th-90th bin, etc. Thus, even when two experiments have widely different median chip intensities or intensity distributions, the top 10 percent (decile) should include the same genes amongst the two chips if they are measuring the same distribution of underlying expression. When two experiments measure the same tissue, but a gene's expression is significantly changed in one due to a disease state, then expression will be attributed to different results-bins. The light to intense color depth of each bin is an indication of the total number of experiments reporting expression in that data bin. To readily differentiate data from disease states and normal tissue, the bin gains a highlighted border in those bins where the majority of data is derived from physiologically normal tissue. Bringing these visualization strategies together provides a *top-down* overview of expression of a single gene across those physiological backgrounds in which its expression has been measured. This can, in turn, be used to map changes in expression arising from disease states, or conversely to mine those tissues that would likely be impacted by a significant shift in a gene's expression.

Clicking on the expression chart at any point displays the supporting evidence for expression in that tissue at that level. As shown in Figure 13, this is highlighted in a popup box. Experiment titles are provided for each contributing result, together with a link to the experiment description and GEO expression value. Since the expression profile of any given gene may be derived from numerous reporter sequences, one link allows the viewer to redisplay the gene's expression solely on the basis of evidence from that reporter. Figure 14, shows the expression profile from just one reporter of GH1. It is not surprising to observe that the reporter is sensitive to both genes since the pituitary specific GH1 gene is homologous to the placental specific growth hormone 2 gene (GH2). The expression of GH1/GH2 in placenta and pituitary is amongst the top 1% of gene expression in these tissues, a less significant observation can also be made in tongue at the ~91st percentile. As indicated by the arrows, expression from these tissues is cascaded to parent tissues, of placenta. This proceeds serially from, female genitals, genitals, urogenital and then to the base class, anatomical system. The functionality to query individual reporters is a powerful aid to understanding the contributions from multiple, particularly 3', splice variants.

The viewer offers several additional mechanisms for exploring gene expression with respect to technology and model organism. The choice of array platform can be selected from a list of those in the GEO database, excluding custom platforms. However, it is strongly recommended that the user restrict their choice to single platforms. When a platform reports a present or absent call for a gene based upon the background and stability of expression readings, results can be filtered to include or assessed by reinitiating the search after selecting another species from the dropdown list. This feature confers the option to assess if a specific animal model provides a reasonable analogue of human expression. Genes can also be manually entered for exploration. Since this input accepts Boolean queries, the expression from entire loci can be displayed by specifying the appropriate query, e.g., the expression *'GH1 OR GH2 OR CSH1 OR CSH2 OR CSHL1'*. This will return the net expression from the growth hormone locus.

For the academic or industrial researcher looking to extend their research into animal models, the system links into catalogues of gene knockout mice strains,

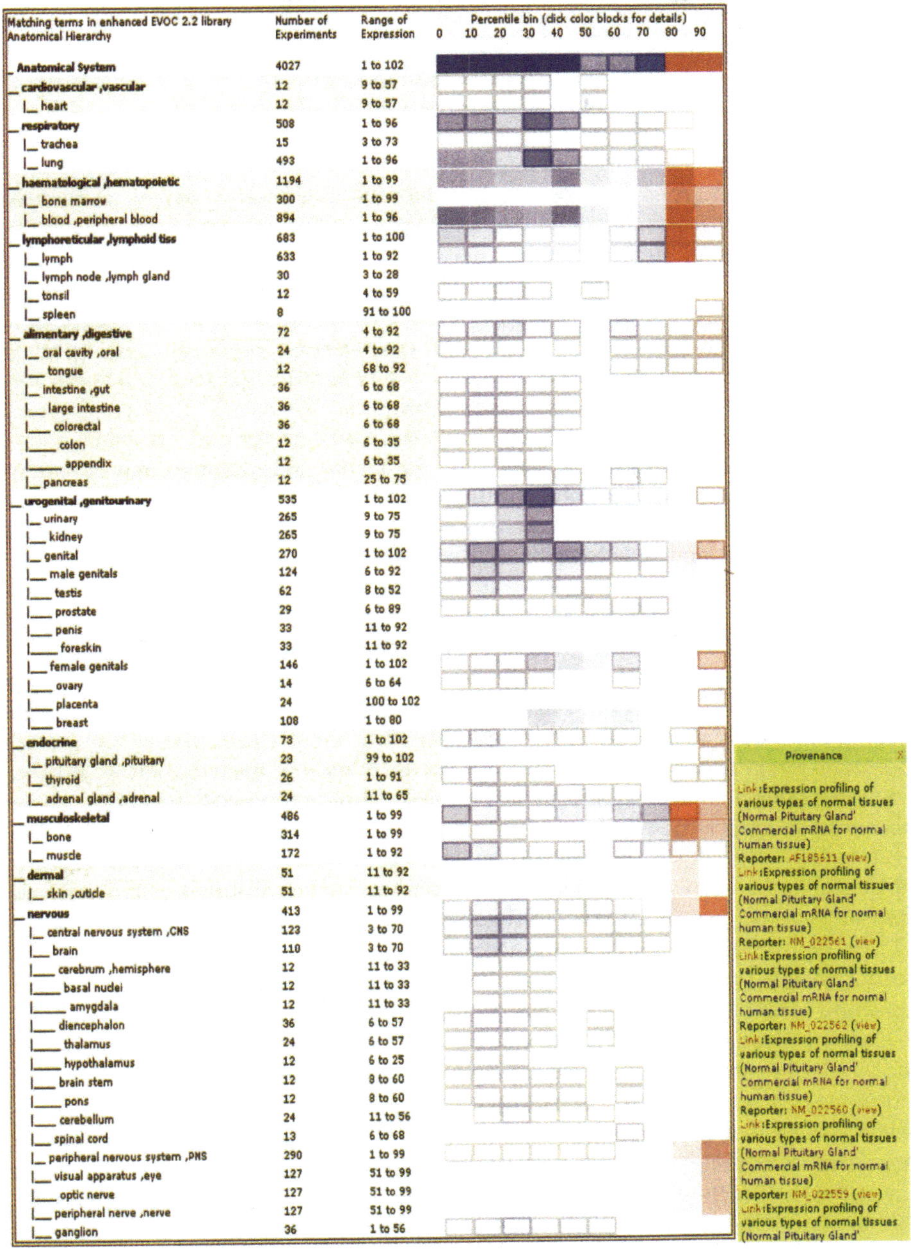

Fig. 13. GEO expression viewer. The distribution of GH1 reporters across tissues identified as present (P) on the Affymetrix platform. The popup box on the right shows the evidence for data in the top percentiles of the 90th-100th percentile-bin of pituitary.

allowing the user to identify which strains would allow the effect of the gene's disruption to be assessed. Currently the system links both the GeneTrap consortia's (IGTC) database and the corporate database at Bay Genomics.

6. APPLICATION

The utility of the system is demonstrated in the following hypothetical clinical scenarios derived from the literature. Consider a clinical study of congenital dwarfism in a small isolated fishing population. A group of distantly related patients within this population are suspected of presenting limited Type I pituitary dysfunction. In addition they exhibit a slight predisposition to thyroid papillary carcinoma, although this may be an environmental factor. Standard trypsin-giemsa staining is employed to reveal a subtle anomaly between 17q22 and 17q24 among those exhibiting dwarfism. Given the opportunity for directed study, the task is to validate the appropriate genetic segments to test.

In this case we select a single patient as the study group and cytoband region between 17q22 and 17q24 as the identifier. UniGenes and chromosomal variations are selected as our area of interest limiting the search to a subset of genes associated with the pituitary gland. The result returns 42 chromosomal variations and 113 genes with significant developmental abnormalities. This includes severe growth retardation in an infant with a substantial deletion between 17q23.2 and 17q24.3. If we sort the results on pituitary, the most specific pituitary expressed gene in this region is human growth hormone 1 (GH1) at 17q24.2 closely followed by the collocated GH2 gene. Both of these genes lie within the region identified by our karyotyping and also overlap that region identified in the OMIM record as leading to severely retarded growth. Following the map viewer link, the entire growth hormone locus that contains both adult and embryonic growth hormone genes located within our region of interest is revealed. WSU expression (Figure 12) shows GH1 to be essentially pituitary specific while GH2 is a placenta-specific variant. GEO expression shows that expression of GH1 is in the top 1% of the genes in normal pituitary tissue, but also shows expression across a range of other tissues (Figure 13). Following the links to the individual reporters specifies greater than a dozen different reporters. Some are tissue specific and some are also clearly reporting GH2, i.e., NM_000515 shown in full in Figure 14.

Examining the expression profiles of the other sequences reported by this search shows that several are expressed sequences without a known protein pedigree or ontological categorization. These candidates are excluded from immediate further analysis. Others are known to be linked to a range of phenotypes not observed amongst our population and can thus be discounted as potential causative agents. Given this evidence we choose to map the growth hormone locus in detail and are able to identify a deletion spanning GH1. Further analysis using OMIM might raise the issue of the genomic stability of the region following deletion or chimerism. For example, PRKAR1A that is linked to thyroid cancer as a chimeric oncogene is located in a neighboring chromosomal segment. Since our hypothetical population also shows a predisposition towards thyroid cancer this association may warrant further investigation within this population.

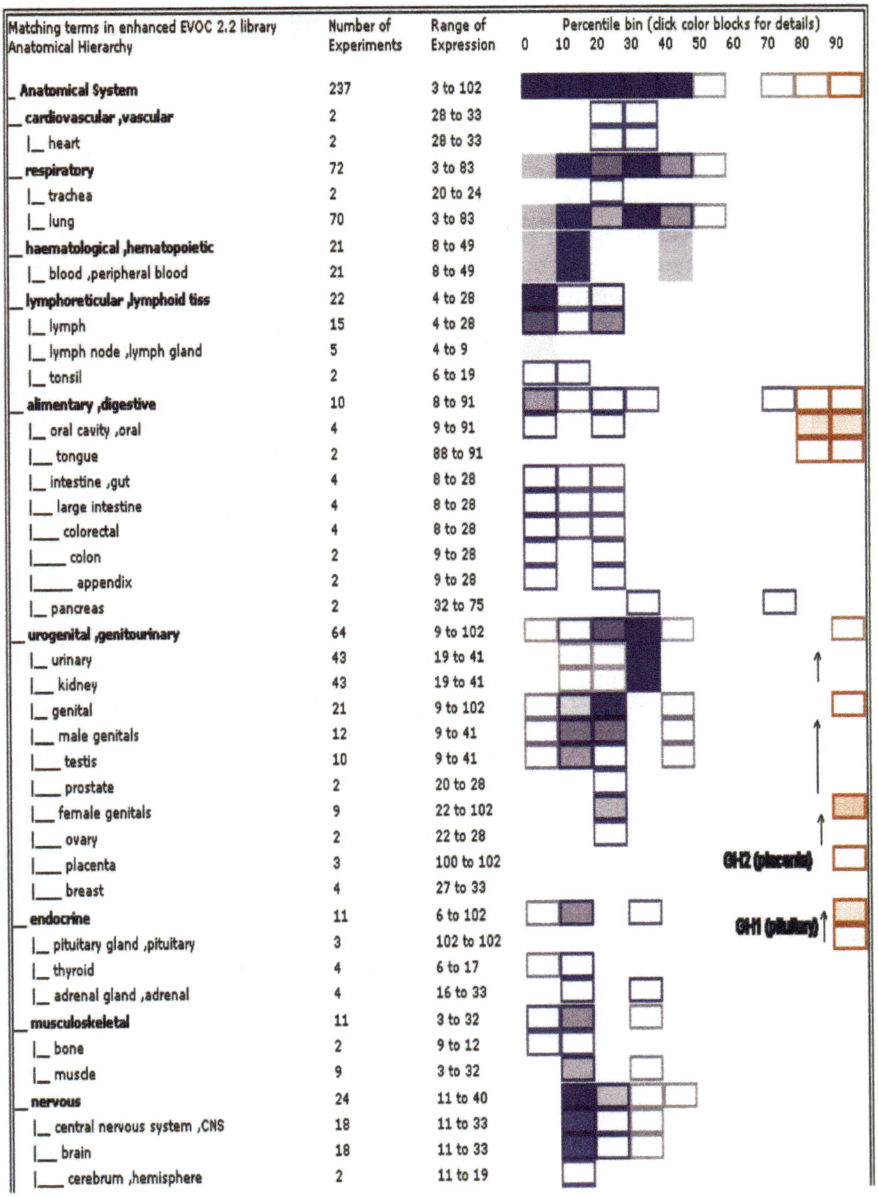

Matching terms in enhanced EVOC 2.2 library Anatomical Hierarchy	Number of Experiments	Range of Expression
_ **Anatomical System**	237	3 to 102
_ **cardiovascular ,vascular**	2	28 to 33
|_ heart	2	28 to 33
_ **respiratory**	72	3 to 83
|_ trachea	2	20 to 24
|_ lung	70	3 to 83
_ **haematological ,hematopoietic**	21	8 to 49
|_ blood ,peripheral blood	21	8 to 49
_ **lymphoreticular ,lymphoid tiss**	22	4 to 28
|_ lymph	15	4 to 28
|_ lymph node ,lymph gland	5	4 to 9
|_ tonsil	2	6 to 19
_ **alimentary ,digestive**	10	8 to 91
|_ oral cavity ,oral	4	9 to 91
|__ tongue	2	88 to 91
|_ intestine ,gut	4	8 to 28
|__ large intestine	4	8 to 28
|___ colorectal	4	8 to 28
|____ colon	2	9 to 28
|_____ appendix	2	9 to 28
|_ pancreas	2	32 to 75
_ **urogenital ,genitourinary**	64	9 to 102
|_ urinary	43	19 to 41
|__ kidney	43	19 to 41
|_ genital	21	9 to 102
|__ male genitals	12	9 to 41
|___ testis	10	9 to 41
|___ prostate	2	20 to 28
|__ female genitals	9	22 to 102
|___ ovary	2	22 to 28
|___ placenta	3	100 to 102
|___ breast	4	27 to 33
_ **endocrine**	11	6 to 102
|_ pituitary gland ,pituitary	3	102 to 102
|_ thyroid	4	6 to 17
|_ adrenal gland ,adrenal	4	16 to 33
_ **musculoskeletal**	11	3 to 32
|_ bone	2	9 to 12
|_ muscle	9	3 to 32
_ **nervous**	24	11 to 40
|_ central nervous system ,CNS	18	11 to 33
|__ brain	18	11 to 33
|___ cerebrum ,hemisphere	2	11 to 19

Fig. 14. Refining expression. The expression of GH1 can be specified to a single reporter: NM_022561. This reporter shows expression in pituitary at the top percentile. Cross hybridization with GH2 in placenta is also noted. Arrows show the inheritance of tissue expression data by parent tissues in the hierarchy (arrows). Moderately high expression (~91st percentile) in tongue is also observed. This is anomalous relative to its distribution in the EST database.

7. CONCLUSION

The ability to easily derive meaningful connections is the key to LARaLINK and will continue to be an area that is actively pursued. Not only are other emerging and established database resources being investigated for incorporation, but the XML output options of the system are being extended. Its service descriptions will shortly be lodged with one of the nascent bioinformatics web-services registries. For the clinician this will provide further access to the resources arising from different approaches and specialties. Whilst current data covers genetics, for the cytogeneticist, epigenetic data such as methylation status, DNA duplex stability and nuclear attachment will be equally important. As this data emerges from projects such as ENCODE and the Human Epigenome we will seek its integration. For both clinician and researcher alike, the advance of web services should ultimately reduce the frustration of broken weblinks and the need for innumerable pages that lead only to more pages of links. As the community embraces and modifies web services, static links will be joined by more proactive relationships brokered by systems such as LARaLINK for dynamic resources, flexible enough to embrace new data as it becomes available and to remove (and potentially archive) old resources as they become obsolete. The key to success will remain the capability to translate between diverse vocabularies and to relate data through dynamic synonyms, parents and children set in a hierarchical set of ontologies_Acknowledgments: The authors gratefully acknowledge the Michigan Economic Development Corporation and the Michigan Technology Tri-corridor for the support of this program by grant 085P1000819 to SAK and NSF grant 0234806. We would like to thank the open source community for all the tools provided, without which this project would not have been possible.

REFERENCES

Barrett T, Suzek TO, Troup DB, Wilhite SE, Ngau W, Ledoux P, Rudnev D, Lash AE, Fujibuchi W and Edgar R (2005) NCBI GEO: mining millions of expression profiles. Nucleic Acids Research. Vol. 33, Database issue D562-D566

Boguski MS, Lowe TM and Tolstoshev CM (1993) dbEST--database for "expressed sequence tags" Nat Genet. 4(4):332-3.

Borgaonkar, D S (1997) Chromosomal Variation in Man, 8th ed. John Wiley/Liss, NY and Chromosomal Variation in Man online database.

Catalog of Human Cancer Genes : McKusick's Mendelian Inheritance in Man for Clinical and Research Oncologists (Onco-MIM). The Johns Hopkins University Press (April 1, 1999)

Diehn M, Sherlock G, Binkley G, Jin H, Matese JC, Hernandez-Boussard T, Rees CA, Cherry JM, Botstein

D, Brown PO and Alizadeh AA (2003) SOURCE: A Unified Genomic Resource of Functional Annotations, Ontologies, and Gene Expression Data. Nucleic Acids Research. Vol. 31 (1) 219-223.

Fayz B, Moldenhauer JS, Wang D, Zhao C, Yao B, Liu D, Weinsheimer S, Gardner L, Johnson A Womble DD, Krawetz SA (2005) LARaLINK: a web application for cytogenetic linkage analysis. Clinical Genetics Vol 67: 314-318

Kelso J, Visagie J, Theiler G, Christoffels A, Bardien-Kruger S, Smedley D, Otgaar D, Greyling G, Jongeneel V, McCarthy M, Hide T and Hide W (2003) eVOC: A Controlled Vocabulary for Gene Expression Data. Genome Research 13: 1222-1230.

Mitelman F. Chromosomal and molecular genetic aberrations of tumor cells. In: Cancer Cytogenetics. New York: Wiley-Liss; 1995

Platts AE, Moldenhauer JS, Fayz B, Wang D, Borgaonkar DS and Krawetz SA (2005) LARaLINK 2.0: A Comprehensive Aid to Basic and Clinical Cytogenetic Research. Genetic Testing [In Press]

Pontius JU, Wagner L and Schuler GD (2003) UniGene: a unified view of the transcriptome. In: The NCBI Handbook. Bethesda (MD): National Center for Biotechnology Information.
Schinzel A (2002) Human Cytogenetics Database 2.0, Oxford University Press
Baraitser M and Winter RM (2001) London Dysmorphology Database, London Neurogenetics Database &
Dysmorphology Photo Library. Oxford: Oxford University Press
Sijmons RH and Burger GTN (2003) The use of a diagnostic database in clinical oncogenetics. Hereditary Cancer in Clinical Practice):1: 31-33,
van Gent DC, Hoeijmakers JH and Kanaar R (2001) Chromosomal stability and the DNA double-stranded break connection. Nat Rev Genet (3):196-206.

ELSEVIER

Applied Mycology
and Biotechnology
An International Series
Volume 6. Bioinformatics

Sequence-Based Analysis of Fungal Secretomes

Nicholas O'Toole[1], Xiang Jia Min[1], Gregory Butler[1,2], Reginald Storms[1,3] and Adrian Tsang[1,3]
[1]Centre for Structural and Functional Genomics, Concordia University, Montreal, Quebec, H4B 1R6, Canada; [2]Department of Computer Science, Concordia University, Montreal, Quebec, H3G 1M8, Canada; [3]Department of Biology, Concordia University, Montreal, Quebec, H4B 1R6, Canada (tsang@vax2.concordia.ca)

Secreted proteins play critical biological roles in fungal species. Here we review and assess computational protocols for the identification of secreted proteins using their amino acid sequences. Protein sequences are screened for the presence of secretory signal peptides and the lack of features that prevent the delivery of proteins to the extracellular space, such as transmembrane segments, C-terminal ER-retention signals, or glycosylphosphatidylinositol (GPI) anchors. We apply such techniques to the complete genomes of 10 fungal species, identifying their putative complete sets of secreted proteins (their *secretomes*). Particular attention is given to predictions for the yeast secretome, which can be validated using the curated subcellular localizations of proteins from yeast. We make distinctions between the soluble and non-soluble portions of secretomes, discussing the roles of putative GPI proteins in the fungal cell wall.

1. INTRODUCTION

Fungi use secreted proteins to accomplish their diverse lifestyles. Most fungi adopt a nutritional strategy in which they secrete extracellular enzymes to break down potential food sources and then transport the resulting products into the cells. To accommodate this way of life, fungi have evolved the most effective and comprehensive arrays of catalytic activities. The ability to breakdown complex substrates, especially lignocellulose, has made fungi the major decomposers of the biosphere and an important driver of the Earth's carbon cycle (Frankland et al. 1982; Carroll and Wicklow 1994). Extracellular proteins form the cornerstones of these roles.

Fungal species, including *Aspergillus niger, Aspergillus oryzae, Aspergillus aculeatus, Aspergillus japonicus, Rizopus oryzae, Trichoderma reesei, Talaromyces emersonii,*

Corresponding author: Adrian Tsang

Kluyveromyces marxianus and *Saccharomyces cerevisiae*, have a long history of use in the food industry (Mattey 1992; Nevalainen et al. 1994; Jeenes et al. 1991; Polizeli et al. 2005) and are assigned a generally recognized as safe (GRAS) status by the United States Food and Drug Administration (http://www.enzymetechnicalassoc.org/). Several fungi, particularly those from the genus *Aspergillus* (*A. niger*, *A. oryzae A. aculeatus* and *A. japonicus*) and *T. reesei*, can secrete very high levels of some proteins into the culture medium. The ability to secrete some proteins at high levels has been exploited to develop strains for industrial scale enzyme production. For example, some commercial strains of *A. niger* and *T. reesei* secrete homologous glucoamylases and cellulases at more than 30g/l (Finkelstein et al. 1989; Durand et al. 1988).

The above *Aspergillus* strains are also proficient at performing eukaryotic post-translational modifications, including proteolytic processing and protein glycosylation (Archer and Peberdy 1997). This makes them attractive hosts for the production of heterologous proteins derived from other eukaryotic sources including humans. These attributes together with well-developed methods for the production of genetically modified strains and performing large-scale fermentations have been exploited commercially to produce organic acids (Mattey 1992) and a variety of native and foreign proteins (Gouka et al. 1997). Enzymes expressed in these fungi have found a wide range of applications in the food (Camacho and Aguilar 2003), feed (Pandey et al. 2000), pulp and paper (Araújo et al. 1999), bioethanol (Bothast and Schlicher 2005) and textile (Csiszár et al. 2001) industries. Remarkably, in the food applications area over 100 different fungal-produced enzyme preparations are marketed commercially.

White rot fungi, particularly the well-studied *Phanerochaete chrysosporium*, have an extraordinary ability to degrade and mineralize a broad spectrum of chemical pollutants (Bumpus and Aust 1987; Paszczynski and Crawford 1995; Reddy 1995). Mineralization of these compounds is initiated by extracellular peroxidases followed by internalization where degradation apparently involves a diverse set of intracellular P450 monooxygenases (Doddapaneni et al. 2005).

Although the hydrolytic capabilities of fungi are well documented, the repertoires of secreted proteins that define their capabilities remain to be fully elucidated. Furthermore, it is often difficult to obtain the high levels of expression required for commercial purposes for some homologous proteins (Gouka et al. 1997) and most heterologous proteins (Berka et al. 1997; Jeenes et al.1994; Ruiz-Duenas et al. 1999). A comprehensive account of all fungal secreted proteins would be helpful in identifying additional enzymes with improved and novel activities for commercial and environmental applications.

Almost all secreted proteins are sorted to the endoplasmic reticulum (ER) during synthesis. They contain a signal peptide at the N-terminus (Blobel and Dobberstein 1975). This peptide sequence directs the ribosomes that are synthesizing the secreted proteins to the rough ER. Once bound, the polypeptide crosses the ER membrane as it is being synthesized. This method of entering the ER is referred to as the co-translational pathway. After the transfer into the ER is completed, the nascent protein is folded and moves through the ER and the Golgi complex. Eventually, the secreted proteins are placed in transport vesicles and shipped to the cell exterior.

Export via the ER/Golgi is often called the classical secretory pathway. Secreted proteins make up only a fraction of the proteins that enter the ER. Proteins that contain the ER signal peptide and enter the ER include residents of the rough ER, smooth ER, Golgi complex, lysosomes, endosomes, and plasma membrane.

Some proteins enter the ER secretory pathway after they are completely synthesized in the cyotosol and released from the ribosomes, the post-translational pathway. Among them are a group of membrane proteins that use a transmembrane signal at the C-terminus to enter the ER (Abell et al. 2003). Mammalian cells are thought to use primarily the co-translational pathway whereas the yeast *Saccharomyces cerevisiae* uses both the co-translational and post-translational pathways (Corsi and Schekman 1996; Kalies and Hartmann 1998).

While many proteins without an N-terminal signal peptide can be found in the ER and the Golgi, over 90% of human secreted proteins contain classical N-terminal signal peptides (Scott et al. 2004). However, a few well-characterized, secreted proteins of mammalian cells do not possess the classical signal peptide and they are exported outside the cell by mechanisms independent of the ER/Golgi secretory pathway (Nickel 2003). A recently developed algorithm, SecretomeP (Bendsten et al. 2004b), attempts to identify these nonclassical mammalian secreted proteins. There are examples of non-classically secreted proteins in fungi, including the *S. cerevisiae* mating pheromone a-factor (Chen et al. 1997) and two galectins from *Coprinus cinerius* (Boulianne et al. 2000), but it is believed that the vast majority of extracellular fungal proteins are processed by the classical secretory pathway.

In this paper we review and assess computational tools for the identification of fungal classically secreted proteins using their amino acid sequences. In section 2, past large-scale studies aimed at the identification of secreted proteins from all kingdoms of life are reviewed. Section 3 is a comprehensive summary of bioinformatics tools that have been developed for identification of secreted proteins. In section 4, we describe the screening protocols used in this work to identify secreted proteins, which are used in section 5 on the proteome of *S. cerevisiae* to identify the set of secreted yeast proteins (the yeast *secretome*). The accuracy of the methods is assessed with the manually curated subcellular localization information available for the yeast proteome. Finally, in section 6, the methodology is extended to 9 other fungal proteomes. Trends within the set of secretomes are discussed and we pay particular attention to putative cell wall proteins, the non-soluble component of fungal secretomes.

2. SEQUENCE-BASED SECRETOME ANALYSES

Proteins targeted to the extracellular space by the classical secretory pathway possess an N-terminal signal peptide, composed of a central hydrophobic core surrounded by N- and C- terminal hydrophilic regions. The signal peptide is recognized by the secretory machinery of the cell and cleaved upon export to the cell membrane in prokaryotes or the endoplasmic reticulum (ER) in eukaryotes. It is important to note that not all proteins processed by the classical secretory pathway will be delivered to the extracellular space, particularly in eukaryotes, where they can finally reside in a number of other cellular compartments (e.g. the Golgi apparatus, lysosomes). Proteins in the secretory pathway that remain within or anchored to the cell often contain additional sequence features responsible for their retention, such as transmembrane helices, endoplasmic reticulum (ER) retention signals or glycosylphosphatidylinositol (GPI) anchors, although in fungi a GPI attachment signal may indicate targeting to the cell wall (de Groot et al. 2005). Therefore fungal GPI proteins are potentially extracellular proteins.

The term *secretome* was introduced by Tjalsma et al. (2000) to denote the complete set of proteins in an organism processed by the secretory pathway. However, it has more recently been used to refer to the extracellular portion of the proteome (e.g, Greenbaum et al. 2001) and it is in this context that we use the term in this work. Putative secretomes for several organisms have recently been identified computationally using a variety of bioinformatics tools on large sets of genomic data. For example, Lee et al. (2003) analyzed the 6165 predicted open reading frames (ORFs) from the genome sequence of *Candida albicans* for the presence of a signal peptide with the SignalP program and the absence of transmembrane helices, GPI anchors or mitochondrial targeting peptides with the TMHMM, Big-PI and TargetP programs respectively. Protein sequences fulfilling these criteria were designated secreted and the final *C. albicans* secretome was predicted to contain 283 ORFs. Similarly, Grimmond et al. (2003) identified full-length cDNA clones from the RIKEN Mouse Gene Encyclopedia project encoding proteins with signal peptides and no transmembrane helices according to SignalP and TMHMM. The mouse secretome identified by these authors contained 2033 potentially secreted proteins. Similar techniques were employed with human and puffer fish genomic data (Klee et al. 2004). These authors used the TargetP algorithm alone to identify signal peptides and thus the study was confined to proteins in the classical secretory pathway. EST data from the nematode *Nippostrongylus brasiliensis* were analyzed by Harcus et al. (2004) with the SignalP program to identify potentially secreted proteins. The over-representation of genes with unknown function among the putative secreted proteins led these authors to claim that secreted proteins are undergoing accelerated evolution in this parasitic organism, presumably because of changing demands on secreted proteins responsible for host-pathogen interactions. Klotz et al. (2005) also used the SignalP algorithm to analyze EST sequences from the protozoan parasite *Eimeria tenella* and found 51 putative secreted proteins. A functional complementation system in yeast was able to experimentally verify that 5 of these were indeed secreted proteins. Trost et al. (2005) have used seven different bioinformatic tools to identify and compare predicted secretomes of pathogenic and nonpathogenic *Listeria* species. Most recently, Wymelenberg et al. (2005) analysed the genome sequence of the rot basidiomycete *P. chrysosporium* with the SignalP, TargetP and TMHMM algorithms to identify 268 potentially secreted proteins. A mass-spectrometric analysis was also performed to experimentally identify 50 secreted proteins, of which 24 were among the subset of genes predicted to be secreted computationally.

3. DATA SOURCES AND TOOLS

3.1 Data Sources

3.1.1 NCBI RefSeq collection

The protein database in NCBI contains sequence data from the translated regions of cDNA sequences and predicted gene models from genomes in GenBank, EMBL and DDBJ as well as protein sequences submitted to PIR, SWISS-PROT, PRF, PDB (Protein Data Bank). The Reference Sequence (RefSeq) collection in NCBI provides a comprehensive, integrated, non-redundant set of sequences, including genomic

DNA, transcript (RNA), and protein products, for major research organisms (Pruitt et al. 2005; http://www.ncbi.nlm.nih.gov/RefSeq/). RefSeq standards serve as the basis for medical, functional, and diversity studies. RefSeqs are used as a reagent for the functional annotation of some genome sequencing projects, including those of human and mouse. Thus, the RefSeq protein sequences are a good resource for secretome analysis. The protein sequences including RefSeqs are accessible via BLAST, Entrez, and the NCBI FTP site. Information is also available in Entrez Genomes and Entrez Gene, and for some genomes additional information is available in the Map Viewer. Specific information about the NCBI sequence resources for fungi is available at http://www.ncbi.nlm.nih.gov/genomes/FUNGI/funtab.html.

3.1.2 The Saccharomyces genome database

The Saccharomyces Genome Database (SGD, http://www.yeastgenome.org) is the most comprehensive resource for yeast molecular biology and genetics. Importantly, the SGD contains manually curated gene ontology (GO, http://www.geneontology.org) terms for all known yeast gene products (Dwight et al. 2002). For the analysis described in this chapter, all GO molecular function, biological process and cellular component terms plus amino acid sequences were downloaded from the SGD website and associated with yeast RefSeq sequences via a strict BLAST alignment (at least 95% pairwise sequence identity). For association of yeast genes with subcellular localization, GO-Slim terms were used.

3.2 Tools

In this subsection we describe and review the major software tools used to predict subcellular localization based on the amino acid sequences of proteins, some of which are used later in the present work.

3.2.1 SignalP

SignalP consists of two different predictors based on neural network and hidden Markov model algorithms for predicting the presence and location of signal peptide cleavage sites in amino acid sequences from different organisms including Gram-positive prokaryotes, Gram-negative prokaryotes, and eukaryotes (Nielsen et al. 1997; Bendtsen et al. 2004a). This tool is particularly useful for EST and genome projects aiming at finding genes encoding secreted proteins. A web server for the current version, SignalP 3.0, with improved prediction accuracy, is available at http://www.cbs.dtu.dk/services/SignalP/.

3.2.2 TMHMM

TMHMM (Krogh et al. 2001) predicts transmembrane helical topology of proteins with a hidden Markov model, discriminating between membrane and soluble proteins with sensitivity and specificity better than 99%. An independent study found that TMHMM was the best performing transmembrane prediction program (Moller et al. 2001) and a recent study by Cuthbertson et al. (2005) showed that TMHMM was among the best four performing transmembrane topology predictors. Krogh et al. (2001) noted, however, that the accuracy of TMHMM is lower when a signal peptide is present in the sequence.

3.2.3 Phobius

Phobius is a combined transmembrane protein topology and signal peptide predictor. The predictor is based on a hidden Markov model (HMM) that models the different sequence regions of a signal peptide and the different regions of a transmembrane protein in a series of interconnected states (Kall et al. 2004). Compared to TMHMM and SignalP, errors coming from cross-prediction between transmembrane segments and signal peptides were reduced substantially by Phobius. False classifications of signal peptides were reduced from 26.1% to 3.9% and false classifications of transmembrane helices were reduced from 19.0% to 7.7%. Phobius is well suited for whole-genome annotation of signal peptides and transmembrane regions. The method is available at http://phobius.cgb.ki.se/.

3.2.4 TargetP

TargetP is a neural network-based tool for large-scale subcellular location prediction of proteins (Emanuelsson et al. 2000). Using N-terminal sequence information only, it discriminates between proteins destined for the mitochondrion, the chloroplast, the secretory pathway, and "other" localizations with a success rate of 85% (plant) or 90% (non-plant) on redundancy-reduced test sets. TargetP is available as a web-server at http://www.cbs.dtu.dk/services/TargetP/.

3.2.5 The PSORT family

The PSORT family server (http://psort.nibb.ac.jp/; http://psort.org/) contains several variant tools for the prediction of protein localization sites in cells. It receives the information of an amino acid sequence and its source origin, *e.g.*, Gram-negative bacteria, as inputs. Then, it analyzes the input sequence by applying the stored rules for various sequence features of known protein sorting signals. Finally, it reports the possibility for the input protein to be localized at each candidate site with additional information. The family includes the following variant tools: (1) PSORT: an old version for plants and bacteria; (2) PSORT II: recommended for yeast and animal; (3) iPSORT: for N-terminal sorting signals for plants or non-plants; (4) PSORT-B: recommended for Gram-negative bacteria; (5) WoLF PSORT: recommended for animal, plant and fungi.

3.2.6 Big-PI fungal predictor

The Big-PI Fungal Predictor (Eisenhaber et al. 2004) is a program to detect glycosylphosphatidylinositol (GPI) anchoring signals in fungi. The program is based on a learning set of preprotein sequences exclusively from fungi and has a sensitivity close to 90% with a false positive prediction rate of approximately 0.1%, which compares extremely favourably with older methods of GPI anchor prediction. The program is available via a web server at http://mendel.imp.univie.ac.at/gpi/fungi_server.html.

3.2.7 SecretomeP

The SecretomeP server (http://www.cbs.dtu.dk/services/SecretomeP/) is trained for prediction of mammalian secretory proteins targeted by the non-classical secretory pathway, i.e. proteins without an N-terminal signal peptide (Bendtsen et al. 2004b). The method used in the server is also capable of predicting signal peptide-containing secretory proteins where only the mature part of the protein has been annotated, or cases where the signal peptide remains uncleaved. However, the method was trained only with mammalian proteins, its potential applications to other organisms remain to be tested. Another limitation is that the server only allows 50 sequences per submission.

3.2.8 LOCSVMPSI

LOCSVMPSI is a web server for the prediction of subcellular localization of eukaryotic proteins using the support vector machine (SVM) and the position-specific scoring matrix generated from profiles of PSI-BLAST. LOCSVMPSI performed better than some widely used prediction methods, such as PSORTII, TargetP and LOCnet (Xie et al. 2005). An online web server can be accessed at http://Bioinformatics.ustc.edu.cn/LOCSVMPSI/LOCSVMPSI.php.

3.2.9 Protein prowler

The Protein Prowler server predicts subcellular localization using sequence-biased recurrent networks (Boden and Hawkins 2005). It was demonstrated that recurrent networks improve the overall prediction performance. Compared to the original results reported for TargetP, the accuracy was increased by 6% and 5% on non-plant and plant data, respectively. The Protein Prowler is available online at http://pprowler.imb.uq.edu.au/.

3.2.10 Locfind

Locfind is based on bidirectional recurrent neural networks trained to read sequentially the amino acid sequence and produce localization information along the sequence for the prediction of the localization of eukaryotic proteins (Reczko and Hatzigerrorgiou 2004). Systematic variation of the network architecture in combination with an efficient learning algorithm leads to a 91% correct localization prediction for novel proteins in fivefold cross-validation. The Locfind system is available at http://139.91.72.10/blstm/blstm.html.

3.2.11 SubLoc

SubLoc uses the support vector machine (SVM) to predict the subcellular localization of proteins from their amino acid compositions (Hua and Sun 2001). The total prediction accuracies reach 91.4% for three subcellular locations in prokaryotic organisms and 79.4% for four locations in eukaryotic organisms. It is available at http://www.bioinfo.tsinghua.edu.cn/SubLoc/.

3.2.12 Other transmembrane protein predictors

Numerous programs have been developed for the prediction of transmembrane segments in proteins. They are briefly listed here:

ConPred II (http://bioinfo.si.hirosaki-u.ac.jp/~ConPred2/; Arai et al. 2004).

DAS (http://www.sbc.su.se/~miklos/DAS/; Cserzo et al. 1997).

DAS-TMfilter (http://mendel.imp.univie.ac.at/sat/DAS/DAS.html; Cserzo et al. 2004).

HMMTOP (http://www.enzim.hu/hmmtop/index.html; Tusnady and Simon, 2001).
kPROT (http://bioinformatics.weizmann.ac.il/kPROT/; Pilpel et al. 1999).
MEMSAT (http://saier-144-37.ucsd.edu/memsat.html; Jones et al. 1994).
OrienTM (http://biophysics.biol.uoa.gr/OrienTM/; Liakopoulos et al. 2001).
PHDhtm (http://cubic.bioc.columbia.edu/predictprotein/; Rost et al. 1996).
PSIPRED (http://bioinf.cs.ucl.ac.uk/psipred/; McGuffin et al. 2000).
SOSUI (http://sosui.proteome.bio.tuat.ac.jp/sosuiframe0.html; Hirokawa et al. 1998).
SVMtm (http://genet.imb.uq.edu.au/predictors/SVMtm; Yuan et al. 2004).
THUMBUP (http://theory.med.buffalo.edu/; Zhou and Zhou, 2003).
TMAP (http://bioweb.pasteur.fr/seqanal/interfaces/tmap.html; Persson and Argos, 1994).
TMBETA-NET (http://psfs.cbrc.jp/tmbeta-net/; Gromiha et al. 2004).
TMFinder (http://www.ccb.sickkids.ca/tools/tmfinder/html/login.html; Deber et al. 2001).
TMMOD (http://liao.cis.udel.edu/website/servers/TMMOD/; Kahsay et al. 2005).
TMpred (http://www.ch.embnet.org/software/TMPRED_form.html; Hoffman and Stoffel, 1993).
TopPred (http://bioweb.pasteur.fr/seqanal/interfaces/toppred.html; Claros and von Heijne, 1994).
waveTM (http://bioinformatics.biol.uoa.gr/waveTM/input.html; Pashou et al. 2004).

4. SCREENING PROTOCOLS FOR SECRETED PROTEINS

Past sequence-based bioinformatic studies of secretomes have relied on a combination of programs to predict signal peptide and transmembrane helices in proteins (usually SignalP and TMHMM respectively), to predict whether a protein contains a secretory signal peptide and no transmembrane regions, and is thus potentially an extracellular protein. However, the hydrophobic stretch of amino acids in the core of a signal peptide is often mistaken for the hydrophobic residues in a transmembrane helix by programs that predict transmembrane helices. Conversely, the signal peptide predictors can mistake an N-terminal transmembrane helix for a signal peptide. The implications of this ambiguity for the prediction of transmembrane topologies in proteins have been explored by Lao et al. (2002a,b). In a comprehensive comparative analysis of 27 different methods for the prediction of transmembrane helices, Chen et al. (2002) found that few existing algorithms were able to reliably distinguish between transmembrane helices and signal peptides. Efficient algorithms for the differentiation of signal peptides and transmembrane helices are in development (e.g. Yuan et al. 2003).

Due to these difficulties a certain degree of inaccuracy is unavoidable when using the SignalP and TMHMM programs in tandem for the prediction of secreted proteins. To overcome the ambiguity of simultaneous predictions of a signal peptide and an N-terminal transmembrane helix, Lee et al. (2003) and Trost et al. (2005) ignore predicted transmembrane helices if they end within 40 amino acids of the N terminus. On the other hand, Wymelenberg et al. (2005) and de Groot et al. (2003), in their study of fungal GPI proteins, exclude proteins with one or more predicted transmembrane helices from their list of secreted proteins, so in these studies the signal peptide prediction is ignored. Naturally the sensitivity and specificity of tests for secreted proteins will be affected by the protocol chosen. We suspect that the former choice would result in fewer false negative predictions of secreted proteins and thus increase the sensitivity of the predictions algorithm at the expense of its specificity.

The *Phobius* program (Käll et al. 2004), a *joint* signal peptide/transmembrane helix predictor, was developed recently in order to resolve the difficulties inherent in combining SignalP and TMHMM or analogous methods. Compared to TMHMM and SignalP, errors coming from cross-prediction between transmembrane segments and signal peptides were reduced substantially by Phobius. False classifications of signal peptides were reduced from 26.1% to 3.9% and false classifications of transmembrane helices were reduced from 19.0% to 7.7%. We have utilized the Phobius software in the present secretome analysis because of its computational efficiency and its potential for improvements in prediction accuracy.

The first step in our sequenced-based screening protocol for secreted proteins is the identification of those proteins with a signal peptide and no transmembrane helices. This data is immediately extractable from the results of the Phobius program. For completeness, in this section on the yeast secretome we also apply the SignalP v.3.0 and TMHMM programs, using the old protocol of Lee et al. (2003) and Trost et al. (2005), i.e. a single transmembrane helix predicted by TMHMM is ignored if a signal peptide is predicted. A signal peptide is considered detected by the SignalP program if both the neural network and hidden Markov model parts of SignalP agree that there is a signal peptide.

Following the Phobius or SignalP/TMHMM step, we identify soluble proteins with a signal peptide that remain localized in the endoplasmic reticulum due to an extension of the C-terminal KDEL motif, first discovered by Munro and Pelham (1987). Recently Wrzeszczynski and Rost (2004) have analyzed the occurrences of protein sorting motifs among proteins from different cellular compartments and inspecting their data we find that the C-terminal [KHR][DENQ]EL pattern (in PROSITE notation) is sufficiently sensitive and specific to screen for retention in the ER. Therefore proteins containing this motif are considered soluble ER proteins and are not included in the secretome.

It is well recognized that the similarity between secretory signal peptides and mitochondrial targeting peptides remains a serious barrier to accurate prediction of subcellular localization. The final step of the screening protocol is the use of the TargetP (Emanuelsson et al. 2000) algorithm to exclude proteins with predicted mitochondrial target peptides from the predicted secretomes. We apply the default, winner-takes-all cutoffs.

5. THE YEAST SECRETOME

5.1 Secretome Lists

The secreted protein screening protocol was applied to all 5866 *Saccharomyces cerevisiae* protein sequences from the NCBI RefSeq resource (see Section 3.1). The final putative "secretomes" using the Phobius and SignalP/TMHMM screenings contained 226 and 198 proteins respectively. Figure 1 illustrates the number of proteins removed at each stage of both screening protocols. The two lists of secreted proteins share 164 members in common.

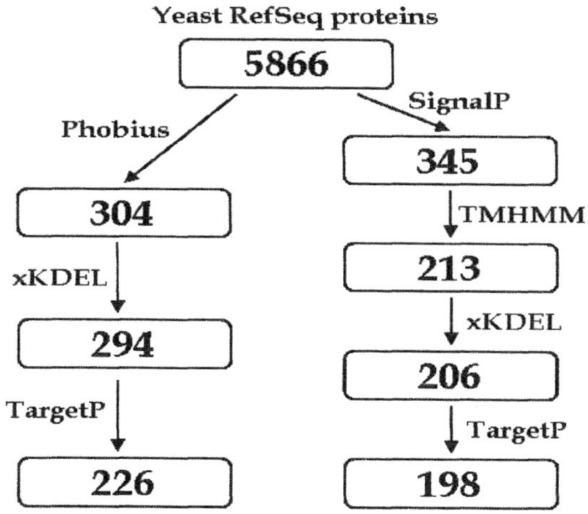

Fig. 1. Flowchart showing the screening of yeast proteins for secretome analysis. "xKDEL" refers to the extended KDEL motif [KHR][DENQ]EL.

5.2 Subcellular Localization of Secretome Proteins

Each member of the yeast secretome lists was associated with a GO term from the Saccharomyces Genome Database. Figure 2a displays the breakdown of the predicted secretomes by curated subcellular localization. In order to generate meaningful divisions in this figure, cellular component GO-Slim terms were used. These were downloaded from the SGD FTP site ftp://genome-ftp.stanford.edu/pub/yeast/data_download/literature_curation/. Under GO-Slim mapping regular GO cellular compartment terms are mapped to one or more of the following 24 higher level subcellular compartments: bud, cell cortex, cell wall, cellular component unknown, chromosome, cytoplasm, cytoplasmic membrane-bound vesicle, cytoskeleton, endomembrane system, endoplasmic reticulum, extracellular region, Golgi apparatus, membrane, membrane fraction, microtubule organizing center, mitochondrial membrane, mitochondrion, nucleolus, nucleus, peroxisome, plasma membrane, ribosome, site of polarized growth and vacuole. Additional information on GO-Slim terms is available at http://www.yeastgenome.org/help/goslimhelp.html. For comparison purposes, the proportions of total yeast genes in the same divisions of GO-Slim terms are displayed in Fig. 2b.

The two screening methods do not produce significantly different results. The large over-representation of proteins in the cell wall, extracellular, vacuole and ER compartments in Fig. 2a compared to Fig. 2b indicates that as expected, the screening protocol detects proteins in the classical secretory pathway. Perhaps surprising at first glance is the smaller than expected portions of the secretomes with an annotated extracellular location. However, there are only 19 ORFs in the SGD with a

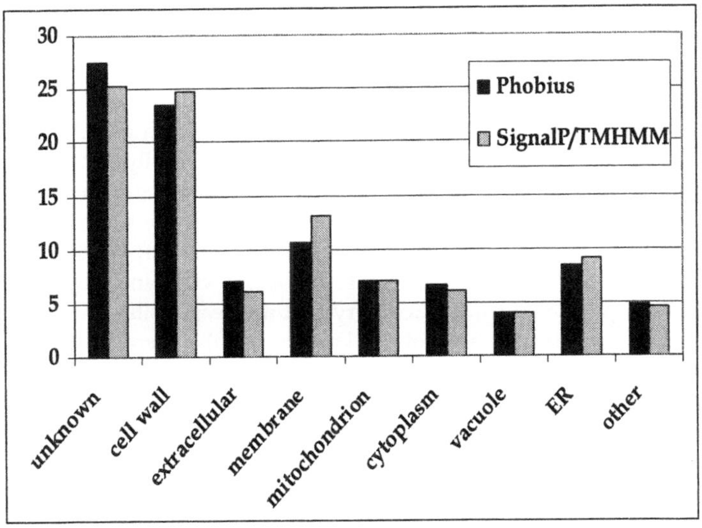

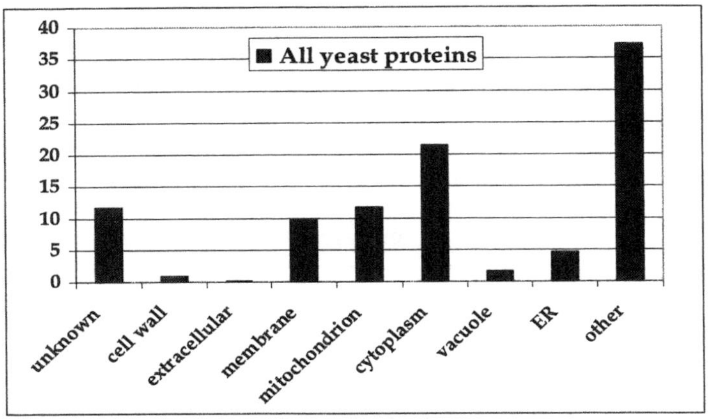

Fig. 2. a) Proportions of the predicted yeast secretome in different subcellular compartments based on GO-Slim annotation from the Saccharomyces Genome Database. The "other" column contains proteins from the "nucleus", "Golgi apparatus" and "ribosome" cellular compartments Go-Slim terms (there were no more than 5 proteins in any one of these compartments using both screening protocols). b) Proportions of all yeast proteins in same subcellular divisions as part a).

cellular compartment GO term of "extracellular region". The Phobius and SignalP/TMHMM methods detect 16 and 12 of the 19 extracellular ORFs in the SGD respectively. Two of the three extracellular ORFs that were not detected by the Phobius protocol are the mating pheromone a-factor genes MFA1 and MFA2, which are known not to be secreted by the classical pathway (Chen et al. 1997). The other extracellular ORF not detected is an unusual secreted RNase RNY1 (Macintosh et al.

2001) whose signal peptide is only detected by the neural network method of SignalP. Excluding the two non-classically secreted proteins from the analysis, the Phobius and SignalP/TMHMM methods have sensitivities of 94% (16/17) and 71% (12/17) for extracellular yeast proteins respectively. However, sensitivity alone is a meaningless statistic because trivially labeling every yeast protein as extracellular yields a sensitivity of 100%. More meaningful statistics on the secreted protein screenings are calculated below.

Another striking feature in Fig. 2a is the large proportion of cell wall proteins found in the secretomes predicted by both methods. Naturally, one would expect cell wall proteins to be processed by the secretory pathway and the detection of a large number of them is partial evidence of the success of the secretome screening. Similarly, our screening *should* detect vacuolar and soluble ER proteins without an extended KDEL motif. Presumably these proteins contain sequence features recognized by the protein sorting machinery that are responsible for their retention within the cell. These features are not yet known, but the identification of proteins that possess them, by methods such as ours, should hasten their discovery. Proteins in the membrane, mitochondrial and cytoplasmic cellular compartments should not have been detected by the secretome screening, assuming their GO annotations are accurate, and these represent a failure of the Phobius/TMHMM, TargetP and Phobius/SignalP programs respectively. There are a larger number of membrane proteins detected by the SignalP/TMHMM method than the Phobius one and this may be due to its protocol of ignoring an N-terminal transmembrane helix when a signal peptide is present. There are a slightly larger number of cytoplasmic proteins detected by the Phobius screening and this, as well as anecdotal evidence gathered by manual inspection of the predictions, suggests that the Phobius criterion for detecting a signal peptide is quite a bit looser than that of SignalP. Note that proteins in the "other" division of Fig. 2a represent a very small proportion of the predicted secretome, whereas they represent over a third of all yeast proteins. (Most of these proteins are nuclear and ribosomal.)

The GO-Slim cellular component terms for yeast proteins can also be used for a quick albeit imperfect calculation of the sensitivity and specificity measures of the testing for secreted proteins (Loong, 2003). A list of the yeast proteins with the GO-Slim cellular component terms "cell wall" and "extracellular region" was extracted. This "positive" list contains 91 proteins (72 cell-wall and, as mentioned before, 19 extracellular). All of these proteins were among the original set of RefSeq sequences that were tested. The Phobius and SignalP/TMHMM methods detected 69 and 61 of these proteins respectively. Therefore the sensitivities of the Phobius and SignalP/TMHMM screenings are 76% and 67% respectively.

Next, a list of yeast proteins which did *not* possess the GO-Slim cellular component terms "cell wall" and "extracellular region" and "cellular component unknown" was extracted. This list contained 4726 unique proteins, but after removing proteins that were not among the original set of RefSeq proteins that were tested, the "negative" list for analysis contained 4666 proteins. 4618 and 4624 of the proteins that were predicted not to be secreted by the Phobius and SignalP/TMHMM methods were in this "negative list". Therefore the specificities of the Phobius and SignalP/TMHMM screenings are both 99%. There does not appear to be a great difference in the power of the two methods for screening. One can apply different criteria to increase the sensitivity of the testing, such as by using an either SignalP neural network *or* SignalP HMM *or* Phobius hit criterion to indicate a signal peptide, but such techniques greatly reduce the specificity of testing, resulting in putative "secretomes" which contain many hundreds of proteins, most of which are false positives. It should also be remembered that the generation of these "positive" and "negative" lists above was quite arbitrary. One might have included ER proteins without a retention signal and Golgi proteins in the "positive" list because, as mentioned above, our screening does not claim to remove soluble proteins from these compartments.

5.3 Yeast GPI Proteins

The results of the previous section suggest that a sizable proportion (indeed, the majority) of the yeast secretome is composed of insoluble proteins attached to the cell wall. It would be desirable to be able to computationally identify the portion of secretomes that are localized to the cell wall. In fungi, GPI modified proteins are the most abundant cell wall proteins (de Groot et al. 2005). In higher eukaryotes a GPI signal indicates that the mature protein is anchored to the plasma membrane, but fungal GPI cell wall proteins are linked to the cell wall via a remnant of the GPI anchor through a largely unknown mechanism. The identification of sequence features that determine whether a fungal GPI protein is cell wall or plasma membrane associated is an active area of research (e.g. Frieman and Cormack, 2004). Certainly, the identification of proteins within a predicted fungal secretome which have GPI attachment signals would go some way towards separating the secretome into its soluble vs. insoluble portions. The amino acid sequences of the yeast proteins predicted to be secreted by the Phobius method in the previous section were input to the Big-PI fungal predictor (Eisenhaber et al. 2004). Of the 226 proteins in the putative yeast secretome, 28 were found to possess a GPI modification site by the Big-PI predictor. These proteins and their associated Biological Process and Cellular Component GO terms from the SGD are listed in Table 1. Almost every protein in the table has a role in the cell wall according to the GO annotations and it appears that the screening for GPI anchors does have the potential to identify the portion of a fungal secretome associated with the cell wall.

Returning to an analysis of the screening methods used in this study, we carefully examined the 21 cell wall proteins in SGD according to the GO-Slim annotation that were not correctly identified by the Phobius algorithm and found that for 10 genes (SPO19, YOR214C, GAS4, AGA1, KRE1, FLO10, UTH1, SAG1, PRY3, FLO5, SPS2, EXG2, FIG1, FLO1 and FLO9), Phobius predicts a small, single transmembrane helix at the extreme C-terminal end of the protein, in addition to a signal peptide. Subsequent analysis with the Big-PI server and inspection of the annotations for these genes shows that all contain a GPI modification site. This suggests that Phobius mistakes the hydrophobic portion of the GPI signal sequence for the string of hydrophobic residues in a transmembrane helix, in an analogous way to the well-known cross prediction of secretory signal peptides and transmembrane helices. We found that the TMHMM program performed better in these cases, where only 3 of the 10 proteins above were designated as having an

Table 1. Predicted GPI proteins in yeast. An asterisk next to the GO term indicates there is more than one term for that gene.

Gene	Name	Biological Process	Cellular Component
YOR383C	FIT3	siderophore transport	cell wall (sensu Fungi)
YOR382W	FIT2	siderophore transport	cell wall (sensu Fungi)
YOR010C	TIR2	response to stress	cell wall (sensu Fungi)
YOR009W	TIR4	biological process unknown	cell wall (sensu Fungi)
YOL030W	GAS5	biological process unknown	cell wall (sensu Fungi)
YOL052C-A	DDR2	response to stress	cytoplasm*
YNL190W		response to dessication	cell wall (sensu Fungi)
YNL300W		biological process unknown	cell wall (sensu Fungi)
YMR251W-A	HOR7	response to stress	endoplasmic reticulum*
YLR390W-A	CCW14	cell wall organization and biogenesis	cell wall (sensu Fungi)*
YLR194C		biological process unknown	cell wall (sensu Fungi)
YLR110C	CCW12	cell wall organization and biogenesis*	cell wall (sensu Fungi)
YLR042C		biological process unknown	cell wall (sensu Fungi)
YLR040C		biological process unknown	cell wall (sensu Fungi)
YKL096W	CWP1	cell wall organization and biogenesis	cell wall (sensu Fungi)
YKL096W-A	CWP2	cell wall organization and biogenesis*	cell wall (sensu Fungi)
YJR150C	DAN1	sterol transport	cell wall (sensu Fungi)
YIL011W	TIR3	biological process unknown	cell wall (sensu Fungi)
YHR143W	DSE2	cell wall organization and biogenesis*	cell wall (sensu Fungi)*
YHR126C		biological process unknown	cellular component unknown
YGR189C	CRH1	biological process unknown	cell wall (sensu Fungi)*
YER150W	SPI1	biological process unknown	cell wall (sensu Fungi)
YER011W	TIR1	response to stress	cell wall (sensu Fungi)
YEL040W	UTR2	cell wall organization and biogenesis	cell wall (sensu Fungi)*
YDR077W	SED1	cell wall organization and biogenesis*	cell wall (sensu Fungi)*
YDR055W	PST1	cell wall organization and biogenesis	cell wall (sensu Fungi)
YCL048W	SPS22	cell wall organization and biogenesis*	plasma membrane
YBR067C	TIP1	cell wall organization and biogenesis	cell wall (sensu Fungi)

extreme C-terminal transmembrane helix by TMHMM. These results indicate that accurate computational discrimination between GPI and transmembrane helices should lead to better screenings protocols for fungal secreted proteins.

6. OTHER FUNGAL SECRETOMES

6.1 Predicted Secretomes

The Phobius/xKDEL/TargetP screening protocol was applied to the RefSeq protein sequences of 9 other fungal proteomes: *Aspergillus nidulans, Candida albicans, Cryptococcus neoformans, Eremothecium gossypii, Magnaporthe grisea, Neurospora crassa, Saccharomyces pombe, Ustilago maydis and Yarrowia lipolytica.* Table 2 lists the number of proteins in these organisms plus *S. cerevisiae* and the number of proteins predicted to be secreted.

Presumably there will be similar inaccuracies among each of these secretomes as was found for yeast. Therefore the variation in the percentage of the proteomes predicted to be secreted is striking. The rice blast fungus *Magnaporthe grisea* has a very large 12.6 percent of its proteins predicted to be secreted, and it and the other filamentous fungi tend to have larger secretomes.

Table 2. Secreted proteins in fungal proteomes.

SPECIES	TOTAL REFSEQ ORFS	SECRETED ORFS	% SECRETED
A. nidulans	18951	1678	8.9
C. albicans	13685	740	5.4
C. neoformans	6606	261	4.0
E. gossipyi	4718	158	3.3
M. grisea	11109	1396	12.6
N. crassa	10085	682	6.8
S. cerevisiae	5866	226	3.9
S. pombe	5045	149	3.0
U. maydis	6522	527	8.1
Y. lipolytica	6521	388	6.0

6.2 GPI Protein Content of Secretomes

GPI proteins in the predicted fungal secretomes above should, as for yeast, play roles in the cell wall. The Big-PI fungal predictor was again run on all protein sequences predicted to be secreted and the numbers of GPI proteins detected is displayed in Table 3.

Table 3. GPI proteins in fungal secretomes.

SPECIES	GPI PROTEINS	% OF SECRETOME	% OF PROTEOME
A. nidulans	97	5.8	0.5
C. albicans	70	9.5	0.5
C. neoformans	15	5.7	0.2
E. gossipyi	11	7.0	0.2
M. grisea	71	5.0	0.6
N. crassa	47	6.9	0.5
S. cerevisiae	28	12.4	0.5
S. pombe	16	10.7	0.3
U. maydis	15	2.8	0.2
Y. lipolytica	52	13.4	0.8

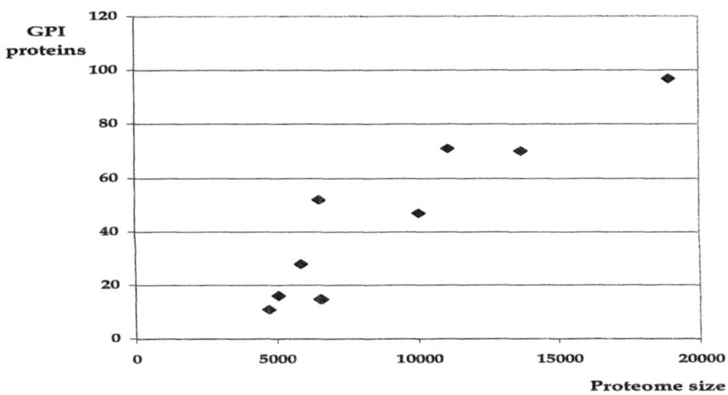

Fig. 3. Number of predicted GPI proteins vs. proteome size for the 10 fungal organisms studied.

In Table 3 we see that those organisms with smaller proportional predicted secretomes (e.g. *S. cerevisiae*) have a greater content of GPI proteins in their secretomes. As a result, the percentage of the proteomes given to GPI proteins is relatively uniform. The correlation between the number of GPI proteins and proteome size is displayed in Fig. 3.

One can speculate that since the great majority of fungal GPI proteins play roles in the cell wall and that there is a relatively good correlation between the number of GPI proteins and proteome size, the amount of proteins required for cell wall maintenance is also proportional to the proteome size. However there is certainly not as strong a correlation between the sizes of predicted secretomes and the sizes of proteomes in Table 2. What is the source of the wide variations in the sizes of secretomes seen in Table 2? Presumably the large secretomes of the filamentous fungi are made up of greater numbers of *soluble* extracellular proteins that are responsible for the essential interaction of the organism with its natural substrate or host. Further analysis of the content of both predicted and experimental fungal secretomes should verify this hypothesis.

7. CONCLUSION

Our investigation has highlighted the challenges in the accurate identification of secretomes using sequence data. It has also made clear that the secretomes of fungi contain a large suite of insoluble cell wall proteins which can be identified computationally via the detection of GPI attachment signals. This work grew out of the bioinformatics requirements of a fungal functional genomics project (https://fungalgenomics.concordia.ca/home/index.php) aimed at the discovery of novel extracellular fungal enzymes with environmental and industrial applications. Putative open reading frames from thousands of assembled EST sequences are routinely processed with most of the tools used in this study, in order to optimally generate a small pool of potentially interesting extracellular proteins for further experimental investigation. Projects such as these will continue to benefit from the ongoing development of computational algorithms that identify subcellular localization using sequence data.

Acknowledgements: The authors wish to thank Lukas Käll for providing a local version of the Phobius software, Søren Brunak for providing a local version of the SignalP software, and Birgit and Frank Eisenhaber for processing data with the Big-PI fungal predictor. This work was financially supported by Genome Canada and Génome Québec.

REFERENCES

Abell BM, Jung M, Oliver JD, Knight BC, Tyedmers J, Zimmermann R and High S (2003). Tail-anchored and signal-anchored proteins utilize overlapping pathways during membrane insertion. J Biol Chem 278:5669-78.

Arai M, Mitsuke H, Ikeda M, Xia JX, Kikuchi T, Satake M and Shimizu T (2004). ConPred II: a consensus prediction method for obtaining transmembrane topology models with high reliability. Nucleic Acids Res 32:W390-W393.

Araújo JHB, Moraes FF and Zanin GM (1999). Bleaching of kraft pulp with commercial xylanases. Appl Biochem Biotechnol 77–79:713–722.

Archer DB and Peberdy JF (1997). The molecular biology of secreted enzyme production by fungi. Crit Rev Biotechnol 17:273-306.

Bendtsen JD, Jensen LJ, Blom N, Von Heijne G and Brunak S (2004b). Feature-based prediction of non-classical and leaderless protein secretion. Protein Eng Des Sel. 17:349-356.

Bendtsen JD, Nielsen H, von Heijne G and Brunak S (2004a). Improved prediction of signal peptides: SignalP 3.0. J Mol Biol. 340:783-795.

Berka RM, Schneider P, Golightly EJ, Brown SH, Madden M, Brown KM, Halkier T, Mondorf K and Xu F (1997). Characterization of the gene encoding an extracellular laccase of Myceliophthora thermophila and analysis of the recombinant enzyme expressed in Aspergillus oryzae. Appl Environ Microbiol 63:3151-3157.

Blobel G and Dobberstein B (1975). Transfer of proteins across membranes. I. Presence of proteolytically processed and unprocessed nascent immunoglobulin light chains on membrane-bound ribosomes of murine myeloma. J Cell Biol. 67:835-51.

Boden M and Hawkins J (2005). Prediction of subcellular localization using sequence-biased recurrent networks. Bioinformatics 21:2279-2286.

Bothast RJ and Schlicher MA (2005). Biotechnological processes for conversion of corn into ethanol. Appl Microbiol Biotechnol 67:19-25.

Bumpus JA and Aust SD (1987). Biodegradation of environmental pollutants by the white rot fungus Phanerochaete chrysosporium: involvement of the lignin degrading system. BioEssays 6:166–170

Camacho NA and Aguilar OG (2003). Production, purification and characterization of a low molecular mass xylanase from Aspergillus sp. and its application in bakery. Appl Biochem Biotechnol 104:159–172.

Carroll GC and Wicklow SC (1994). The Fungal Community. Its Organization and Role in the Ecosystem. 2nd ed. New York: Marcel Dekker.

Chen CP, Kernytsky A and Rost B (2002). Transmembrane helix predictions revisited. Protein Sci 11:2774-2791.

Chen P, Sapperstein SK, Choi JD and Michaelis S (1997). Biogenesis of the Saccharomyces cerevisiae mating pheromone a-factor. J Cell Biol 136:251-269.

Claros MG and von Heijne G (1994). TopPred II: an improved software for membrane protein structure predictions. Comput Appl Biosci 10:685-686.

Corsi AK and Schekman R (1996). Mechanism of polypeptide translocation into the endoplasmic reticulum. J Biol Chem 271:30299-30302.

Cserzo M, Eisenhaber F, Eisenhaber B and Simon I (2004). TM or not TM: transmembrane protein prediction with low false positive rate using DAS-TMfilter. Bioinformatics 20:136-137.

Cserzo M, Wallin E, Simon I, von Heijne G and Elofsson A (1997). Prediction of transmembrane alpha-helices in prokaryotic membrane proteins: the dense alignment surface method. Protein Eng. 10:673-676.

Csiszár E, Urbánszki K and Szakás G (2001). Biotreatment of desized cotton fabric by commercial cellulase and xylanase enzymes. J Mol Catal B Enzym 11:1065–1072.

Cuthbertson JM, Doyle DA and Sansom MS (2005). Transmembrane helix prediction: a comparative evaluation and analysis. Protein Eng Des Sel 18:295-308.

De Groot PW, Hellingwerf KJ and Klis FM (2003). Genome-wide identification of fungal GPI proteins. Yeast 20:781-796.

De Groot PW, Ram AF and Klis FM (2005). Features and functions of covalently linked proteins in fungal cell walls. Fungal Genet Biol. 42:657-675.

Deber CM, Wang C, Liu LP, Prior AS, Agrawal S, Muskat BL and Cuticchia AJ (2001). TM Finder: a prediction program for transmembrane protein segments using a combination of hydrophobicity and nonpolar phase helicity scales. Protein Sci 10:212-219.

Doddapaneni H, Chakraborty R and Yadav JS. (2005). Genome-wide structural and evolutionary analysis of the P450 monooxygenase genes (P450ome) in the white rot fungus Phanerochaete chrysosporium: evidence for gene duplications and extensive gene clustering. BMC Genomics 6:92.

Durand H, Clanet M and Tiraby G (1988). Genetic improvement of Trichoderma reesei for large scale cellulase production. Enzyme Microb. Technol. 10: 341-345.

Dwight SS, Harris MA, Dolinski K, Ball CA, Binkley G, Christie KR, Fisk DG, Issel-Tarver L, Schroeder M, Sherlock G, Sethuraman A, Weng S, Botstein D and Cherry JM (2002). Saccharomyces Genome Database (SGD) provides secondary gene annotation using the Gene Ontology (GO). Nucleic Acids Res. 30:69-72.

Eisenhaber B, Schneider G, Wildpaner M and Eisenhaber F (2004). A sensitive predictor for potential GPI lipid modification sites in fungal protein sequences and its application to genome-wide

studies for Aspergillus nidulans, Candida albicans, Neurospora crassa, Saccharomyces cerevisiae and Schizosaccharomyces pombe. J Mol Biol. 337:243-253.
Emanuelsson O, Nielsen H, Brunak S and von Heijne G (2000). Predicting subcellular localization of proteins based on their N-terminal amino acid sequence. J Mol Biol. 300:1005-16.
Finkelstein DB, Rambosek J, Crawford MS, Soliday CL, McAda PC and Leach J (1989). Protein secretion in Aspergillus niger. In: Genetics and Molecular Biology of Industrial Microorganisms (eds Hershberger CL, Queener SW, Hegeman G) American Society for Microbiology, Washington, DC, pp 295-300.
Frankland JC, Hedger NH and Swift JJ (1982). Decomposer Basidiomycetes: Their Biology and Ecology. Cambridge, United Kingdom: Cambridge University Press.
Frieman MB and Cormack BP (2004). Multiple sequence signals determine the distribution of glycosylphosphatidylinositol proteins between the plasma membrane and cell wall in Saccharomyces cerevisiae. Microbiology. 150:3105-3114.
Gouka RJ, Punt PJ and van den Hondel CA (1997). Efficient production of secreted proteins by Aspergillus: progress, limitations and prospects. Appl Microbiol Biotechnol. 47:1-11.
Greenbaum D, Luscombe NM, Jansen R, Qian J and Gerstein M (2001). Interrelating different types of genomic data, from proteome to secretome: 'oming in on function. Genome Res, 11: 1463–1468.
Grimmond SM, Miranda KC, Yuan Z, Davis MJ, Hume DA, Yagi K, Tominaga N, Bono H, Hayashizaki Y, Okazaki Y and Teasdale RD (2003). The mouse secretome: functional classification of the proteins secreted into the extracellular environment. Genome Res, 13:1350–1359.
Gromiha MM, Ahmad S and Suwa M (2004). Neural network-based prediction of transmembrane beta-strand segments in outer membrane proteins. J Comput Chem. 25:762-767.
Harcus YM, Parkinson J, Fernandez C, Daub J, Selkirk ME, Blaxter ML and Maizels RM (2004). Signal sequence analysis of expressed sequence tags from the nematode Nippostrongylus brasiliensis and the evolution of secreted proteins in parasites. Genome Biol 5:R39.
Hirokawa T, Boon-Chieng S and Mitaku S (1998). SOSUI: classification and secondary structure prediction system for membrane proteins. Bioinformatics 14:378-379.
Hoffman K and Stoffel W (1993). TMbase - A database of membrane spanning protein segments. Biol. Chem. Hoppe-Seyler 374:166.
Hua S and Sun Z (2001). Support vector machine approach for protein subcellular localization prediction. Bioinformatics 17:721-728.
Jeenes DJ, Mackenzie DA and Archer DB (1994). Transcriptional and post-transcriptional events affect the production of secreted hen egg white lysozyme by Aspergillus niger. Transgenic Res. 1994 3:297-303.
Jeenes DJ, Mackenzie DA, Roberts IN and Archer DB (1991). Heterologous protein production by filamentous fungi. Biotechnol Genet Eng Rev 9:327-367.
Jones DT, Taylor WR and Thornton J M (1994) A model recognition approach to the prediction of all-helical membrane protein structure and topology. Biochemistry 33:3038-3049.
Kahsay RY, Gao G and Liao L (2005). An improved hidden Markov model for transmembrane protein detection and topology prediction and its applications to complete genomes. Bioinformatics. 21:1853-1858.
Kalies KU and Hartmann E (1998). Protein translocation into the endoplasmic reticulum (ER)-two similar routes with different modes. Eur J Biochem. 254:1-5.
Käll L, Krogh A and Sonnhammer EL (2004). A combined transmembrane topology and signal peptide prediction method. J Mol Biol. 338:1027-1036.
Klee EW, Carlson DF, Fahrenkrug SC, Ekker SC and Ellis LB (2004) Identifying secretomes in people, pufferfish and pigs. Nucleic Acids Res, 32:1414–1421.
Klotz C, Marhafer RJ, Selzer PM, Lucius R and Pogonka T (2005). Eimeria tenella: Identification of secretory and surface proteins from expressed sequence tags. Exp Parasitol, in press.
Krogh A, Larsson B, von Heijne G and Sonnhammer EL (2001). Predicting transmembrane protein topology with a hidden Markov model: application to complete genomes. J Mol Biol 305:567-580.
Lao DM, Arai M, Ikeda M and Shimizu T (2002b). The presence of signal peptide significantly affects transmembrane topology prediction. Bioinformatics 18:1562-1566.
Lao DM, Okuno T and Shimizu T (2002a). Evaluating transmembrane topology prediction methods for the effect of signal peptide in topology prediction. In Silico Biol 2:485-494.

Lee SA, Wormsley S, Kamoun S, Lee AF, Joiner K and Wong B (2003). An analysis of the Candida albicans genome database for soluble secreted proteins using computer-based prediction algorithms. Yeast. 20:595-610.

Lee SA, Wormsley S, Kamoun S, Lee AF, Joiner K and Wong B (2003). An analysis of the Candida albicans genome database for soluble secreted proteins using computer-based prediction algorithms. Yeast, 20: 595–610.

Liakopoulos TD, Pasquier C and Hamodrakas SJ (2001). A novel tool for the prediction of transmembrane protein topology based on a statistical analysis of the SwissProt database: the OrienTM algorithm. Protein Eng. 14:387-390.

Loong TW (2003). Understanding sensitivity and specificity with the right side of the brain. BMJ. 327:716-719.

MacIntosh GC, Bariola PA, Newbigin E and Green PJ (2001). Characterization of Rny1, the Saccharomyces cerevisiae member of the T2 RNase family of RNases: unexpected functions for ancient enzymes? Proc Natl Acad Sci U S A. 98:1018-1023.

Mattey M (1992). The production of organic acids. Crit Rev Biotechnol 12:87-132.

McGuffin LJ, Bryson K and Jones DT (2000). The PSIPRED protein structure prediction server. Bioinformatics 16:404-405.

Moller S, Croning MD and Apweiler R (2002). Evaluation of methods for the prediction of membrane spanning regions. Bioinformatics. 2001 17:646-653.

Munro S and Pelham HR (1987). A C-terminal signal prevents secretion of luminal ER proteins. Cell. 48:899-907.

Nevalainen H, Suominen P and Taimisto K (1994). On the safety of Trichoderma reesei. Biotechnol. 37:193-200.

Nickel W (2003). The mystery of nonclassical protein secretion. A current view on cargo proteins and potential export routes. Eur J Biochem. 270:2109-2119.

Nielsen H, Engelbrecht J, Brunak S and von Heijne G (1997). Identification of prokaryotic and eukaryotic signal peptides and prediction of their cleavage sites. Protein Eng 10:1-6.

Pandey A, Soccol CR, Nigam P and Soccol VT (2000). Biotechnological potential of agro-industrial residues. I, sugarcane bagasse. Bioresour Technol. 74:69–80.

Pashou EE, Litou ZI, Liakopoulos TD and Hamodrakas SJ (2004). waveTM: wavelet-based transmembrane segment prediction. In Silico Biol4:127-131.

Paszczynski A and Crawford RL (1995). Potential for bioremediation of xenobiotic compounds by the white rot fungus Phanerochaete chrysosporium. Biotechnol Prog 11:368–379.

Persson B and Argos P (1994). Prediction of transmembrane segments in proteins utilising multiple sequence alignments. J Mol Biol 237:182-192.

Pilpel Y, Ben-Tal N and Lancet D (1999). kPROT: a knowledge-based scale for the propensity of residue orientation in transmembrane segments. Application to membrane protein structure prediction. J Mol Biol 294:921-935.

Polizeli ML, Rizzatti AC, Monti R, Terenzi HF, Jorge JA and Amorim DS (2005). Xylanases from fungi: properties and industrial applications . Appl Microbiol Biotechnol 67:577-91.

Pruitt KD, Tatusova T and Maglott DR (2005). NCBI Reference Sequence (RefSeq): a curated non-redundant sequence database of genomes, transcripts and proteins. Nucleic Acids Res. 33:D501-D504.

Reczko M and Hatzigerrorgiou A (2004). Prediction of the subcellular localization of eukaryotic proteins using sequence signals and composition. Proteomics 4:1591-1596.

Reddy CA (1995). The potential of white rot fungi for the treatment of pollutants. Curr Opin Biotechnol. 6:320–328.

Rost B, Fariselli P and Casadio R (1996). Topology prediction for helical transmembrane proteins at 86% accuracy. Protein Sci. 5:1704-1718.

Ruiz-Duenas FJ, Martinez MJ and Martinez AT (1999). Heterologous expression of Pleurotus eryngii peroxidase confirms its ability to oxidize Mn(2+) and different aromatic substrates. Appl Environ Microbiol 65:4705-4707.

Scott M, Lu G, Hallett M and Thomas DY (2004). The Hera database and its use in the characterization of endoplasmic reticulum proteins. Bioinformatics 20:937-944.

Trost M, Wehmhoner D, Karst U, Dieterich G, Wehland J and Jansch L (2005). Comparative proteome analysis of secretory proteins from pathogenic and nonpathogenic Listeria species. Proteomics 5:1544-1557.

Tusnady GE and Simon I (2001). The HMMTOP transmembrane topology prediction server. Bioinformatics 17:849-850.
Wrzeszczynski KO and Rost B (2004). Annotating proteins from endoplasmic reticulum and Golgi apparatus in eukaryotic proteomes. Cell Mol Life Sci 61:1341-1353.
Wymelenberg AV, Sabat G, Martinez D, Rajangam AS, Teeri TT, Gaskell J, Kersten PJ and Cullen D (2005). The Phanerochaete chrysosporium secretome: database predictions and initial mass spectrometry peptide identifications in cellulose-grown medium. J Biotechnol. 118:17-34.
Xie D, Li A, Wang M, Fan Z and Feng H (2005). LOCSVMPSI: a web server for subcellular localization of eukaryotic proteins using SVM and profile of PSI-BLAST. Nucleic Acids Res 33:W105-110.
Yuan Z, Davis MJ, Zhang F and Teasdale RD (2003). Computational differentiation of N-terminal signal peptides and transmembrane helices. Biochem Biophys Res Commun. 312:1278-83.
Yuan Z, Mattick JS and Teasdale RD (2004). SVMtm: support vector machines to predict transmembrane segments. J Comput Chem. 25:632-636.
Zhou H and Zhou Y (2003). Predicting the topology of transmembrane helical proteins using mean burial propensity and a hidden-Markov-model-based method. Protein Sci. 12:1547-1555.

Applied Mycology
and Biotechnology
An International Series
Volume 6. Bioinformatics

ELSEVIER

Using Web Agents for Data Mining of Fungal Genomes

Audrius Meškauskas
Alte Gfennstr. 22, CH-8600 Dubendorf, Switzerland (AudriusA@Bioinformatics.org).

We created an application called Sight, a Java™-based package that provides a user-friendly interface to generate and connect agents for automatic genomic data mining without requiring programming skills from the user. Sight was originally developed to automate analysis of the human genome and attempts to generate web agents for fungus-related Internet resources revealed that some of those resources use new methods of representing the information they report, and some servers returned multiple intermediate pages leading towards their response, which created difficulties for automated recovery of results. Consequently, it was not possible to use effectively the old version of Sight so this version of the application was adapted with a little additional programming, creating a new version for which these features of the fungal genome servers do not represent a problem. The new version of Sight (v. 3.0.0) that is tailored to servers carrying fungal databases is freely available for download from the project website at these URLs: http://bioinformatics.org/jSight/ and http://jsight.sourceforge.net/index_SF.htm.

1. INTRODUCTION

As more genomes are sequenced and added to the publicly available databases the opportunities for data mining expand in terms of both the range of organisms available and the biological systems open to analysis. For the most part, these opportunities are still enjoyed by the relatively few people (or teams) who have the necessary combination of the biological expertise to frame the research questions and the programming skills to take full advantage of the database server resources that are available. No computer programming can yet supply the biological expertise, but many of the operations involved in data mining server databases are mechanical and repetitive. These are open to management by appropriate programming techniques to produce routines that carry out the data mining on behalf of the researcher and consequently use the available computer power to extend the research reach of the individual in time and space. Importantly, these advantages are not limited to those who have the programming skills to create those routines in the first place. Today's programming languages enable the development of applications that make creation of programming routines a matter of assembling a sequence of preprogrammed modules, that application itself writing the code that will eventually carry out the job. In much the same way that presentation applications allow the user to assemble a multimedia presentation without need of programming skills to bring together text, images, video and sound, so these data mining applications allow the non-programmer to assemble a sequence of routines that retrieve sequences, query databases, retrieve results and then, potentially formulate new searches on the basis of responses to earlier queries. These applications produce 'web agents', which are computer programs that search the Internet on behalf of their author for data that the author requires.

Existing systems for automated access to the Internet resources are usually either specialized for a single given task or written as a general purpose tools. By their levels of complexity, these can be grouped in the following categories: (1) Single task, single step web agents work with a single web service only. For example, WebBlast (Ferlanti et al. 1999) stores the search requests in the local query and schedules submissions to the server (for example, during the night time). BioQuery (Brundege and Dubay, 2003) periodically searches NCBI database (Wheeler et al., 2003) using the previously stored queries. As a rule, such systems automate a submission only. The returned response is analyzed by the user. (2) Multiple task, single step web agent systems are collections of the single task web agents, having shared user interface and shared mechanisms for integrating the new web services. For example, Sewer (Basu, 2001) is a JavaScript based system of web pages containing web forms for accessing various bioinformatics web services. These forms replace the original forms of the corresponding web pages, creating a more effective environment for the user. The Sewer assistance ends after the request is sent. Differently from Sewer, Proteomix (Chikayama et al. 2004) needs specialized software on the server side, communicating via a more advanced protocol called SOAP (Simple Object Access Protocol). This tool is more specialized, analyzing the given protein sequence in a variety of ways.

1. Multiple step, fixed workflow systems connect several agents into a workflow. Such systems are frequently used for automated genome annotations. First, some gene prediction service must assemble the sequenced clones and predict the genes for the given DNA sequence. Next, one (or, more often several) other services provide some conclusions about the predicted protein (for example, that there are known proteins to which it is similar, that known functional domains are present, how many transmembrane regions it may contain, and so on). The results are usually stored in a database. Examples of such systems could be Genotator (Harris, 2000), Pedant (Frishman et al. 2001) or Fountain (Buerstedde and Prill, 2001).
2. Changeable workflow, fixed agent set systems are used for flexible workflows, where the agents for the workflow are picked from a fixed list. The integration of new services still needs serious programming effort in these cases. Several such systems were suggested quite a long time ago, usually adapting already existing frameworks for the needs of bioinformatics

analysis. In this category we could mention Kleisli (Kolatkar et al. 1998), GAIA (Bailey et al. 1998) or Tambis (Stevens et al. 2000).

3. Changeable workflow, extendable agent set systems allows the end user to add new agents without significant programming effort. Depending on the method by which the new agents are added, this group falls into the following subgroups: (a) Systems in which the originators take pains to create a simple way for adding new agents written by user proficient at programming. Many projects (for example EDITtoTrEMBL) support integration of the user-written code, and this possibility usually remains in more advanced systems like Sight or Taverna; (b) Systems that automatically generate code templates that must be finished by the user with a little additional programming. In this case the system itself usually focuses on handling the communication between agents. A good example of such a skeleton generator is Decaf (Graham et al., 2003); (c) The new agent is generated using a graphical user environment, which is the approach employed by the Sight program discussed in more detail below. In these cases the operator provides all the necessary additional information, but the program writes the code; (d) The web server that provides the web service the user wishes to access may also offer additional web document(s) that can be used to create web agent(s) for the service. This document defines the formats of the possible requests and responses. It is usually written in WSDL, the machine-readable language created for describing web services. The server must also be connected using a specific protocol, different from the protocol used to submit a web form. The most advanced experimental applications like Taverna (Oinn et al., 2004) rely exclusively on this new approach and cannot integrate the ordinary web services. We think that a more useful approach could be to add WSDL support without discarding the ability to create agents for work with the classic web forms.

We have created an application called Sight, which is a Java™-based package that provides a user-friendly interface to generate and connect agents for automatic genomic data mining, allowing the user without programming skills to tailor web agents for his/her individual requirements (Meškauskas et al., 2004). Sight allows the assembly of an arbitrary flowchart-like workflow. Using a Web form the user chooses agents from a built-in library and connects the response data fields of one agent to the request data fields of another. This user-interactive agent generator produces agents that can be connected for sequential tasks using the application. To minimize Internet connections for trivial tasks, Sight incorporates the algorithms for several functions such as pattern searches, protein translations or simple sequence manipulations have been included in the application for local use.

There are several other facilities available in the application, but Sight's built-in web agent generators have never been tested on the fungal-specific Internet resources. Web pages devoted to such resources were designed later than pages devoted to plant or human sequences. As a result, they make intensive use of advanced features like JavaScript language, unusual (often nested) tables, multiple pages per response, and so on. Such features provide a serious challenge for web agent applications that must still be able to extract a clear data structure from a complicated multi-page server response. The purpose of this work was to test and adapt the Sight system, enabling easier creation of fungal-related workflows. The list of web resources used to tailor Sight to fungal genome servers (Table 1) was taken from the chapter by Moore et al. (this volume).

2. RESULTS

A Sight web agent is essentially an active flow chart in which each element is a working preprogrammed routine. The user assembles the flow chart according to the task s/he wishes to perform. I will outline here the specific changes that have to be made, either to the program or to its implementation, in order to apply Sight to data mining of fungal genomes.

2.1 Main Conceptions of the Sight Workflow

Sight architecture has been significantly modified since the program was first briefly described in the literature (Meškauskas et al., 2004), by the addition of loops, confluences (convergences), and so on. These new features will be described below.

2.1.1 Sight agent

The reusable elementary unit (or agent) executes a single remote or local algorithm. For this we need two data structures, defining the submitted request and the received response.

Sight request consists of the multiple named items (fields), each storing a string value. They normally correspond to the fields, checkboxes and other controls in the web form that was used as the initial data to generate the agent. In contrast to the request, the result of the bioinformatical web agent often needs to be an array of records. For example, a similarity search service may return multiple hits to the sequences in the database; a gene prediction program may find multiple genes in a DNA sequence; a program for predicting transmembrane segments may detect multiple transmembrane helixes, etc. Hence, Sight agent response is programmed as an array of records. These records also consist of multiple named fields.

As the request and response format differs for each agent, the agents also contain explanatory data structures defining these formats. For each request or response field they define its type, name and arbitrary comment. The request fields can also have a default value and a list of other possible values.

Table 1. The fungus-related Internet resources supported by the most recent version of the Sight web agent application (version 3.0.0).

Organism	URL	Resources
Aspergillus fumigatus	http://www.tigr.org/tdb/e2k1/afu1/	Similarity search (nucleotide sequences only)
Cryptococcus neoformans	http://www.tigr.org/tdb/e2k1/cna1/	Similarity search (protein and nucleotide sequences)
Phanerochaete chrysosporium	http://genome.jgi-psf.org/whiterot1/whiterot1.home.html	Similarity search (nucleotide sequences only).

2.1.2 Sight Workflow

For any workflow, connection between two agents it is only possible if the result of the master agent can be converted to the request for the slave agent. Some systems require that these two data structures should be identical or (like Decaf) leaves the solution of the problem to the programming user. On the other hand the Sight application generator produces Java™ code to create a slave request from the master response. More exactly, the request is created using the full hierarchy of response and request (the master response, master request, the master of master response, the request that was sent to the master of master and so on) right up to the level of workflow input data (Fig. 1).

The code for such a workflow is generated automatically; the user simply connects the required request and response fields. Fig. 2 illustrates the simple case of a branching, tree-like workflow. As shown in this example, the system can potentially generate a large number of requests, especially for the agents standing lower in the hierarchy. For example, if the master agent has returned 7 records, these will represent 7 requests for the slave agent. If this slave agent has returned, for example, 5 records for each request and has its own slave agent, the number of requests for this slave-of-slave will already be 30. The inbuilt Sight security system limits the number of parallel submissions to the same web service to prevent server overload.

Previous Sight versions supported linear and tree like workflows only. However, the current version (v. 3.0.0), as presented in this manuscript, also supports loops and confluences (Figs 3 & 4). The confluence arises when two or more branches of the tree workflow must join together again to provide data for a shared agent (Fig. 3). Our solution for the type conversions for confluences is to process all possible combinations of the records in the two master agent responses. For example, if the workflow has branched and two slave agents have the shared slave-of-slave agent, and one of these two agents has returned 5 and another 3 records in response, it is possible to combine 15 different requests for the shared slave-of-slave agent.

2.1.3 Loops

Sight 3.0.0 supports circular workflows. Such algorithms are used, for example, in building sequence similarity networks or in reconstruction of metabolic pathways. The loops are realised with a pair of two communicating specialised agents: loop starter and loop closer (Fig. 4). The loop starter just passes all its requests through. When the loop closer receives the request, it communicates the loop starter, initiating the additional "virtual request". This "virtual request" is processed by the agents between the loop starter and loop closer and may initiate the subsequent new virtual requests. The l oop is terminated when one of the agents between the starter and closer returns the empty response (no records) or when the maximal number of iterations is exceeded.

2.1.4 Storing the Results

The results of running the workflow must be stored for subsequent viewing or analysis by the user. For systems with a fixed workflow the results are usually stored in the database. However each user-defined workflow usually needs a new database structure. It is difficult to implement a user-friendly interface for accessing these multiple different databases. The older versions of Sight stored the results in the html documents.

Taverna tried another approach, creating a complicated folder and subfolder structure on the local file system. In the new Sight version we implemented the ability to store the results in the form of network. The agent responsible for storing

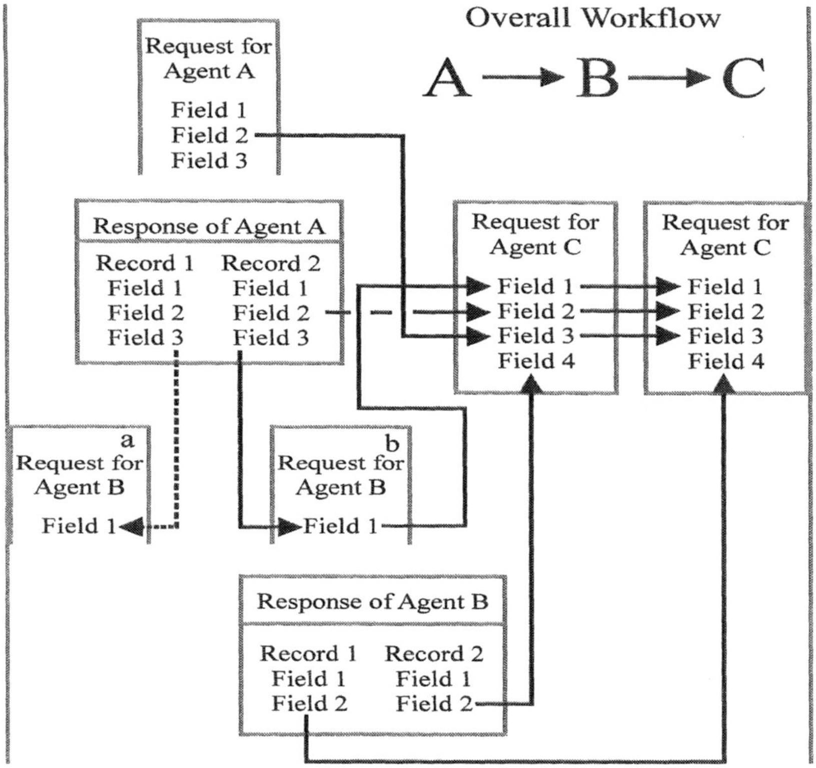

Fig. 1. An illustration of data flow during data type conversion for the case of the simple linear workflow employing three Sight agents A, B and C. The initial request for the agent A consists of 3 fields. This agent returns a request from two records. Each of them also have 3 fields. The request of agent B consists of one field, and the value for this is taken from field 3 in the agent A response (= results) record. As agent A returns two records, two independent requests (**a** and **b**) for agent B will be created (shown as dashed and solid lines). Let us suppose that for one of these two requests agent B returned two records, each having 2 fields. Now, finally, the request is made to agent C consisting of 4 fields that must be filled by various values from the workflow. Field 1 is identical to the field from the agent B request, field 2 is identical to the field 2 from the agent B results record, field 3 takes its value from field 2 in the initial workflow request for agent A, and finally field 4 takes its value from field 2 in the agent B response record. As agent B has returned two records to the request **b**, two requests for agent C will be created for this branch. However as agent B also has another request (**a**), the total number of requests to agent C depends on the number of records in the agent B response to its request **a**. If this response contains, for example, 3 records, the total number of the requests to agent C will be 3 + 2 = 5.

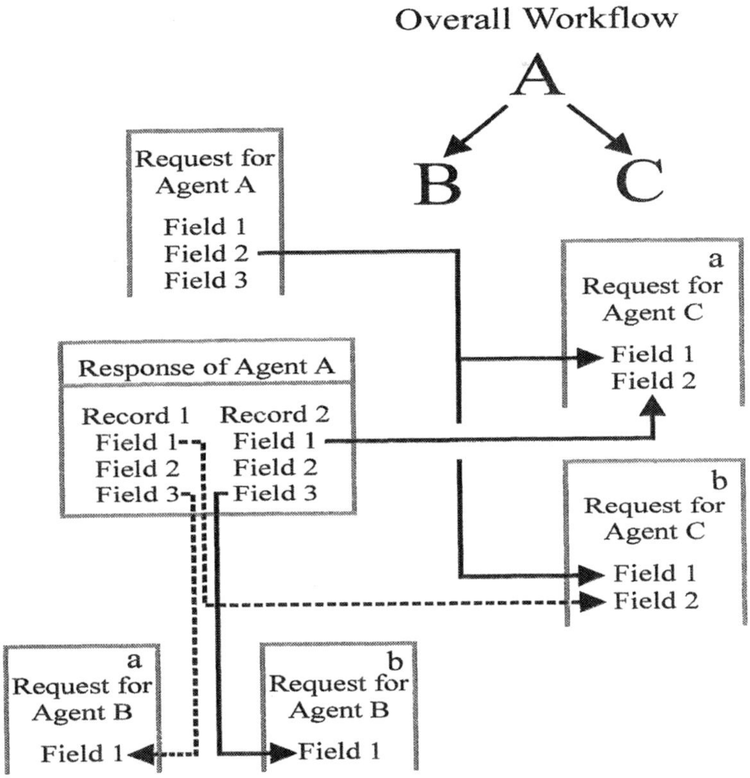

Fig. 2. Illustration of data flow during data type conversion for the case of a branching (tree-like) workflow. Here, agent B requires the value held in field 3 from the master response record. Agent C requires field 1 from the master response record, but it additionally needs the data in field 2 from the master request. As the master A has returned two records in response, both slave agents (B and C) receive the requests (**a** and **b**).

the network takes the names of the two nodes that must be connected. As the agent receives more and more requests, the number of currently existing nodes and connections increases. The created network can be viewed with the free bioinformatical graph viewer CytoScape (Shannon et al., 2003). Sight also has a specialised group of agents (loggers) that just append the requests to the local files. In this way the interesting information can be logged separately in FASTA or some other format.

2.1.5 Filters, conditional analysis and annotation events

Even systems that annotate all input sequences benefit from the possibility to abandon some workflow branches if the results of previous analysis does not satisfy the given conditions (Moller et al., 1999). This strategy is especially effective when the purpose is to find data structures that match the a specified set of search criteria. For example, Sight was used to find and analyze gene sequences similar to the

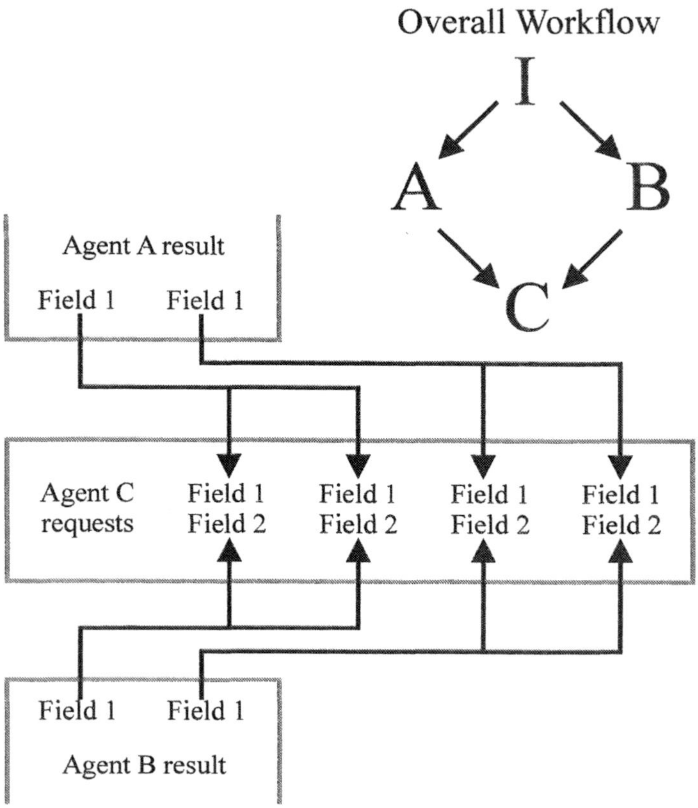

Fig 3. The simple case of confluence. In this workflow, the agent I is a master for two agents A and B. Agent C is a shared slave agent for A and B. If agent I returns a single record result, agents A and B both receive a single request (not shown). However, both A and B results contains two records, and the shared slave needs fields from both master agents, this workflow generates four requests for agent C. Confluences are only supported in the new version of the program Sight 3.0.0 alpha.

sequences of membrane ionic channels. That Sight application submitted the given sequence to the BLAST similarity search service. The headers of the returned similarity hits were then scanned for specific keywords (for example, 'channel'). Only if these keywords were found was the initial sequence allowed to become the subject of much more detailed analysis. Apart from accelerating the computation, this made the report files much shorter and easier to read.

In Sight, we use the concept of a filter agent to perform this task. The filter scans its input data for the user-defined regular expressions (wildcards). If the search conditions are satisfied, the filter returns a non-zero result that is passed to its slave agents. This causes the slave agent, which has access to the results and requests of all the master agents in the workflow hierarchy, to continue the analysis using the workflow data. However, if the filter agent returns a zero result (no matches with

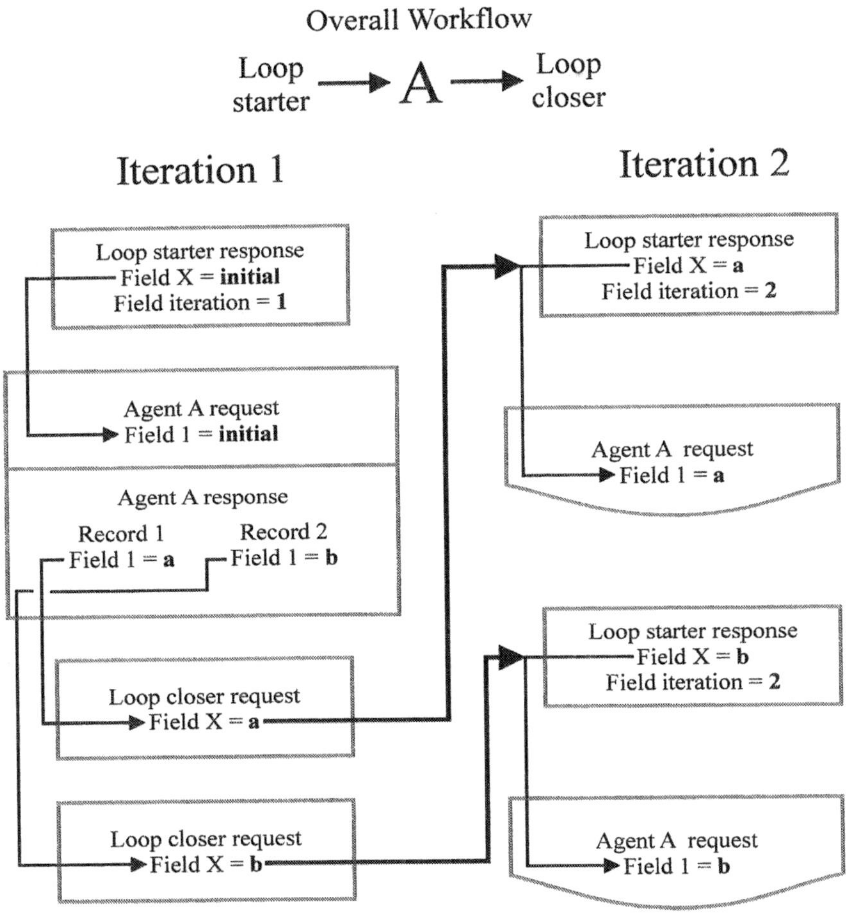

Fig 4. The concept of processing loops in Sight. The figure illustrates a simple loop, where agent A is placed between the loop starter and loop closer. During the first iteration the loop starter send one request to its slave agent A. As agent A returned two records in its result, the loop closer (slave agent for A) receives two requests and during the next iteration produces two virtual requests for the loop starter. The words "**initial**", "**a**" and "**b**" are sample values and illustrate how the structures are converted during iterations. Loop agents can handle up to 3 loop variables (X, Y and Z). Loops are supported in Sight 3.0.0 alpha and higher.

the user defined expressions), its slave agents do not receive any further requests and the conditional analysis is not performed. The regular expression filter has two sets of wildcards. To satisfy the search conditions, one of them must be present, and the other must be absent in the filter request. Sight also has a numeric filter that checks that a number which is returned falls with a certain, user-defined, interval. The most complicated cases require the Weka data mining engine (Witten and Frank, 1999). Weka filter uses machine learning algorithms and needs two additional input data sets (negatives and positives).

Filters can be combined with annotation event listeners. This group of agents writes their input data into separate files. For example, if the input data are records in Swissprot format, the subsequent filter may scan them to pick proteins from organisms belonging to some user-defined taxonomic group. The output of the filter may be connected to the agent that writes its input in Swissprot format again. Only entries that have passed the filter will be written, so that the agents output a database of proteins belonging to the given taxonomic group. In a similar way, Sight can pick proteins that are located in a given cellular compartment, that belong to a given protein category, and so on. The words 'possible', 'probable', 'by similarity', and so on, can also be detected and, if needed, taken into consideration.

One of the available Sight annotation listeners writes its output in CytoScape format (xxx.sif). CytoScape is a tool for creating networks rather than annotation tables. The agent's input consists of the two node names that must be connected. The same names can appear in the later input records, forming the network structure that can be later opened and viewed by CytoScape.

2.2 Testing the Web form Submission Module

To create a web agent we must first submit a biologically relevant request; if a similarity search is the main interest then the request must be for a protein or nucleotide sequence that the user expects to be found in the server database. For easier preparation of test requests, the web agent generation system needs some built-in example sequences. As Sight was written as a tool for analysis of the human genome, the program was initially equipped with sample protein, DNA and RNA sequences taken from human sources. Such sequences have no analogues in fungal genomes, so the corresponding similarity searches return no hits. To provide realistic queries, I have extended the built-in example set by adding the RNA and protein sequences of ribonucleotide reductase M2. A similarity search test operation will now find similar sequences in any fungal genomes the user cares to test.

2.3 Testing and Extending the Redirection System

In the simplest case a web server accepts the web query form the user has completed on-line, performs the requested search, and returns a web page that contains the results of the bioinformatics analysis. The duration of analysis is usually limited by automatic breaks in the connection after several minutes caused either at the client or server side. Recent experience shows that increasing numbers of web services are returning an intermediate page. This page confirms that the request has been accepted and is being dealt with, and contains the hyperlink that must be followed to retrieve results. If this hyperlink is followed before the server has completed the task, another page is returned suggesting the enquirer might like to be more patient and wait for some further time. Essentially, this strategy allows the server an unlimited amount of time to complete the task. However, most of the fungus-related web services go even further. Instead of returning the complete result, they tend to provide only a synopsis or general description. For example, the result page of a similarity search over the *P. chrysosporium* genome does not contain the score for the similarity search. Such important details can only be retrieved by following additional links from the response page. If the web agent is to extract the user's results effectively, then it must be able to deal automatically with these intermediate pages returned by the database server. Sight already supported the most trivial cases, but we found that the redirection problem required rather more attention. The first difficulty is that the number of intermediate pages may not be known in advance and they may need different algorithms to find the hyperlink to follow. Another problem is that in some cases the user is expected to follow several links from a single page, which creates a task related to web crawling for the web agent. A new version of Sight has been produced that replaces such a sophisticated agent with a mini-workflow consisting of specialised sub agents that find the hyperlinks the user is expected to follow to get all of the results. Mini-workflows are assembled and tested like the ordinary Sight workflows, but after generation, they join the final Sight application as main agents. Mini-workflows cannot be realised simply as part of the ordinary Sight workflow because a different caching strategy is required. The intermediate hyperlinks that have to be followed must not be cached. Instead, the final result must be cached using the initial request to compute the caching key.

2.4 Testing and Extending Agent Algorithms

It is typical to illustrate the results of searches with graphics. However, the services for which Sight was originally designed report the same information in text form and so until now Sight's web agents did not need to analyze these figures. Unfortunately, fungal genome servers are different. For example, the result of a similarity search of the *P. chrysosporium* genome comprises just one large image illustrating the position(s) of the hit(s) that have been found in the total DNA sequence. To get more information, it is necessary for the user to mouse-click on the image. From the point of view of a web agent, it is necessary to follow those hyperlinks defined in the "hot spots" in the image, but the graphic contains several "hot spots" and only some of them lead to pages with useful additional information. The links that are needed cannot always be recognized from the surrounding data by an automatic routine, but in many cases they can be recognized from the Internet address itself. In the example quoted the link must contain the substring "getAlignment". Hence, we have produced a new Sight version that includes the previously missing feature of selecting data fields by content rather than by context.

Experience with fungal genome servers also showed that the automated table analysis of the original version of Sight also needed serious improvement. Details of search results of the *P. chrysosporium* genome are presented in the form of multiple tables, there being one table per search hit. Instead of working with the single table generated by human genome servers, the fungal web needs to collect results from several very similar tables. Interestingly, these tables themselves are cells of an even larger "supertable". To solve this problem in Sight, we have now implemented a concept of the table "domain", defining the table position in the nested supertables (Fig. 5).

2.5. Discussion and Prospects

The version of Sight which has now been tailored to use fungal genome servers will be extremely useful to mycologists involved in data mining. There are still some remaining problems, though. The most challenging among these are servers that return work results to the enquirer by E-mail (for example, the Whitehead Institute servers). It is not difficult to program automated E-mail checking, but the issue that arises is that in complicated workflows there may be several agents issuing requests and there is a problem in deciding which agent in the workflow sent a specific request which is answered by a specific E-mail. For example, if two agents in a workflow submit requests to the same server (say, related searches on different sequences), then how can the E-mail messages be correctly sorted automatically when the server returns them? Evidently, the server response must contain some kind of "submitter name" which the agents can use for sorting. Many servers do provide such support, allowing the user to specify, for example, the sequence header. That header may be used subsequently by the web agents for their own orientation. This is not a trivial problem and, for the moment, this part of the application is not sufficiently reliable, and we are still not ready to release it for wider use.

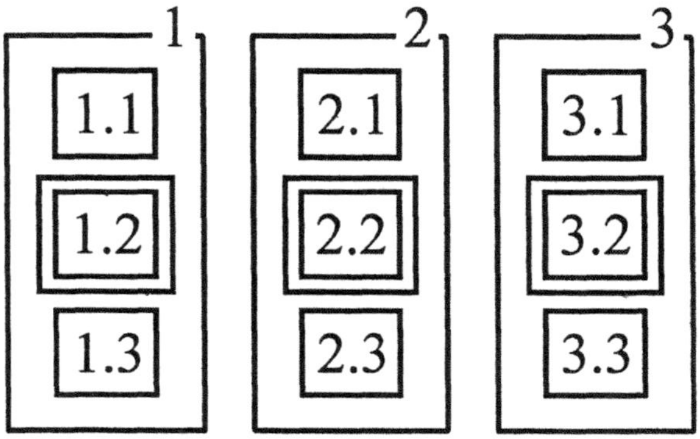

Fig. 5. Illustration of the concept of the table domain on the results web page returned by the genome server in response to a query. In this example, the required data are located in the table in the group of cells shown as 1.2, 2.2 and 3.2 and the web agent identifies these by their content.

3. CONCLUSION

During attempts to generate web agents for fungus-related Internet resources (Table 1) it was recognized that some of those resources use new methods of representing the information they report. In some cases it was not possible obtain search details with the previously available version of Sight (v. 2.1.2), and some servers returned multiple intermediate pages leading towards their response which created difficulties for automated recovery of results. Despite these problems, it was possible to use Sight to create web agents that are able to automate searches of fungal genomes. The previous version of the application was adapted with a little additional programming, creating a new version for which these features of the fungal genome servers do not represent a problem. The new version of Sight (v. 3.0.0) that is tailored to servers carrying fungal databases is freely available for download from the project website at these URLs: http://bioinformatics.org/jSight/ and http://jsight.sourceforge.net/index_SF.htm.

Acknowledgement:: I thank Dr David Moore (School of Biological Sciences, University of Manchester) for making the original suggestion that I should test the Sight application on fungal resources and implement the adaptations that were needed.

REFERENCES

Bailey LC, Fischer S, Schug J, Crabtree J, Gibson M and Overton GC (1998). GAIA: framework annotation of genomic sequence. Genome Research 8:234-250.

Basu MK (2001). SeWeR: a customizable and integrated dynamic HTML interface to bioinformatics services. Bioinformatics 17: 577-578.

Buerstedde JM and Prill F (2001). FOUNTAIN: A JAVA open-source package to assist large sequencing projects. BMC.Bioinformatics: 2: 6-7.

Brundege JM and Dubay C (2003). BioQuery: an object framework for building queries to biomedical databases. Bioinformatics 19:901–902.

Chikayama E, Kurotani A, Kuroda Y and Yokoyama S (2004). ProteoMix: an integrated and flexible system for interactively analyzing large numbers of protein sequences. Bioinformatics: (in press).

Ferlanti ES, Ryan JF, Makalowska I and Baxevanis AD (1999). WebBLAST 2.0: an integrated solution for organizing and analyzing sequence data. Bioinformatics 15:422-423.

Frishman D, Albermann K, Hani J, Heumann K, Metanomski A, Zollner A and Mewes HW (2001). Functional and structural genomics using PEDANT, Bioinformatics 17:44-57.

Graham J, Decker K.S and Mersic M (2003). Decaf - a flexible multi-agent system architecture Autonomous Agents and Multi-Agent Systems. 7(1): 7-27.

Harris NL (2000). Annotating sequence data using Genotator. Molecular.Biotechnology 16:221-232.

Kolatkar PR, Sakharkar MK, Tse CR, Kiong BK, Wong L, Tan TW and Subbiah S (1998). Development of software tools at BioInformatics Centre (BIC) at the National University of Singapore (NUS). Pacific Symposium on Biocomputing 735-746.

Meškauskas A, Lehmann-Horn F and Jurkat-Rott K (2004). Sight: automating genomic data-mining without programming skills. Bioinformatics 20:1718-1720.

Moller S, Lesser U, Fleischmann W and Apweiler R. (1999). EDITtoTrEMBL: a distributed approach to high-quality automated protein sequence annotation. Bioinformatics 15: 219-227.

Oinn T, Addis M, Ferris J, Marvin D, Greenwood M, Carver T, Pocock MR, Wipat A and Li P (2004). Taverna: a tool for the composition and enactment of bioinformatics workflows. Bioinformatics in press.

Shannon P, Markiel A, Ozier O, Baliga NS, Wang JT, Ramage D, Amin N, Schwikowski B and Ideker T (2003). Cytoscape: a software environment for integrated models of biomolecular interaction networks. Genome Res 13: 2498-2504.

Stevens R, Baker P, Bechhofer, S, Ng G, Jacoby A, Paton NW, Goble CA and Brass A (2000). TAMBIS: transparent access to multiple bioinformatics information sources. Bioinformatics 16:184-185.

Wheeler DL, Church DM, Federhen S, Lash AE, Madden TL, Pontius JU, Schuler GD, Schriml LM, Sequeira E, Tatusova TA and Wagner L (2003). Database resources of the National Center for Biotechnology. Nucleic Acids Res. 31: 28-33.

Witten IH and Frank E (1999). Data Mining: Practical Machine Learning Tools and Techniques with Java Implementations. Morgan Kaufmann Publishers, San Francisco.

Applied Mycology
and Biotechnology
An International Series
Volume 6. Bioinformatics

ELSEVIER

Searching Biological Databases Using Biolinguistic Methods

Gautam B. Singh
Center for Bioinformatics, Department of Computer Science & Engineering, Oakland University, Rochester, MI 48309, USA (singh@oakland.edu).

Existing methods based on homology rely on current research in genome analysis using *n-grams* (i.e. breaking the genome up into "words or "syllables"), protein motifs, and other bio-linguistic techniques have shown promise. In particular, as new protein structures and functions are identified, these bio-linguistic approaches can reach across multiple genomes to identify similar genes, elucidating their functions. Likewise new genes or disease gene variations identified through sequencing of individuals can be compared to known genes for identification of changes to their "normal" functions. In this review, we describe algorithms for searching biological databases using the *n-gram* analysis. Our results demonstrate that these algorithms are more sensitive than those currently available for both genomics and proteomics analysis, allowing a more accurate portrayal of similarity of gene function. The algorithm's capabilities extend to the comparison of biological sequences using phylogenetic and bio-chemical properties that enable the results to be significant from perspective of structure and function of genomic and proteomic data analysis. Recent years have seen an explosive growth in the speed and capacity of data collection and storage devices. The biological databases are experiencing an unprecedented growth where they are doubling every fifteen months. The algorithms described are amenable to parallelization with effective domain database partitioning. This makes them an attractive alternative for searching protein databases by developing high-speed functionally partitioned searches.

1. INTRODUCTION

At the cornerstone of bioinformatics research is the task of comparing or aligning sequences. The power of sequence alignment stems from the empirical finding that biological sequence similarity generally implies similarity in biological functions and a common ancestry. This is one of the main reasons that scientists search biological databases - their objective is to find other sequences similar to their query. Given that thefunction and the ancestry of the neighbors found in the database might have been characterized previously, similarity searches help scientists understand the functions of sequence under investigation. Furthermore, as the genomic databases are continuing their exponential growth rate, and doubling approximately every 14 months, the significance of *sensitive* searches of databases is becoming an important issue to address. Currently, the techniques/algorithms of similarity detection can be divided into three categories:

- *Homology-based Methods* such as trees and clustering. This type of methodology is utilized in Blast and Smith-Waterman search algorithms.
- *Structure-based Methods* such as threading and fold recognitions,

and

Nonhomology Methods such as domain fusions, phylogenetic profiles, correlated expression, and conserved gene position.

Among these three categories, the first two are based on sequence or structure similarity; and the third one is based on properties shared by functionally related proteins to identify protein-protein relationships. Although homology-based methods have been widely used, they have disadvantages. Homology-based methods are based on the assumption that similar sequences will share similar functions. This assumption does not hold true in many cases where similar sequences are structurally and functionally diverse. For example, about 25-40% of the genes in a bacterial genome usually do not find matches with known genes. It is also generally accepted that in several cases, where the evolutionary relationships between the sequences is distant, the occurrences of common *motifs* amongst the sequences is indicative of relatedness computing sequence similarities that aim at finding common sub-strings between the query and each sequence in the database. In this method, the probability that the query and the database sequences are related is proportional to the extent and the number of the shared subsequences. One limitation results from the fact that since the value of these similarity scores is dependent upon the length of the common subsequence, a rank-ordering of all scores is required in order to establish the significance of a given score. When close homologues do not exist, non-homology methods come to the rescue. Unlike homology-based or structure-based methods, non-homology based methods work like experimental genetic methods. They work by identifying the context that the protein operates in. Genomes contain considerable information about the functions and relationships between genes and proteins. This functional information is encoded in forms such as patterns of gene fusion, conservation of gene position, patterns of gene co-inheritance and other sorts of evolutionary information (Marcotte 2000). Some specific examples of the different types of information utilized by the non-homology based methods are described below:

Phylogenetic profiles method: This method is based on the assumption that proteins that function together in a pathway or structural complex are likely to evolve in a correlated fashion. During evolution, all such functionally linked proteins tend to be either preserved or eliminated in a new species (Pellegrini et al. 1999). In this method, a gene is described by its phylogenetic profile - a string that encodes the presence or absence of a homologous gene in other genomes. This string is then used to search for other genes with similar profiles. Liberles et al. 2002 discuss inverse phylogenetic profiles and non-exact profiles using phylogenetic trees. As a standalone method, this method seems to perform quite well (Marcotte et al. 1999).

Inductive Logic Programming(ILP): Muggleton introduced ILP by combining inductive learning and logic programming (Muggleton S. and Raedt L. D. 1994). ILP integrates first-order relational logic with background knowledge. There are three main elements in an ILP learning system: examples (E), background knowledge (B), and hypothesis (H). The elements can be separated into positive and negative examples. Each element of ILP is a logic program. Learning is achieved through prediction of the simplest hypothesis (H) given the background knowledge (B) and examples (E). ILP is an expressive learning approach and the rules generated by this approach can be understood by humans. The disadvantage of ILP is the lack of probability in its rules. Since biogenetic problems are characterized by a high degree of uncertainty, the hypotheses will have a higher descriptive power if they incorporate stochastic measurements.

Rosetta Stone or Domain fusion method: This method is based on searching for gene fusion events when comparing genomes. Gene fusion appears to be common and driven at least partially by the process of sub-functionalization (Lynch and Force 2000). The Rosetta Stone method is one of several methods introduced to go beyond sequence similarity when predicting the function of a gene (Marcotte 1999; Enright et al. 1999).

Conserved gene order: Genes that are found as neighbors in multiple organisms often encode interacting proteins and could be used to predict protein function (Tamames 2001). Dandekar et. al. (1998) compared nine bacterial and archaeal genomes and found that numbers of gene pairs are conserved. The proteins encoded by conserved gene pairs seems to interact physically.

A common misconception is that protein sequences that are homologous share similarity near the functional sites or regions of other active domains within the protein (Pearson W. 1997). This partly accounts for the popularity of the databases of sequence motifs, such as PROSITE (Bairoch et al. 2002), which tabulate amino acid patterns as regular expressions. As features that result from the convergence to a common property, sequence motifs are very informative. This is observed for example, in sites for glycosylation and phosphorylation. While most sequences that share statistically significant similarity ($E( ) < 0.02$) are homologous, many distantly related homologous sequences do not share significant *sequence* homology. Homologous sequences share a common ancestor, and thus a common *protein structure*. Depending upon the evolutionary distance and divergence path, two or more homologous sequences may have very few absolutely conserved residues. It is observed that albeit protein sequences may share a high level of similarity ($0.02 < E() < 20$), it may not be considered significant from the standpoint of retrieving related sequence from the ever-growing sequence databases, particularly with the rank-ordering needed to establish relatedness. Furthermore, it has been successfully shown that in several such cases, the occurrences of common *motifs* amongst the sequences are indicative of relatedness (Pearson 1997).

As illustrated in Fig. 1 biologists solve the problem of sequence interpretation using one of the two methodologies (Kanehisa M. 2000), namely sequence homology similarity and motif based similarity. In *sequence similarity* search, also known as the homology search, a query sequence is compared with each of the database sequences. The homology search is like searching against all the known sentences written in the DNA language, and when matching sentences are found the meaning in the precedents is used to interpret the new sentence. Alternatively, the *motif search* based approach is akin to having a dictionary and possibly utilizes the knowledge of grammar, if applicable, on the use of motifs and their functional significance. The query sequence is checked against the motif dictionary and the presence and absence of motifs is used for association of functional significance.

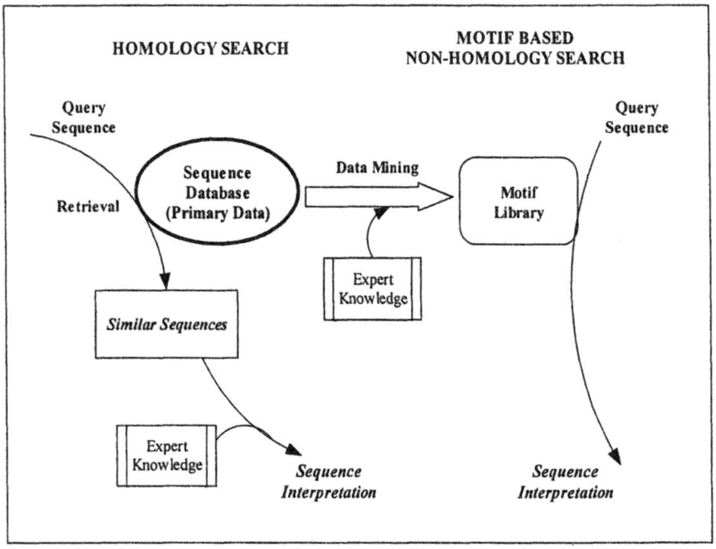

Fig. 1. Comparison of *similarity* and *motif-based* search process for functional interpretation.

The compilation of a dictionary of motifs and other empirical rules is a challenge by itself. Computer scientists, biological chemists and statisticians are collaborating to build computational tools and methods to better understand the function of proteins in terms of language theory. As in languages, where sequences of letters determine patterns of words and sentences, sequences of amino acids in proteins determine protein structure and the dynamics of its functions. Thus, motifs and higher level patterns can be thought of as syllables or words that have particular properties and functional significance. Scientists can thus predict the geometrical structure and functional dynamics based on the similarity of the series of motifs found in the sequences.

In this paper we present some additional evidence substantiating and validating the *k-mer* profiles (sometimes referred to as *n-gram*) algorithm for developing a data-mining system. The sequence retrieval system described earlier utilizes linguistic properties of biological sequences as its basis for establishing relatedness. As described in more detail in the following sections, the information contained in a profile of the various words occurring in the sequences is used for this purpose. In this manner, a biological sequence is considered to be a carrier of tokens. Unique words of specific sizes are considered to be these tokens and the sequence similarity is defined in terms of the average frequencies of the tokens occurring in the sequences compared. A measure for similarity of word frequencies is defined using information theoretical considerations. The similarity measure, called the *divergence* metric, is an absolute measure that does not require its significance to be evaluated relative to a population of similarity scores.

2. SEQUENCE REPRESENTATION

A word frequency profile based method aims at establishing that genomes from different organisms have characteristic frequency profiles, similar to each physical material having its unique resonant frequency. By observing the frequency spectrum of an unknown material, one can determine its ingredients. Protein sequences are considered to be made up of some "words" (n-grams). Each different organism has its own word set, where a given word that is frequently found in one organism may be totally absent in another. It has been shown that these frequently used phrases are not due to random variation (Blaisdell 1996; Brocchieri and Karlin 1995; Ganapathiraju et al. 2002). The *n-grams* of the genome of natural organisms and that of artificial genomes generated by Monte Carlo simulation were compared (Sigrist et al. 2002). It was found that the frequencies of natural genomes are well above the baseline variation due to chance sampling. This indicates that n-gram profiles of the query sequence may help in retrieving related sequences from databases.

The sequences being compared are represented by the profile of their *n-gram* word frequencies. While it is possible to construct exhaustive profiles for DNA alphabet (example, a *4-mer* profile for a DNA sequence will have 256 words). The scale-space representation at lower resolutions (smaller word size) is an integration of the representation at higher resolutions. For example, the frequency f_{AAA} of the tri-nucleotide AAA in a sequence is the sum of the frequencies of f_{AAAA}, f_{AAAC}, f_{AAAG} and f_{AAAT}. Thus, a comparison at lower resolution provides the aggregation of statistical information at higher resolution. This enables the search using lower-resolution profiles to become more sensitive.

The compilation of a dictionary of motifs and other empirical rules is a challenge by itself. Computer scientists, biological chemists and statisticians are collaborating to build computational tools and methods to better understand the function of proteins in terms of language theory. As in languages, where sequences of letters determine patterns of words and sentences, sequences of amino acids in proteins determine protein structure and the dynamics of its functions. Thus, motifs and higher level patterns can be thought of as syllables or words that have particular properties and functional significance. Scientists can thus predict the geometrical structure and functional dynamics based on the similarity of the series of motifs found in the sequences.

In this paper we present some additional evidence substantiating and validating the *k-mer* profiles (a.k.a. *n-gram*) algorithm for functional proteomics. As described below, the information contained in a profile of the "words" occurring in the sequences is used for this purpose. In our model, a biological sequence is considered to be a carrier of tokens.

Unique words of specific sizes are considered to be these tokens and the sequence similarity is defined in terms of the average frequencies of the tokens occurring in the sequences compared. A measure for similarity of word frequencies is defined using information theoretical considerations. The similarity measure using *L-Divergence* based on information theoretical foundations is used for comparing sequence profiles.

Profiles based method aims at modeling the biological sequence as a mosaic of patterns. The n-gram profiles may be viewed as a stochastic language model where the mapping from genome sequence to functions is the conceptual analog to mapping words to the semantics. Furthermore, the low-resolution profile may be as much as 10 times more compact than the high-resolution profile and corresponds to a higher level of sensitivity, as in general, a larger number of short words will be shared by any two given sequences. In contrast, the higher-resolution profile is based on commonality of longer words. The profile representation at lower resolutions (smaller word size) is an integration of the representation at higher resolutions. The *n-gram* profile analysis may be utilized in an adaptive manner, increasing the value of *n* when the algorithm is desired to be more selective. Moreover, higher weights may be assigned to specific words that have been previously documented as known motifs with known functional significance. To overcome the above shortcomings, the modeling approach described in this paper may be extended to multi-mer model or functional profile. In a multi-mer model words, having different lengths, and are concatenated without overlapping; each word being derived from a lexicon by using the well-known pattern databases such as TRANSFAC (Attwood et al. 2002) or PRINTS (Needleman and Wunsh 1970) or by using linear programming approaches such as the expectation-maximization to construct our own dictionary. The advantages of the former approach are that it is simple and the meaning of each word is known. The advantage of the latter is that by constructing our own dictionary, new patterns may be discovered.

3. SEQUENCE COMPARISON MEASURES

The section on sequence representation demonstrated that the relative word frequencies, expressed as a probability density functions, do adequately capture the functional differences between sequences. However, we need to have the ability to quantify differences between the sequences that, for example, might belong to the same category. Albeit the metric D computed in Kolmogorov-Smirnov test may be used for this purpose, we have utilized a metric from informational theory for capturing the divergence between the sequence profiles. One advantage of using this approach is that the result of comparing two sequences falls between 0 through 1. This enables us to use divergence measures for the comparison of these *pdfs*. Two such measures are the *variation distance*, $V(p_1,p_2)$ and the *L-divergence*, $L(p_1,p_2)$. Either of these may be used for measuring the divergence between two probability density functions p_1 and p_2:

$$V(p_1, p_2) = \frac{1}{2} \sum_{x \in X} |p_1(x) - p_2(x)|$$
$$I(p_1, p_2) = \sum_{x \in X} p_1(x) \log \frac{p_1(x)}{p_2(x)} \tag{1}$$
$$J(p_1, p_2) = I(p_1, p_2) + I(p_2, p_1)$$

$$K(p_1, p_2) = \sum_{x \in X} p_1(x) \log \frac{p_1(x)}{\frac{p_1(x) + p_2(x)}{2}} \tag{2}$$
$$L(p_1, p_2) = K(p_1, p_2) + K(p_2, p_1)$$

$$L(p_1, p_2) \leq V(p_1, p_2) \tag{3}$$

Experimentations on random bit strings were performed in order to better understand the characteristics of the *L-divergence* measure. The results of these experiments are summarized in Figure 2. The experimental results are based on studying the effect of mutating thousands of synthesized sequences of varying compositions. Each of these sequences is mutated by a specific fraction, and the divergence between the original and the mutated sequence is measured and plotted. Sequence composition is plotted along the *x*-axis, mutation level along the *y*-axis, and corresponding divergence value along the z-axis.

It is interesting to point out that when the sequence composition is random, mutation of the original sequence does not cause the sequences to diverge further. This is evident in the sequence with a composition of *0.5*, i.e. a sequence with equal number of *0*'s and *1*'s. We observe that no amount of mutation causes this sequence to diverge further from its parent sequence. This is in accordance with our expectation. Thus, if a sequence is random to start with any level of random mutations will preserve this randomness in accordance with our expectation. Also, if a sequence is random to start with, any level of random mutations will preserve this randomness. And, the mutated sequence continues to remain similar to the original sequence by the virtue of the fact that both are random. It may also be noted that the sequences that have a definite distribution of characters, such as those comprised only of *0*'s or *1*'s, are the ones that diverge the most when mutated.

Figure 3 shows experimental comparison with four commonly used distance metrics. About 1700 HIV envelope protein sequences are employed and retrieval sequences by different distance metrics are compared in term of coverage from top Needleman-Wunsch fit scores (Karlin S, Blaisdell E, and Brendell 1990). This sequence comparison method is a standard for establishing global alignment based sequence homology. To evaluate retrieval performance, coverage is defined as the spread of the sequences retrieved by a method in the set of sequences retrieved using Needleman-Wunsch. For example, we established that the top 1% of sequences retrieved using the Bhattacharyya method has coverage of 50% within the list of sequences retrieved using the Needleman-Wunsch method. This means that the top scoring sequences using Bhattacharyya fall within the top 50% of the sequence ranking created using Needleman-Wunsch.

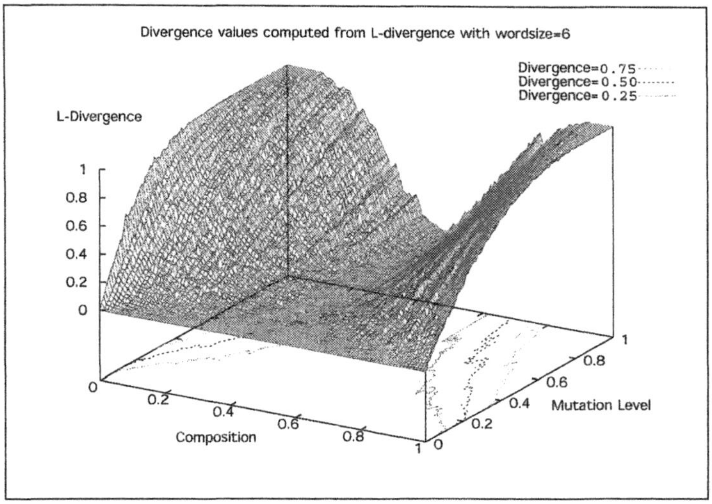

Fig. 2: Divergence Surface Plot depicts the variation of divergence values as a function of difference in sequences. A given sequence is randomly mutated and the divergence between the original and the mutated sequences is plotted. Interestingly, when the composition of the sequence has the highest entropy to begin with (composition = 0.5), the level of mutation keeps the divergence values unchanged

Generally, a lower coverage value indicates a higher consistency of the scoring method to the global alignment scoring system - which is the present standard. Thus, the scoring system based on L-Divergence generally agrees with the scores achieved through the conventional methods.

For computing the divergence between protein sequences we have utilized multiple profiles. The average divergence between two protein sequences represented using profiles p_i^f and p_j^f on functional space f (example, hydrophobic properties, residue charge properties, residue aromatic properties, etc.) is defined as a weighted sum as follows:

$$d(p_i, p_j) = \sum_f w_f L(p_i^f, p_j^f) \qquad \text{with} \quad \sum_f w_f = 1 \tag{4}$$

In this manner sequences can be compared using an aggregation of independent comparisons on a wide range of functional spaces.

Karlin's group has reported statistical comparison of genomes (Blaisdell E. Campbell A. and Karlin S., 1996; Karlin, S., Blaisdell E., and Brendell V., 1990; Karlin S. and Lagunda I. 1994).

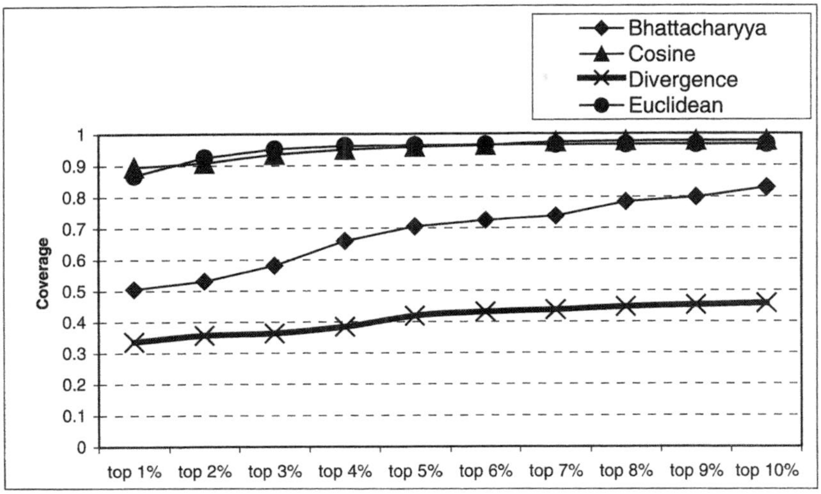

Fig. 3: Comparison of the Divergence based scoring for n-gram profile similarity based on divergence measure and other distance metrics.

The comparisons within and between species sample sequences are based on the biases observed in the usage of di-, tri- and tetra-nucleotides within the whole genomes. It was discovered that viral genomes have distinctive signature of short oligonucleotide abundance that pervades throughout the genome and distinguishes it from other genomes. Their studies were based on computing the di-nucleotide biases as the odds ratio $r_{XY} = f_{XY}/f_X f_Y$ where the frequency of nucleotide X is f_X and f_{XY} is the frequency of di-nucleotide *XY*. They further utilized the relative abundance values for estimating the distance $d(f,g)$ between two genomes *f* and *g*. This is defined as:

$$d(f,g) = \frac{1}{16}\sum_{ij} | r_{ij}(f) - r_{ij}(g) | \qquad (5)$$

The metric used for relative dinucleotide abundance defined in Eq. (5) is similar to the variational distance $V(p_1, p_2)$ defined in Eq. **Error! Reference source not found.**. The variational distance forms an upper bound to the L-divergence metric. Their experimental results have shown that the closely related organisms have highly similar profiles (or genome signatures) than do distantly related organisms, and thus also lays down strong biological foundations for the biological sequence retrieval algorithm that we are proposing to build.

4. DNA SEQUENCE RETRIEVAL

Eight sequences were randomly selected each from humans, bacteria, Arabidopsis, and yeast. The lengths of the randomly selected query sequences were approximately

100 bp, 500 bp, 1000 bp, 5 kb, 10 kb, 20 kb, 50 kb and 100 kb. The divergence scores for these sequences from the four species were computed against the Primate section of GenBank (GB-PRI) using the n-gram profile method. It was observed, that as expected the human sequences were closest to the primates on an average, for sequences for all sizes. Generally, the gap between humans and primates diminishes as sequence length became larger. The results shown in Fig. 4 illustrate that the average divergence based on the topmost 5% matches places humans to be closest neighbors of primates. Furthermore, this proximity increases as sequence size increases (i.e. more evidence of words and syllables used is gathered).

From Fig. 4 it is evident that the bacterial sequences are generally the farthest away from the primate section of Gene Bank. This is in accordance to the evolutionary distance between these two species. The plot on the right shows the ratio of the average score for bacteria's top 1% neighbors found in GB-PRI and human's top 1% neighbors. This clearly shows that the divergence scores for humans tend to an order of magnitude smaller than those of bacteria. In this manner, the profile based genome retrieval algorithm demonstrates the power of this method to discriminate the sequence neighbors.

This is expected to be in contrast to methods based on local sequence similarities where the chances of finding substring matches of a large query sequence to the database would generally increase with the increase in query sequence size.

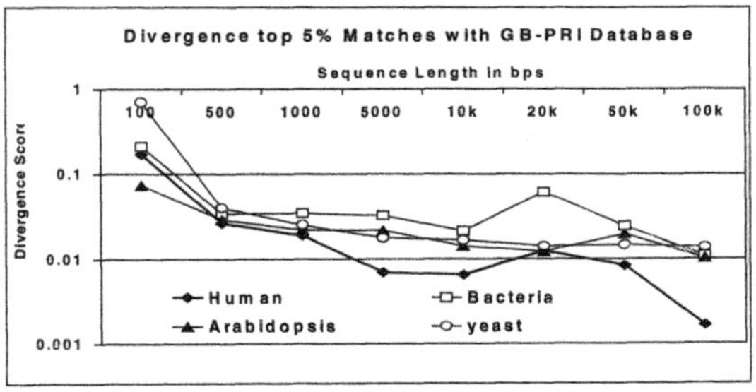

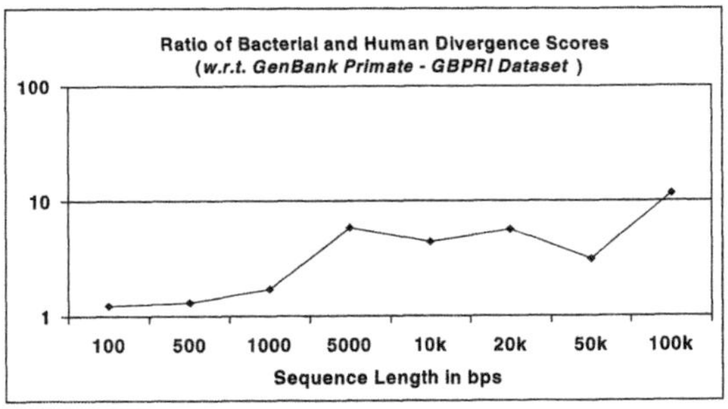

Fig. 4. Divergence Statistics. *Top*: Average divergence scores between four species studied and primate dataset in GenBank (GB-PRI). *Bottom:* Human sequences are generally an order of magnitude closer to the primates as compared to the bacterial sequences.

By increasing the sensitivity of genomic database searches, we expect that the set of sequences retrieved will also include those that are *remotely related* to a query sequence. We believe that such an approach will help scientists uncover distant relationships amongst genes. As illustrated in the previous section, composition-based genomic sequence retrieval algorithm shows promising results. Further investigation into phylogenetic reconstruction using an example data set for phylogenetic reconstruction is shown in Figure 5. Phylogenetic reconstruction, also known as the determination of the evolutionary "tree of life", is considered to be a major problem in computational sciences.

Research has been on the presentation of the phylogenetic problem as a problem in optimal construction of a binary tree, representing the evolution of life. The problem of counting the number of possible trees and determining the optimal tree is an NP-hard problem, and several algorithms for tree visualization have been defined (Blanchette et al. 2002).

The distance based clustering methods, as well as neighborhood joining algorithms for tree constructions have been used in research, with each method deserving of its own merits.

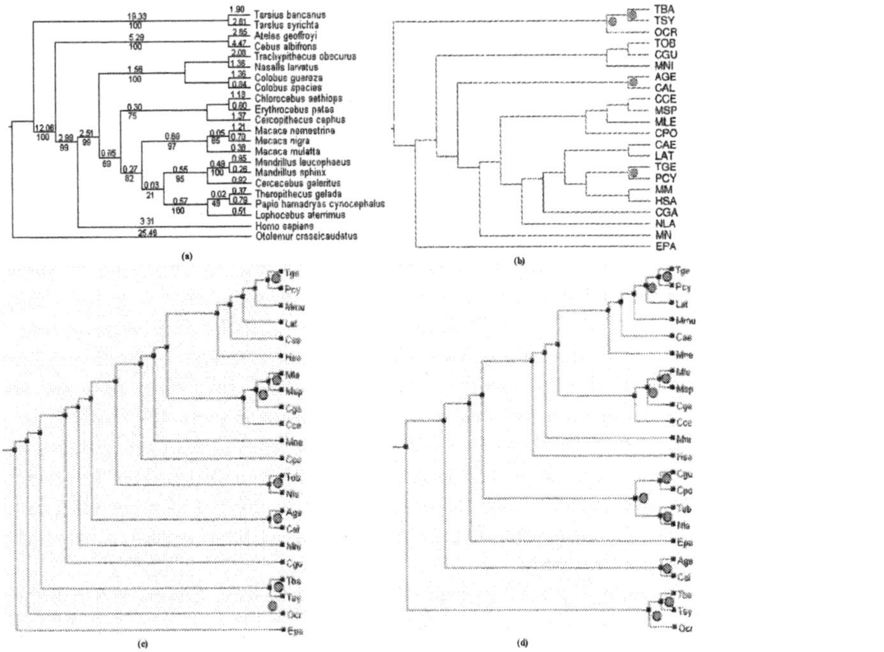

Fig. 5: (a) The phylogenetic tree representing evolutionary relationships between the Old World Monkeys. (b) Phylogenetic tree using BLASTN score as the basis for distance in UPGMA. Phylogenetic trees using

UPGMA with sequence distances using the *n-gram* distance measure with (c) word size =6 and (d) word size=9. Dots indicate biologically correct taxonomy. It is clear that phylogeny using *n-gram* biolinguistic method yields biologically correct results.

The algorithmic basis for probabilistic tree construction and optimality of trees is often defined in terms of minimization of Hamming distance in parsimony and Bayesian inference in Maximum Likelihood Estimation for tree reconstruction (Bonet, Steel, Warnow and Yooseph 1998; Yang 1996; Golding and Felsenstein 1990; Felsenstein 1981; Liu and Singh 2002).

5. Extension to Protein Sequence

Proteins serve both the structural and functional components inside the cell. They are chemically different from DNA and RNA as they are composed of amino acids. The protein alphabet is comprised of 20 symbols: A, C, D, E, F, G, H, I, K, L, M, N, P, Q, R, S, T, V, W, and Y. The canonical structure of an amino acid is shown below:

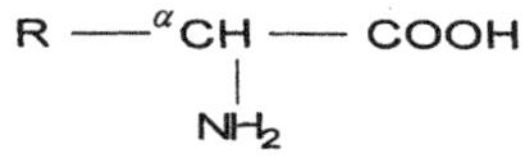

The properties of the amino acids are characterized by the nature of the side chain group, denoted as "R", associated with the amino acid. Each amino acid has a unique side group attached to the α-carbon atom. Amino acids are categorized into eight classes based upon the bio-chemical properties of the side chain. These classes are, (i) simple amino acids that contain a functional group as the side chain, for example, glycine, alanine, valine, leucine, isoleucine; (ii) Hydroxy amino acid that contain a hydroxyl group as the side chain, for example, serine, and threonine; (iii) Sulfur containing amino acids, example, cysteine and methionine; (iv) Acidic amino acids that contain carboxyl group in its side chain, for example, aspartic and glutamic acid; (v) Amino acid amides - amino acids in which the carboxyl group has been transformed into an amide group (-CO.NH_2) example, asparagine and glutamine; (vi) Basic amino acids contain an amino group (NH_2) as a part of its side chain, for example, lysine and arginine; (vii) Heterocyclic amino acids have a side chain rings with at least one non-carbon atom in the ring, for example, tryptophan, histidine, and proline; (viii) Aromatic amino acids have a benzene ring as a part of the side chain, as in the case of phenylalanine and tyrosine. Furthermore, linear chains of amino acids fold into specific three-dimensional structures that are solely dependent upon the linear chain of amino acids synthesized from the mRNA. In this manner, the gene and the chain of amino acids, determine the shape of the protein, which in turn establishes its function. A protein may be shaped like a stiff rod for the purposes of providing support (example, collagen, keratin), form prong-based structures to "zip-open" macromolecules (leucine-zippers, and zinc fingers) or form hooks to be used as a part of muscle contraction (myosin).

One issue with protein *n-grams* is that of the profile size. Given the number of characters in the protein alphabet, a 4-mer profile for proteins will be comprised of 20^4 or 160,000 words. However, by using residue properties, such as charge and hydrophobic propensities, these sequences may be represented by an alphabet with a smaller number of symbols, and hence reduce the amount of data needed for comparison. Since the charge and hydrophobic properties of the individual amino acids greatly influences the structure and function of the peptide sequences, such a representation is not only computationally tractable, but may also be functionally more significant.

The mappings utilized for transforming an amino acid sequence into a hydrophobic and charge property sequence are shown in Table 1**Error! Reference source not found.**. For a *charge based transformation*, a peptide sequence is transformed into a new sequence over the alphabet **+**, **-** , and **0**. Analogously, for a *polarity based transformation*, the hydrophobic properties of individual residues is used to construct a sequence over the alphabet **B, L,** and **A**, with the characters **B** replacing a hydrophobic residue, **L** replacing a hydrophilic residue, and an **A** denoting an ambivalent residue. In this manner we are able to compare the occurrences of the proximal and distal amino acids as described in Brocchieri and Karlin (1995).

Table 1: Amino Acid (a) Hydrophobic Profile, and (b) Charge Profile Mapping

(a) Hydrophobic Properties of the amino acids	
Hydrophobic Amino Acids	Alanine (A), Phenylalanine (F), Isoleucine (I), Leucine (L), Methionine (M), Proline (P), Valine (V), Tryptophan (W), Tyrosine (Y)
Hydrophilic Amino Acids	Aspargarine (N), Glutamine (Q), Serine (S), Threonine (T)
Neutral	Cystine (C), Aspartate (D), Glutamate (E), Glycine (G), Histidine (H), Lysine (K), Arganine (R)
(b) Charge Properties of the amino acids	
Negatively Changed Amino Acids	D, E
Positively Charged Amino Acid	H, K, R
Neutral charge	A, C, F, G, I, L, M, N, P, Q, S, T, V, W, Y

The hydrophobic group's molecules adhere amongst themselves with the elimination of water to form linkages between various segments of a chain or between different chains. This is very much like the coalescence of oil droplets suspended in water. The association of various "R" groups, or the side chains, in this manner leads to a very strong bonding and often leads to bringing together groups that can form hydrogen or ionic bonds in the absence of water. The hydrophobic profiles thus capture the affinity of regions to form hydrophobic core.

The hydrophobic profile for a protein sequence is computed in three simple steps. In the first step, the hydrophobic transformation described **Error! Reference source not found.** is used to generate a sequence on a three character alphabet, namely, {B, L, A}.

Based on the value of *n* in the *n-gram* profile, the frequency of all possible *n-grams* is computed in the second step. In the final step, the frequency histogram is converted to a probability density function. For example, the amino acid sequence "NSDYNKLVY" will be transformed to "LLABLABBB". If *2-meer* profiles are considered, the frequencies for the various dimers will be {(AA: 0), (AL: 0), (AB: 2), (LA: 2), (LL: 1), (LB:0), (BA:0), (BL:1), (BB:2)}. Using Laplace rule, we can compute the probabilities for dimers {AA, AL, AB, LA, LL, LB, BA, BL, BB} as $\{\frac{1}{15}, \frac{1}{15}, \frac{3}{15}, \frac{3}{15}, \frac{2}{15}, \frac{1}{15}, \frac{1}{15}, \frac{2}{15}, \frac{3}{15}\}$ respectively. Hydrophobic profiles from two randomly selected sequences in enzyme category EC 1.1.1.1 (major category *oxidoreductases)* were chosen. These three profiles were found to exhibit a similar shape which is attributed to the fact that enzyme sequences within a given category are responsible for similar functions. The results are shown in Fig 6. Contrast this to the hydrophobic profiles for two randomly selected sequences from categories 3.4.27.20 (accession number P27237) and 6.3.4.11. These differences in profiles are again attributed to the fact that the enzymes belonging to the major category 3 are *hydrolases* while those in category 6 are *ligases,* functions that are complementary.

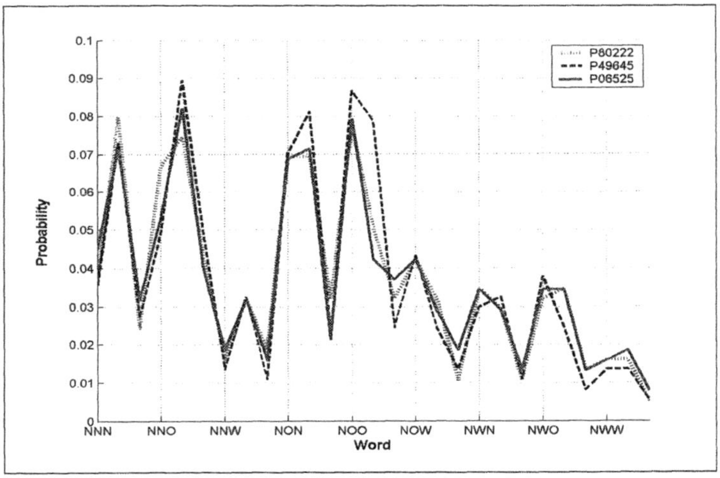

Fig. 6. The hydrophobic property profiles of three randomly selected enzyme sequences from the Enzyme class EC 1.1.1.1. Note the strong conservation of words in the three different sequences that belong to the same functional class.

to each other. Since the distribution of the profile is not characterized, a non-parametric distribution free test such as Kolmogorov-Smirnov test was applied to determine if the difference in the two profiles shown in Fig. 7 is significant. Kolmogorov-Smirnov test tries to determine if two datasets differ significantly based upon the D-statistic designed to measure the maximum distance between the two cumulative distribution functions without requiring us to know the distribution of the data *a-priori*. The cumulative distribution functions were computed and plotted as shown in Fig. 8. The maximum distance between two distributions was measured as D=0.222 for the value of $X \approx 0.315$. This corresponds to α-value of 0.01 and thus we may assert with a 99.99% confidence that the two distributions belong to different populations.

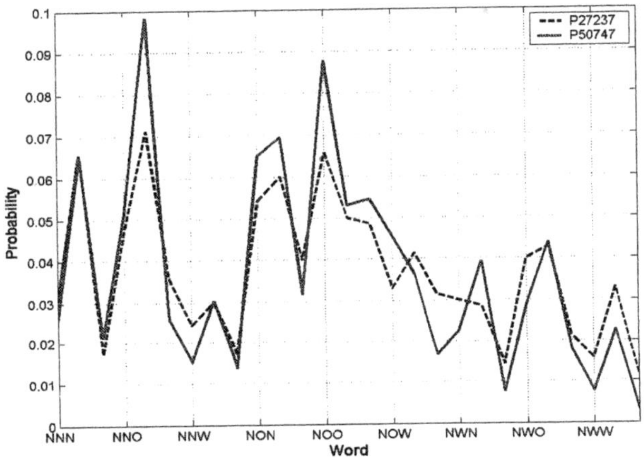

Fig. 7: Hydrophobic property profiles for two randomly selected sequences from class 3.4.27.20 (accession number P27237) and 6.3.4.11 (accession number P50747). The difference in the word probabilities, albeit following a similar trend, exhibits significant variation.

6. PROTEIN CLASSIFICATION

Proteins are of great importance in biological systems. Some of the functions are as follows. Proteins may act as *catalysts* that enhance the rate of reaction necessary to sustain life, as well as serve as the *components of the tissues* holding the skeletal elements together. Nucleoproteins help in carrying *genetic information* from one generation to the next, while cytokines are a type of protein that facilitate in *cell-to-cell signaling*. Proteins perform *transport functions* through active transport mechanisms using catalysis and adsorption. Various proteins are known *hormones* regulating the growth and controlling physiological functions. Sometimes under the condition of non-digestion and lack of denaturizing, proteins may become *toxic* as in the case of snake venom and insect bites. Proteins are used by living forms for the *treatment of shock*, as in the case of blood plasma.

Our first experiment aims at clustering the enzyme sequences randomly selected from each of the six major enzyme classes, 1 through 6. The enzymes in class 1 are *oxidoreductases* and help bring about the oxidation-reduction reactions between two substrates. The enzymes in class 2 are *transferases* that catalyze the transfer of functional subgroups from one substrate to another. The *hydrolases* are the group of enzymes that catalyze the hydrolysis of substrates and grouped into class 3. The enzymes in class 4 are the *lyases* – a class of enzymes that catalyze the removal of functional groups from substrates using mechanisms other than hydrolysis. The *isomerases* in class 5 catalyze the inter-conversion of optical, geometrical and positional isomers by intra-molecular rearrangements. The enzymes in class 6 are *ligases* that catalyze the linking together of compounds using the ATP as the energy source.

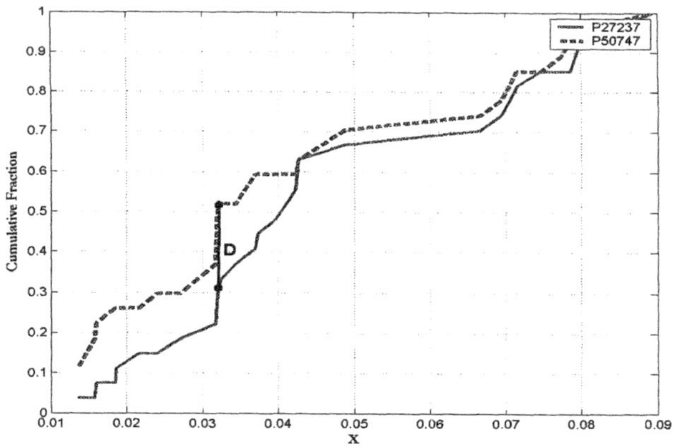

Fig. 8. The Kolmogorov-Smirnov test with the statistic D of 0.2222 establishes with a 99.99% level of confidence (with a = 0.01) that the two profiles in Fig. belong to different populations.

Six sequences were randomly selected from each of the six major enzyme groups described above. The dissimilarity matrix was computed by utilizing the *L-divergence* measure between each sequence pair. The sequences were subsequently clustered using a hierarchical clustering procedure. The clustering results are shown in Fig 9. The results demonstrate that clusters formed correspond reasonably well to the known enzyme classes.

Our second experiment aims at establishing if the profiles reflect the functional classification of a protein. For this experiment, we selected 8-10 sequences from Calcium Channel Protein, Myosin protein, Sodium Channel Protein, Microtubules Associated Proteins, and Thyroglobulin Precursor protein. The shape of a protein determines its biological activity. Proteins are classified according to their biological roles. Calcium channels are proteins that allow calcium ions to flow into cells. This influx of calcium plays a role in many important cell functions including muscle contraction. Myosin is a contractile protein and plays a very important role in cellular movement. Na-Channel protein is a protein embedded in the plasma membrane. It is found in the nerve and muscle cells and is used in the rapid electrical signaling found in these cells. Microtubules are conveyer belts inside the cells and move vesicles, granules, organelles like mitochondria via special attachment proteins. They also serve a cytoskeleton role. Structurally they are linear polymers of tubulin which is a globular protein. Thyroglobulin is a large globular glycoprotein and plays an important role in the biosynthesis of the thyroid hormones and is typically present in the colloid of thyroid gland follicles. Fig 10 shows the results of comparing each of these clusters with each other and representing the result as-*log* (p_{ij}) where the probability is computed to designate the chance similarity between clusters *i and j*. As the function of each of the five protein clusters is sufficiently different from each other, the intra-cluster divergence scores were the lowest. This yields a low probability of chance similarity, and consequently, the value of the measure -*log* (p_{ii}) is the highest. It is important to point out the elevated score of the similarity between the Calcium and the Sodium channel protein clusters. This result is biologically significant as these proteins have similarity attributed to their function as channel proteins.

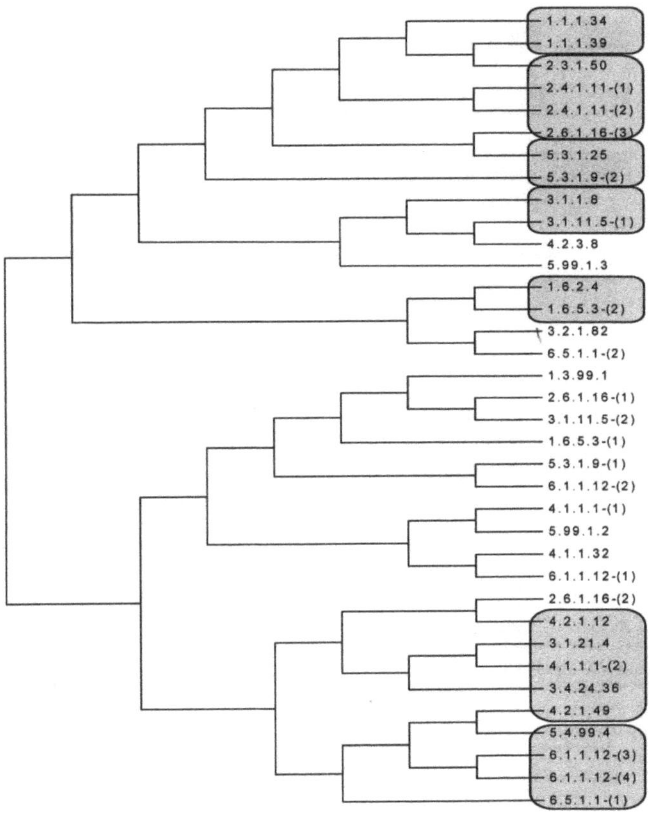

Fig. 9: Clustering of thirty-six randomly selected enzyme sequences, with six sequences selected from each major enzyme class. The clustering is based on divergence score computed using 4-mer hydrophobic- and charge-based profiles of the protein sequences.

7. MULTIPROCESSOR BIO-LINGUISTIC SEARCH ENGINE (BSE)

Our research is also focused on achieving a partitioning of the database such that database retrieval processes may be parallelized and implemented within the Bio-linguistic Search Engine or BSE. As shown in Fig.11, the input query sequence is parsed

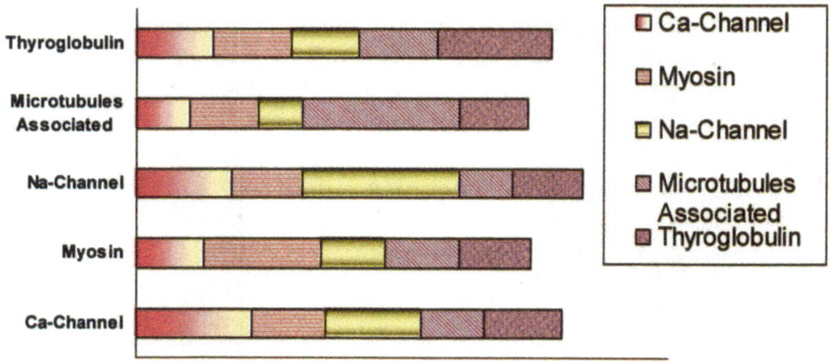

Fig.10. Comparison of inter-cluster divergences for five protein clusters. The two channel protein clusters (i.e. Ca- and Na-channels) are closest to each other.

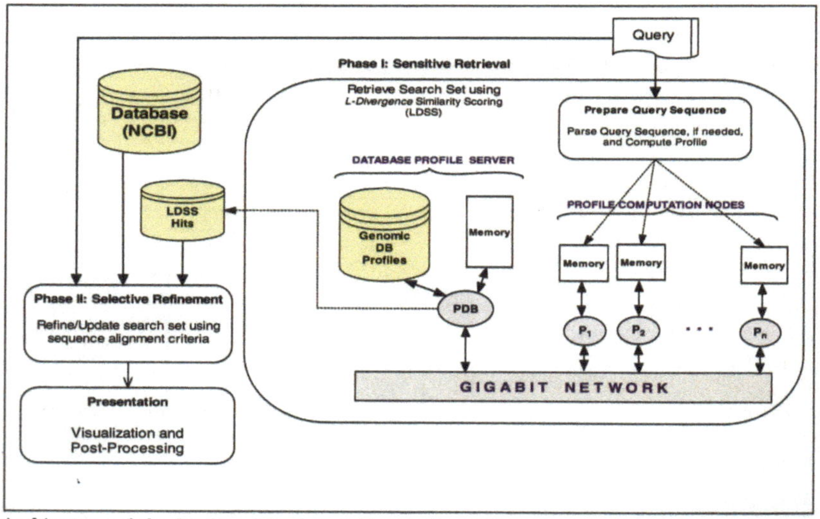

Fig.11: Architecture of the Bio-linguistic Search Engine. In addition to searching the database using the profile based methods, set theoretic operations are applied to contrast the results with NCBI's BLAST and annotation servers.

to evaluate its profile, which is compared to the pre-computed profiles for all the sequences in the database. The search proceeds in a parallel manner by using a number of *Profile Computation Nodes*. The parallel implementation of the profile based searching methodology is easy to implement as the time required to complete the computation is constant and enables the implementation of a static domain partitioning methodology. The implementation utilizes a high speed gigabit networking test-bed for distributing the profiles to the individual nodes, with the *Database Profile Server* used for maintaining and distributing the entire set of genomic database profiles. The resulting set of sequence neighbors is represented by a set of indices of the genomic database sequences that are the query's neighbors. These values are utilized for example to compute the set of database neighbors that have not been reported by NCBI-BLAST server. Maintaining a mapping to the main databases using accession numbers, offers our system the capability to further tap the annotation servers, such as the ENSEMBL mirror deployed at MCBI. In this manner, in addition to providing a phylogenetic tree describing the evolutionary relationships amongst the retrieved sequences, a variety of analysis algorithms for function discovery including multiple sequence alignments, clustering and visualization may be utilized in an incremental manner to better assist in the interpretation of the retrieved sequences within a biological context (Worley et al. 1994).

8. CONCLUSION

In this chapter we presented the basis of sequence database searching using bio-linguistic methods based on *n-gram* sequence representation. We demonstrated as a result of representing the genomic sequences using a profile of individual word frequencies and representing polypeptide chains by their hydrophobic and ionic properties (which have a direct bearing on the secondary and tertiary structures that the proteins forms), we are able to achieve classification and phylogenetic inferences on the data sets that exceed the results currently achievable with the methods based on sub-string comparison. An extension of this concept to a generalized functional space opens up the possibility of creating a number of signatures of a protein sequence capturing other properties such as benzene ring, aromatic group, carboxyl group, or amides on the side chain. The representation of the biological sequences within these spaces allows us to capture the essence of the properties that make them perform the function that they do. We thus characterize such a comparison as a *functional* comparison from sequence data - a new direction in the area of comparative proteomics.

Acknowledgements: This work is partially supported by grants from the National Science Foundation (NSF) Award # EIA-0306064 and by Michigan Center for Biological Information (MCBI), Wayne State University-Node, Detroit, MI.

REFERENCES

Attwood T K, Blythe M J Flower, D R Gaulton, A Mabey, J E Maudling, N McGregor, L. Mitchell, AL Moulton, G Paine, K and Scordis P (Jan.2002). PRINTS and PRINTS-S shed light on protein ancestry, Nucleic Acids Res., vol. 30, no. 1, pp. 239-241.

Bairoch A Bucher P and Hofmann K (1997).The PROSITE database, its status in 1997," *Nucleic Acid Res.*, vol. 25 pp. 217-221.

Blaisdell E Campbell, A and Karlin S (1996). Similarities and Dissimilarities of Phage Genomics, *Proc.Natl.Acad.Sci.USA*, vol. 93 pp. 5854-5859.

Blanchette M, Schwikowski B, and Tompa M, (2002). Algorithms for phylogenetic footprinting," *J.Comput.Biol.*, vol. 9, no. 2, pp. 211-223.

Bonet M, Steel M, Warnow T, and Yooseph S, (1998). Better methods for solving parsimony and compatibility, *J.Comput.Biol.*, vol. 5, no. 3, pp. 391-407.

Brocchieri L and Karlin S (1995). How are Close Residues of Protein Structures Distributed in Primary Sequences?, *Proc.Natl.Acad.Sci.USA*, vol. 92 pp. 12136-12140.

Dandekar T Snel, B Huynen, M and Bork, P (1998).Conservation of gene order: a fingerprint of proteins that physically interact, *Trends Biochem Sci*, vol. 23 pp. 324-328.

Enright A J LLiopoulos I Kyrpides, N C and Ouzounis CA, (1999). Protein interaction maps for complete genomes based on gene fusion events, *Nature*, vol. 402 pp. 86-90.

Felsenstein J (1981). Evolutionary trees from DNA sequences: a maximum likelihood approach, *J.Mol.Evol.* vol. 17, no. 6, pp. 368-376.

Ganapathiraju M, Weisser D, Rosenfeld R, Carbonell J, Reddy R, and Klien-Seetharaman J. (2002). Comparitive n-gram analysis of whole genome protein sequences. 2002. Human Languages Technology Conference.

Galperin M Y and Koonin E V, (June2000). Who's your neighbor? New computational approaches for functional genomics, *Nat.Biotechnol.*, vol. 18, no. 6, pp. 609-613.

Golding B and Felsenstein J, (Dec.1990). A maximum likelihood approach to the detection of selection from a phylogeny, *J.Mol.Evol.*, vol. 31, no. 6, pp. 511-523.

Hide W Burke, J and Davison D.(1994). Biological Evaluation of d^2, an algorithm for high-performance sequence comparison, *J.Comp.Biol*, vol. 1 pp. 199-215.

Javier Tamames, (2001). Evolution of gene order conservation in prokaryotes, *Genome Biology*, vol. 2, no. 6, pp. research0020.1-0020.11.

Kanehisa M (2000). *Post-genome Informatics* Oxford University Press.

Karlin S, Blaisdell E, and Brendell V, (1990). Indentification of significant sequence patterns in proteins, *Methods in Enzymology*, vol. 183 pp. 388-402.

Karlin S and Lagunda I, (1994). Comparison of Eukaryotic Genomic Sequences, *Proc.Natl.Acad.Sci.USA*, vol. 91 pp. 12832-12836.

Liu D and Singh, G (2002). Entropy Based Clustering of High Dimensional Genomic Data Sets. 2002. Proceedings of the 2nd. SIAM International Conference on Data Mining (Workshop on Clustering High Dimensional Data Sets), Arlington, VA.

Liberles D A Thoren, A Heijne, G V and Elofsson A. (2002). The use of phylogenetic profile for gene predictions, *Current Genomics*, vol. 3, no. 3, pp. 131-137.

Lynch M and Force A (2000). The probability of duplicate gene preservation by subfunctionalization, *Genetics*, vol. 154 pp. 459-473.

Matys V, Fricke E, Geffers R, Gossling E, Haubrock M, Hehl R, Hornischer K, Karas D, Kel A E, Kel-Margoulis O V, Kloos D U, Land S, Lewicki-Potapov B, Michael H, Munch R, Reuter I, Rotert S, Saxel H, Scheer M, Thiele S, and Wingender E (Jan.2003). TRANSFAC: transcriptional regulation, from patterns to profiles, *Nucleic Acids Res.*, vol. 31, no. 1, pp. 374-378.

Marcotte EM (2000). Computational genetics: finding protein function by nonhomology methods, Current Opinion in Structural Biology, vol. 10, no. 3, pp. 359-365

Marcotte E M Pellegrini, M Ng H L Rice, D W Yeates, T O and Eisenberg D (1999). A combined algorithm for genomewide prediction of protein function, Nature, vol. 402, no. 6757, pp. 83-86.

Muggleton S and Raedt L D(1994). Inductive logic programming: theory and methods . Journal of Logic Programming. vol. 19, 20 pp. 629-679.

Needleman S and Wunsh C (1970). A general method applicable to search for similaritie s in the amino acid sequences of two proteins. J Mol Biol vol. 48 pp. 443-453.

Pearson W (1997). Identifying distantly related protein sequences, *Comp.Appl.of BioSci.* vol. 13, no. 4, pp. 325-332.

Pellegrini M, Marcotte E M, Thompson, M J, Eisenberg, D and Yeates, T O (1999). Assigning protein functions by comparative genome analysis: protein phylogenetic profiles. Proc Nat Acad Sci, USA. vol. 96 pp. 4285-4288.

Sigrist, C J, Cerutti, L, Hulo, N, Gattiker, A, Falquet, L, Pagni, M, Bairoch, A, and Bucher P (Sept. 2002). PROSITE: a documented database using patterns and profiles as motif descriptors. Brief Bioinform vol. 3, no. 3, pp. 265-274.

Worley K Wiese, B and Smith R (1995). BEAUTY: An enhanced BLAST-based search tool that integrates multiple biological information resources into sequence similarity search results. Genome Research vol. 5 pp. 173-184.

Yang Z (Feb.1996). Phylogenetic analysis using parsimony and likelihood methods. J Mol Evol vol. 42, no. 2, pp. 294-307.

Zhai Y, Tchieu, and Saier M H Jr., (Jan.2002). A web-based Tree View (TV) program for the visualization of phylogenetic trees. J MolMicrobiol Biotechnol vol. 4, no. 1, pp. 69-70.

Keyword Index